HOUSING, CARE AND PSYCHOLOGICAL WELLBEING OF CAPTIVE AND LABORATORY PRIMATES

Benjamin B. Beck

NOYES SERIES
IN
ANIMAL BEHAVIOR, ECOLOGY,
CONSERVATION AND MANAGEMENT

Benjamin B. Beck, Ph.D., Consulting Editor

A series of professional and reference books in ethology devoted to the better understanding of animal behavior, ecology, conservation, and management.

WOLVES OF THE WORLD: Perspectives of Behavior, Ecology, and Conservation.
Edited by *Fred H. Harrington* and *Paul C. Paquet*

IGUANAS OF THE WORLD: Their Behavior, Ecology, and Conservation.
Edited by *Gordon M. Burghardt* and *A. Stanley Rand*

HORSE BEHAVIOR: The Behavioral Traits and Adaptations of Domestic and Wild Horses, Including Ponies.
By *George H. Waring*

GAZELLES AND THEIR RELATIVES: A Study in Territorial Behavior.
By *Fritz R. Walther, Elizabeth Cary Mungall,* and *Gerald A. Grau*

THE BIOLOGY AND MANAGEMENT OF AN EXTINCT SPECIES: *PÈRE DAVID'S DEER.*
Edited by *Benjamin B. Beck* and *Christen Wemmer*

APES OF THE WORLD: Their Social Behavior, Communication, Mentality and Ecology.
By *Russell H. Tuttle*

TIGERS OF THE WORLD: The Biology, Biopolitics, Management, and Conservation of an Endangered Species.
Edited by *Ronald L. Tilson* and *Ulysses S. Seal*

HOUSING, CARE AND PSYCHOLOGICAL WELLBEING OF CAPTIVE AND LABORATORY PRIMATES.
Edited by *Evalyn F. Segal*

Housing, Care and Psychological Wellbeing of Captive and Laboratory Primates

Edited by

Evalyn F. Segal

San Diego State University
San Diego, California

np

NOYES PUBLICATIONS
Park Ridge, New Jersey, U.S.A.

Library of Congress Catalog Card Number: 89-30309
ISBN: 0-8155-1201-5
Printed in the United States

Published in the United States of America by
Noyes Publications
Mill Road, Park Ridge, New Jersey 07656

10 9 8 7 6 5 4 3 2 1

Library of Congress Cataloging-in-Publication Data

Housing, care and psychological well-being of captive and laboratory primates / edited by Evalyn F. Segal.
p. cm.
Bibliography: p.
Includes index.
ISBN 0-8155-1201-5 :
1. Primates as laboratory animals. 2. Primates--Psychology.
3. Primates--Housing. I. Segal, Evalyn F.
599.8'045--dc19 89-30309
CIP

I dedicate this book to Mac and Sarah *(Macaca sylvanus)*; Rachel and Leah *(Macaca mulatta)*; Monkey 1, Monkey 2, Monkey 3, Monkey 4, JoJo, Fred and Cissy *(Saimiri sciureus)*; and to the memory of Christopher Parker *(Homo sapiens)*.

Preface

HOW THIS BOOK CAME TO BE

Legislative History

Congress, in 1985, amended the Animal Welfare Act of 1966 to specify new standards for the care of laboratory animals. One of the new provisions of the amended Act required

> "a physical environment adequate to promote the psychological wellbeing of primates" (United States Code, 7, Section 2143 (a)(2)(B) Supp. IV 1986).

Later that year, the National Institutes of Health responded by revising the NIH *Guide for the care and use of laboratory animals* to require that animals be provided with an environment

> "appropriate for their species," and which "contribute[s] to their health and comfort" (p. 83, 1985 revision of *NIH Guide*).

The *NIH Guide* incorporated the revised *Public Health Service policy on humane care and use of laboratory animals* adopted by the NIH Office for Protection from Research Risks. Principle 4 of the *PHS policy* states:

> "Unless the contrary is established . . . procedures that cause pain or distress in human beings may cause pain or distress in other animals" (*NIH Guide*, 1985 rev.ed., p. 82).

In March, 1987, the Secretary of Agriculture proposed regulations to implement the amended Animal Welfare Act, and followed NIH's lead by defining a painful animal research procedure as

> "any procedure that would reasonably be expected to cause more than slight or momentary pain or distress in a human being . . . " (*Federal Register*, 1987, 52, 10,297).

Responding to public commentary on the 1987 proposals, USDA's Animal and Plant Health Inspection Service published revised regulatory proposals in the *Federal Register* of March 15, 1989 (pp. 10822-10954). Noting that some animal researchers had termed APHIS's 1987 definition of animal pain "anthropomorphic," APHIS nevertheless retained the definition in the latest proposals, on the grounds that there is no satisfactory alternative.

Taken together, these events mark a watershed. As the bioethicist Rebecca Dresser has stated, by these various actions

> "federal officials registered their agreement that the experimental animal's experiences should carry some moral weight" (Dresser, 1988a, p.726). "Moreover, the proposed . . . regulations and the PHS policy explicitly direct scientists and review committees to assume, in the absence of clear evidence to the contrary, that interventions producing pain and distress in humans do so in animals as well. In short, beliefs that were once challenged as anthropomorphic fantasies are now firmly entrenched in the federal research oversight system" (p. 743).

Editorial History

The idea for this book came from a symposium I chaired (Segal, 1986) at the annual meeting of the American Psychological Association. The meeting was in Washington, D.C., and the staff of APA's Committee on Animal Research and Ethics (CARE) arranged for symposium participants to meet over lunch with USDA/APHIS representatives to share thoughts about the difficult task before them of drafting regulations to implement provisions of the new legislation. I had ample opportunity then and since to confirm Rebecca Dresser's observation that

> "the new federal regulatory provision engendering the greatest confusion of all is the Animal Welfare Act's requirement to furnish 'a physical environment adequate to promote the psychological well-being of primates'" (Dresser, 1988a, p. 753).

It was in hopes of diminishing that confusion that I began this book.

I was joined at the start by my departmental colleague Patricia Scollay, an anthropologically-trained primatologist. When a letter inviting contributions to the book went out over both our signatures, it was Patti Scollay's reputation in the anthropology community that induced primatologists to join us. I thank everyone who agreed to contribute to the book on the strength of Patti's invitation, and who then had to entrust their manuscripts to me, after other commitments obliged Patti to withdraw. For me, editing this book has amounted to a short course in primatology, and my first contact with many of its leading practitioners. I thank the authors for an enlightening experience, and I thank Patti for her indispensable help at a critical time.

I also thank the authors for putting up with my high- and heavy-handed editing. Among other things,

* I defied grammatical custom by sometimes (whenever I could) using the personal relative pronoun *who* with nonhuman primates, instead of the impersonal *that*.

* Over authors' protestations, I rounded off behavioral data to no more than two or three significant digits. I think behavioral scientists should not pretend to greater precision of measurement than our methods of observing and recording behavior allow; nor should we apologise for that fact.

* I tried to banish the passive voice. It impedes the flow of thought. In graduate school we learn to equate scientific objectivity with impersonality, and impersonality with the passive voice. But all that's nonsense. Science isn't done by robots; and when science deals with living animals, it is not unusual for scientists to feel a personal interest and concern for their subjects; we needn't apologise for that, either, but neither should we mislead the public and each other by pretending indifference. Our understanding of the biology of intact organisms comes from field studies and laboratory experiments that flesh-and-blood people carry out with flesh-and-blood animals. Rather than closing our eyes to the inevitability of personal interaction between observer and subject, we should acknowledge it and be alert to its manifold effects -- on our subjects' physiology and behavior, on how we observe and record their physiology and behavior, and on how we think and reason about our data.

PSYCHOLOGICAL WELLBEING IN PRIMATES??

> ". . . well-being connotes more than the absence of pain and distress. It implies as well that an individual's physiological, security, and behavioral needs are fulfilled." (Dresser, 1988a, p. 747)

A Personal History

I came to a scientific interest in the behavior of captive primates, and eventually to a concern with their wellbeing, via psychological training in the experimental analysis of learning and behavior. My first exposure to nonhuman primates was in the 1960's, when for about five years, in collaboration with Stephen Holloway, David Oden, and Sam Deadwyler, I studied learning in four squirrel monkeys that we called Monkeys 1, 2, 3, and 4. In the early 1970's, assisted by Roger Dunn, I returned to studying learning in squirrel monkeys, this time with three who had been named by their caretakers: JoJo, Fred, and Cissy. A little later, my late colleague Christopher Parker secured for me a young Barbary macaque from the San Diego Zoo. He was Brandon when he came to us but my master's student, Ralph Burr, liked "Mac" better, so Mac he has been, to me and a long series of student collaborators, including Debbie Wilkie, Bill Murphy, Susie Salstrom, Ray Ryan, Tere Lindquist, and Karen Callaway. Later I obtained a Barbary macaque companion for Mac and named her Sarah. When Sarah died and I could not obtain another Barbary macaque, the Yerkes Primate Research Center sent two rhesus macaques to be Mac's companions; I named them Rachel and Leah. Rachel has moved on to a zoo social group; Mac and Leah are still together half a decade later.

Although I worked with Squirrel Monkeys 1, 2, 3 and 4 for a long time, I felt less personal interest in them than I did in JoJo, Fred, Cissy, Mac, Sarah, Rachel and Leah. It was easy to think of the numbered monkeys simply as data-generators (albeit very good data-generators, Monkey 1 especially), whereas the monkeys who had names stood out thereby as distinct individuals, each with his or her own aptitudes, predilections, and idiosyncrasies. Moreover, once I had become attuned to their individuality, I was surprised to notice resemblances between them and me that I had not been expecting. In situations where, had it been I, I would have felt embarrassed, the monkeys (particularly Mac, the Barbary macaque I came to know best) acted much as I do when I feel embarrassed. In situations where, had it been I, I would have felt proud of what I had accomplished, the monkeys (particularly Mac) acted much as I do when I am feeling proud of what I have accomplished.

Although I was careful to include the right number of subjunctives and counterfactual conditionals in that last sentence, some will still call the shift in viewpoint it reveals "creeping anthropomorphism." I explain it differently. As I came to know the individual characteristics of the personalized monkeys, I saw that Occam's Razor cuts two ways. Though it is folly to impute more of human-like psychological processes to another species than are needed to understand its behavior, it is equally folly to impute fewer. The Rule of Parsimony invites a presumption of homology in species as closely-related as the primates. As recent events show, it is no longer viewed as unscientific, or unduly subjective, to presume that, absent evidence to the contrary, the psychological needs of nonhuman primates are on a continuum with human psychological needs. We do not know very much about either the one or the other, but we can expect that whatever we learn about the psychological needs of one will shed some light on the needs of the other.

Consensus

There is, in fact, a surprising amount of agreement among the views and recommendations appearing in the chapters of this book. Several overlapping themes pervade the book. Chief among them is that **sociality is of the essence of primates. Above all else, primates, captive as well as free, nonhuman as well as human, need companionship.**

Second, behavior and habitat form a co-adapted system. To remove animals from the habitat for which nature has fitted them, and then fail to provide any substitute opportunities for them to express their natural repertoires, is to distort their behavior and physiology so far as to render them dubious subjects for biological research of any kind, whether physiological, medical or behavioral. If we seriously want to understand primate physiology and behavior, we need to house our subjects in physical environments that are complex enough, responsive enough, and changeable enough to allow the development of a fair range of motor and social skills.

Third, primates have a restless intelligence and curiosity that require exercise fully as much as their muscles do. They are born problem-solvers and manipulators, and they are bored silly when they have nothing to challenge them. Then (as we well know) they occupy themselves with scrapping and bickering and picking apart themselves and their environment. (Idle hands do the devil's work.) In the wild, much of the intellectual challenge, as well as motor and social skill development, comes from foraging for food. In captivity, provisions are too easily come by, and the animals have too much time on their hands, and too little to do. There is a growing consensus that we should be a little less free with food provisions, and instead should require animals to work (forage, hunt, learn prescribed tasks, solve problems and puzzles) for at least some of their subsistence. The following chapters are full of ideas about intellectually challenging tasks for captive primates, ranging from foraging for small tidbits in deep woodchips or corncobs, to fishing for goldfish, to diving for raisins, to raking food through a maze, to tracking targets on a video screen -- and much much more.

The politics of primate social life presents other kinds of intellectual challenge. Animals must maneuver for rank, court a mate, make friends, forge coalitions, fend off competitors, rear young; and always, learn to read, and remain ever alert, to the subtle body language that signals companions' likely next moves. The companionship of even one other animal may meet the minimum social needs of captive primates, but life in a larger social group offers so much more.

Another pervasive theme in the book is primates' need to have some control over their physical and social environment and the events of their daily life. To fend off the behavior of despair, they need to be able to make choices -- as to food treats, play activities, work activities, companions, locales. The extent of allowable control and the number of feasible options will vary with the research setting, but even under the most stringent research requirements, introducing options and turning over a modicum of control to the animals will pay dividends in psychologically and physiologically healthier subjects and data of greater reliability and generality.

A New Concern

Some investigators of primate behavior have expressed concern about the impact that new governmental regulations may have on their research and their subjects. They fear that requiring every facility to comply with uniform standards intended to be within reach of all may amount to a step backward for the most forward-looking facilities. There must be room to maneuver, to try out new ways of doing things, even if, sometimes, that means violating one or another minimum standard.

For example, many primates profit from the opportunity to forage for tidbits among deep woodchip litter (see Chamove & Anderson's review in this volume). It would be exorbitantly wasteful of materials and personnel to lay down fresh woodchips every day, and fortunately there is no need to: woodchips become *more* germicidal the longer they are left in place, and cage walls and furnishings stay cleaner when the floor is covered by woodchips, even though that means less frequent hosing. Yet the practice of leaving woodchip litter in place for days or weeks on end, and forgoing daily cage hosing, goes against modern sanitation practices. Placing branches and browse in animals' cages as enrichment raises similar issues for conscientious veterinarians: If it can't be autoclaved, can it be hygienic? We must hope issues of this kind will be resolved in ways that meet veterinarians' standards of hygiene, suit research purposes, and still satisfy the psychological needs of research primates.

THE ETHICS OF ANIMAL RESEARCH: A PERSONAL VIEW

Ethical convictions derive neither from logic nor from science, though both may serve ethics, logic by detecting contradictions among ethical precepts and science by providing the means to ethical ends. We can neither prove nor disprove the assertion that human interests have moral priority over the interests of other species. It does not even make sense to ask for proof. It is simply difficult, given human psychology, to deny the prior claims of our fellows. Like other species, we seem predisposed to place the interests of our own kind above those of other kinds. Surely all species, and not only the human, could be called "speciesist."

My best guess as to the origin of the human sense of ethical obligation to others is the bonds of childhood. Our first obligation is to survive; given our incompetence as primate infants, that requires our elders' assiduous nurturance. From the experiences of nurturing and being nurtured, and the affection those engender, arise ethical bonds of mutual obligation among family members.

Over historical time, the perception of who constitutes "family" (or "human being") -- those to whom ethical obligations are owed -- has gradually expanded, first to neighbors, then to clan, province, nation, religion, race, finally, for some of us, to the whole species. Now some people are urging an even further-expanded bioethics that holds that the interests of other species -- in living out their lives, expressing their behavioral natures, minimizing suffering, and so on -- should weigh in our ethical balance, along with our own interests. No one has any clear answers as to the proportionate ethical weights to assign to the interests of humans and nonhumans, but if an expanded bioethics were to gain widespread support, presumably the first nonhuman species to be included in an enlarged concept of the "humanoid family" would be our nearest relations, the other primates.

Nowhere in this book is it asserted that research with captive primates should not be done. Everywhere it is asserted that it should be done with care, respect, and humanity. A few would add, with love.

FINAL COMMENTS

I am appending below a list of some articles on primate wellbeing that I have lately come across and that for the most part appeared too late for authors to have included in their chapters. I also included a few older references of particular importance and utility (like the Animal Welfare Act). Finally, I offer a short list of addresses of primate wellbeing resources (like the newly-established Animal Welfare Information Center of the National Agricultural Library); I do not pretend that the list is complete.

Several of the recent articles I list below are the work of a group led by Viktor Reinhardt, a veterinarian at the Wisconsin Regional Primate Center. Reinhardt's group are devoting themselves to serving the psychological wellbeing of individually-caged monkeys by seeking ways to provide monkey companions even to some monkeys of long-standing social isolation. Where introducing a companion is out of the question,

Reinhardt *et al.* are attempting other forms of cage enrichment. I regret that I did not know of Reinhardt's important work in time to solicit his contribution for this book.

San Diego, California — Eve Segal
1 April 1989

REFERENCES

Anderson, J. R. & Chamove, A. S. (1986) Infant stumptailed macaques reared with mirrors or peers: Social responsiveness, attachment, and adjustment. *Primates, 27*, 63-82.

Animal and Plant Health Inspection Service, USDA. (1989, March 15). Proposed rules to implement the 1985 Congressional amendments to the Animal Welfare Act. *Federal Register*, 10822-10954.

Animal Welfare Act of 1966, as amended in 1970, 1976, and 1985. *United States Code 7*, §§ 2131-2157 (1982 & Supp. IV 1986).

Barnard, N. (1988, September) Inherent stress: The tough life in lab routine. *Lab Animal*, 21-23, 26-27.

Bayne, K.A.L. (1989, January). Nylon balls re-visited. *Laboratory Primate Newsletter 28:1*, 5-6.

Boccia, M. L. (1989, January). Preliminary report on the use of a natural foraging task to reduce aggression and stereotypies in socially housed pigtail macaques. *Laboratory Primate Newsletter 28:1*, 1-2.

Bramblett, R.D. & Bramblett, C.A. (1988, October). A liquid dispenser for caged primates. *Laboratory Primate Newsletter 27:4*, 16.

Champoux, M., Hempel, M., & Reinhardt, V. (1987) Environmental enrichment with sticks for singly-caged adult rhesus monkeys. *Laboratory Primate Newsletter 26*, 5-7.

Dresser, R. (1988a) Assessing harm and justification in animal research: Federal policy opens the laboratory door. *Rutgers Law Review, 40*, 723-795.

Dresser, R. (1988b) Standards for animal research: Looking at the middle. *Journal of Medicine and Philosophy, 13*, 123-143.

Engler, K. (1989, March). The Animal Welfare Information Center of the National Agricultural Library. *Lab Animal 18:2*, 37-41.

Fajzi, K., Reinhardt, V. & Smith, M.D. (1989, March) A review of environmental enrichment strategies for singly caged nonhuman primates. *Lab Animal 18:2*, 23-29, 31, 33, 35.

Government regulations for animals used in research. (1988). *Scientists Center for Animal Welfare Newsletter, 10*, 1, 4.

Heath, S. J. (1987) Behavioral enrichment for primates: What are the options? *Scientists Center Newsletter 9(1)*, 11-12.

Holden, C. Experts ponder simian well-being. (1988) *Science, 241*, 1753-1755.

Kelley, T.M. & Bramblett, C.A. (1981). Technical note: Urine collection from vervet monkeys by instrumental conditioning. *American Journal of Primatology, 1*, 95-97.

Line, S.W., Clarke, A.S. & Markowitz, H. (1987) Plasma cortisol of female rhesus monkeys in response to acute restraint. *Laboratory Primate Newsletter 26*, 1-4.

NIH Guide for the care and use of laboratory animals (1985 rev. ed.) Office of Protection from Research Risks, National Institutes of Health.

Novak, M. A. & Suomi, S.J. (1988,) Psychological well-being of primates in captivity. *American Psychologist 43*, 765-773.

Petto, A.J. & Devin, M. (1988, July) Food choices in captive common marmosets (*Callithrix jacchus). Laboratory Primate Newsletter 27:3*, 7-9.

Public Health Service policy on humane care and use of laboratory animals. (1986 revision). Office of Protection from Research Risks, National Institutes of Health.

Reinhardt, V. (1987) Advantages of housing rhesus monkeys in compatible pairs. *Scientists Center Newsletter 9(3)*, 3, 5-6.

Reinhardt, V. (1988, October). Preliminary comments on pairing unfamiliar adult male rhesus monkeys for the purpose of environmental enrichment. *Laboratory Primate Newsletter 27:4*, 1-3.

Reinhardt, V., Cowley, D., Eisele, S., Vertein, R. & Houser, D. (1987). Preliminary comments on pairing female rhesus monkeys for the purpose of environmental enrichment. *Laboratory Primate Newsletter 26:2*, 5-8.

Reinhardt, V. & Dodsworth, B. (no date) *Environmental enrichment for individually caged rhesus monkeys at the WRPRC.* A set of slides (the number grows with each availability notice).

--- (no date) *Facilitated socialization of previously singly caged adult rhesus monkeys.* VHS videotape, 20 minutes, in color.

Both the slide set and videotape can be borrowed from Audiovisual Services, Wisconsin Regional Primate Research Center Library, 1223 Capitol Court, Madison WI 53715 (608/263-3512)

Reinhardt, V., Eisele, S. & Houser, D. (1988, April) Environmental enrichment program for caged macaques: A review. *Laboratory Primate Newsletter 27:2*, 5-6.

Reinhardt, V. & Smith, M.D. (1988, July) PVC pipes effectively enrich the environment of caged rhesus monkeys. *Laboratory Primate Newsletter 27:3*, 4- 5.

Rhesus monkey colony management: A 1988 update. (1989, January). *Laboratory Primate Newsletter 28:1*, 8-14. (Transcript of a workshop presented at the June, 1988 meeting of the American Society of Primatologists; M. R. Clarke, Ed.)

Segal, E. F. (Chair) (1986, August 25). *Improving the quality of life of laboratory primates.* Symposium held at the meeting of the American Psychological Association, Washington, D. C.

Wallace, G. (1988, Spring) Improving life for primates. *Caring for Animals. 5:1* (no page number). (Published by the Canadian Federation of Humane Societies, #102-30 Concourse Gate, Nepean, Ontario K2E 7V7).

Watson, L.M. & Petto, A.J. (1988, July) Infant adoption and reintroduction in common marmosets (*Callithrix jacchus). Laboratory Primate Newsletter 27:3*, 1-3.

Wolff, A.V. (1989, January). Polyvinyl chloride piping as perch material for squirrel monkeys. *Laboratory Primate Newsletter 28:1*, 7.

SOME RESOURCES ON LABORATORY ANIMAL WELFARE

Animal and Plant Health Inspection Service (APHIS), U.S. Department of Agriculture, Federal Building, Room G-100, 6505 Belcrest Road, Hyattsville MD 20782

Animal Welfare Information Center (AWIC), Room 301, National Agricultural Library, Beltsville, MD 20705. 301/344-3212.

Caring for Animals. Canadian Federation of Humane Societies, #102-30 Concourse Gate, Nepean, Ontario K2E 7V7.

Laboratory Primate Newsletter (J.E. Schrier, Ed.). Primate Behavior Laboratory, Psychology Department, Brown University, Providence, RI

Office of Protection from Research Risks, National Institutes of Health, Building 31, Room 4809, Bethesda, MD 20892.

Scientists Center for Animal Welfare, 4805 St. Elmo Avenue, Bethesda, MD 20814.

University of Illinois Laboratory Animal Welfare Project, Carol Boast, Principal Investigator, 226 Mumford Hall, 1301 W. Gregory Drive, Urbana IL 61801: preparing *Laboratory animal welfare training resource directory* and *Laboratory animal welfare research guide.* (Letter to EFS of June 14, 1988)

Contributors

Mary Lee Abshire
Central Washington University
Ellensburg, WA

James R. Anderson
Universite Louis Pasteur
Strasbourg, France

Nancy A. Ator
The Johns Hopkins University
School of Medicine
Baltimore, MD

Kathryn Bayne
National Institutes of Health
Bethesda, MD

Cynthia L. Bennett
Santa Barbara Zoological Park
Santa Barbara, CA

William M. Blackmore
University of Southern California
Los Angeles, CA

Mollie A. Bloomsmith
University of Texas M.D. Anderson
Cancer Center
Bastrop, TX

Mark Bodamer
Central Washington University
Ellensburg, WA

Joseph V. Brady
The Johns Hopkins University
School of Medicine
Baltimore, MD

Claud A. Bramblett
University of Texas at Austin
Austin, TX

Arnold S. Chamove
Stirling University
Stirling, Scotland

Roberta L. Cox
University of California, Riverside
Riverside, CA

Karen A. Craven
The Johns Hopkins University
School of Medicine
Baltimore, MD

Jeremy F. Dahl
Yerkes Regional Primate Research Center
Emory University
Atlant, GA

Roger T. Davis
Washington State University
Pullman, WA

Phyllis Dolhinow
University of California
Berkeley, CA

Karla H. Drewsen
University of Massachusetts
Amherst, MA

John E. Fa
Universidad Nacional Autonoma de Mexico
Mexico D.F.

Deborah H. Fouts
Central Washington University
Ellensburg, WA

Roger S. Fouts
Central Washington University
Ellensburg, WA

Jeffrey A. French
University of Nebraska at Omaha
Omaha, NE

Steven G. Gilbert
Health and Welfare Canada
Ottawa, Ontario, Canada

David Marsland Haring
Center for the Study of Primate Biology and History
Duke University
Durham, NC

Mary Kathryn Izard
Center for the Study of Primate Biology and History
Duke University
Durham, NC

James E. King
University of Arizona
Tucson, AZ

Scott Line
University of California
Davis, CA

Rosanne B. Lorden
Eastern Kentucky University
Richmond, KY

Joseph M. Macedonia
Center for the Study of Primate Biology and History
Duke University
Durham, NC

Hal Markowitz
San Francisco State University
San Francisco, CA

Patricia Miller-Schroeder
Northern Frontier Zoo
Earlton, Ontario, Canada

Eileen M. Molzen
University of Nebraska at Omaha
Omaha, NE

Ronald D. Nadler
Yerkes Regional Primate Research Center
Emory University
Atlanta, GA

Vicky R. Norwood
University of Arizona
Tucson, AZ

Melinda A. Novak
University of Massachusetts
Amherst, MA

Peggy O'Neill
National Institute of Child Health and Human Development
Dickerson, MD

James D. Paterson
University of Calgary
Calgary, Alberta, Canada

Michael E. Pereira
Center for the Study of Primate Biology and History
Duke University
Durham, NC

Ramon J. Rhine
University of California, Riverside
Riverside, CA

Duane M. Rumbaugh
Georgia State University and
Yerkes Regional Primate Research Center
Emory University
Atlanta, GA

Eric Salzen
University of Aberdeen
Old Aberdeen, Scotland

Anne Savage
University of Wisconsin
Madison, WI

E. Sue Savage-Rumbaugh
Georgia State University and
Yerkes Regional Primate Research Center
Emory University
Atlanta, GA

Elwyn LaVerne Simons
Center for the Study of Primate Biology and History
Duke University
Durham, NC

Charles T. Snowdon
University of Wisconsin
Madison, WI

Mark A. Taff
University of California
Berkeley, CA

Roger K. Thomas
University of Georgia
Athens, GA

Jaylan S. Turkkan
The Johns Hopkins University School of Medicine
Baltimore, MD

David Washburn
Georgia State University and
Yerkes Regional Primate Research Center
Emory University
Atlanta, GA

Ellen Wrenshall
Health and Welfare Canada
Ottawa, Ontario, Canada

Patricia Chapple Wright
Center for the Study of Primate Biology and History
Duke University
Durham, NC

Contents

I. ORDER: PRIMATES

Family: Hylobatidae (Gibbons)

I. ORDER: PRIMATES

— 1 —
Mental Well-Being in Anthropoids

Claud Bramblett

This chapter has a threefold purpose: 1) to discuss the mental wellbeing of nonhuman primates from the perspective of an anthropologist who specializes in primatology; 2) to make specific recommendations about care and management of primate colonies; and 3) to relate, from 20 years of experience with field and captive primates, how these recommendations can affect a primate colony.

UP FROM THE ANTHROPOID

First to clarify: in this chapter mentalism does not imply anything nonphysical or nonbiological but rather implies the assumption of biological continuity in the way we and our phylogenetically-close relatives process information. We have long been aware of evidence for mental abilities in other species (Griffin 1983). Learning has long been a vital part of the anthropoid lifeway. The ability to devise and revise "mental maps" (Altmann & Altmann 1970) of spatial and temporal distribution of foods may have been of critical importance in mental development (Milton 1981; Richard 1981). The biological importance of learning does not diminish in proportion to the facility to learn; a snail's learning ability may be as important to its survival as memory is to a human.

As behavioral systems began to rely increasingly on learning, social roles began to be transmitted partly by learning, and this allowed something new to be added to the species' organization: values. The special adaptive property of value systems is that they control behavior without waiting for DNA accidents. Transmission of roles and values by means of learning and a primitive communication system presupposes a considerable sophistication of mental processes. Appreciation of context and intention movements is made very powerful by the perceptive aptitudes of the participants.

Primatologists were surprised that nonhuman primates with such comparatively simple communication systems and technologies could prove to be so adept at learning and cognitive tasks in the laboratory. We had underestimated the mental abilities that were part of anthropoid societies and behavioral systems (this is the primary lesson from the ape language experiments). It was their large mental aptitudes that produced such complexities and behavioral plasticity among species with limited communication systems.

Gardner and Gardner (1978) moved against the mainstream when they proposed a search for general laws of behavior rather than accepting surface dissimilarities as evidence of fundamentally different sets of laws controlling behavior in different species. Although they have been cautious in their speculations about mental processes in nonhuman primates, others have not. For example, Kohler (1959, page 236) concluded that "...chimpanzees manifest intelligent behavior of the general kind familiar in human beings." He argued that lack of speech was the chief difference between anthropoids and human beings. The classic description of Vicki as a chimpanzee raised in a human environment concluded that only language abilities distinguished the chimp's intellectual skills from those of a human child (Hayes & Hayes 1951). Premack and Woodruff (1978) speculate that chimpanzees impute mental states to others in a very human fashion. Indeed, their description of chimpanzees as mentalists is a strong (and perceptive) departure from behaviorism. Smith (1987) points out that acts of deception raise the question of awareness and intentionality among some nonhuman species.

One "classic" experiment for self-awareness was the recognition of "self" in a mirror (Gallup Wallnau & Suarez 1980; Platt & Thompson 1984). Vervets in our laboratory who grew up with mirrors in their outdoor cage behaved differenty toward mirror images and real objects. We tested this difference by flashing objects (a peanut hidden in the hand) systematically to see which individuals would react by reaching into the mirror and which would turn immediately to the person holding the peanut. Only one older, wild-born female reached toward the mirror image: all others made the correct choice without hesitation (Zabel 1980).

When we repeated part of Gallup's experiment by sedating vervets and painting facial hair, we too saw no evidence of recognition or interest in the discoloration when viewed from a mirror. However, we carried the experiment a step further by painting the leg where it was easily seen by the subject, but there was no interest there either. If a vervet did not respond to easily visible leg paint, there is no reason to expect it to respond to facial paint visible only in a mirror. In the observation period following recovery from sedation and reintroduction to the group, other vervets paid no attention to easily visible painted marks even without the use of a mirror. Thus, we concluded that our experimental design (and that of Gallup et al. 1980) was flawed. (See a further example under *How can an ethologist help?* below.)

Another type of self-recognition, that of group membership, is demonstrable. Even our mixed-taxa groups readily identified their cagemates and behaved differently toward them than they did toward animals who lived for years in full view but in a different cage. However, the politics of subgroups also require complex differential recognition of context and identity. Cheney Seyfarth and Smuts (1986) have suggested that these complex social relationships are evidence of sophisticated cognitive mechanisms. Furthermore, the adaptive value of these relationships may have stimulated the evolution of intelligence.

At least one special event occurred in the ancestral human lineage: the communication system became progressively free of innate signals, and more and more dependent upon learning. The human vocal repertory progressed toward a speech system instead of a system of innate calls. The shift included both improvements in learning abilities and changes in the anatomy and neuromotor control of the speech apparatus. We tend to think of this as a special and unique process, but it may, in fact, have occurred several times in the geological history of our planet and many of our vertebrate relatives may have moved varying distances down this path.

The development of speech is accompanied by an elaboration of neural apparatus that supports speech abilities. It is quite possible that much of the grammar or syntax that characterizes human speech evolved with this speech apparatus and is not a product of mental processes. If this is accurate, then our scientific preoccupation with language and speech handicaps the development of Griffin's proposed science of cognitive ethology.

The rise of speech is perhaps second in importance only to the previous trend toward increased reliance on learning in communication and behavior. Speech itself is a clumsy way to communicate. It is slow, requires extensive training, and it demands great effort to avoid confusion since it is not naturally redundant unless accompanied by gestures and intention movements. Furthermore, speakers from distant groups encounter dialect confusion. Fortunately, the older mental abilities are still present to judge context, situations, and intentions. The learned content of speech demands special investment in learning and practice.

There were at least two dramatic alterations of adaptive value that became available with fluency in speech. The first was a different kind of possible information content. One could make direct reference to items and details that were displaced in time and space. Even imaginary subjects could acquire a mental reality and utility. The second major adaptive feature was that language permitted the efficient transmission of value systems - systems that could motivate complex behavioral sequences and insist upon consistently adaptive acts. It is not technology that is the height of human achievement (that is too new). It is the efficient way in which we transmit values. Value systems are different from less generalized rule systems in that value systems set goals to work toward or to avoid. They are systems that focus energy and effort. Some values are supported by sets of rules that have grown over time to translate the values (or principles) into behavioral choices. An unusual property of some values is that they are adaptive only as ideals. It could be catastrophic for some idealistic goals actually to be realized.

Human value systems and culture, like other forms of learning, must have their basis in neural tissues that are biologically specialized for that particular function. The individual is primed to acquire value systems at the appropriate time in development. Furthermore, a value system can exist with no reinforcement other than mental processes, and sometimes can survive the most extraordinary negative stimulation.

By selecting value systems that organize behavior in adaptive ways, the same human genotype can exibit numerous behavioral phenotypes and so adapt to a wide variety of niches. Just as gene changes may be too sluggish to readily accommodate rapid changes occurring in human ecosystems, a great danger to present humanity is that our value systems may be too slow to change in the world that we are consuming. The mental wellbeing controversy is a forum that illustrates a human dilemma about value systems.

THE ANTHROPOLOGIST'S DILEMMA

A basic problem when dealing with values is the difference between the way they are perceived and the way they operate to alter behavior. The consequences of a particular "value" do not necessarily correlate with its imagined properties. Indeed, it may work in the opposite manner, or in a way that would be intellectually objectionable if it were perceived. For example, a preference for male offspring may be a vital component of a value system in a culture whose ecology is based upon a gender ratio that contains a surplus of females. [If you tend to stop having children once you have a male, families who have male first children tend to be smaller. Attempts to produce sons if first child is female result in production of more females than males (Lave and March 1975; pages 69-71).]

These cultural attributes also apply to the research community. Imagine the set of values that leads to the proposed changes in federal regulations to ban natural materials from primate housing. Likewise, a primate who is forced to sit all day in its own excrement is perceived as "clean" if the cage can be sanitized once a day.

The predisposition to assume the values of the culture in which we find ourselves at a critical stage of development shapes our behavioral plasticity into a specific array of adaptive responses. This process was surely adaptive at a time in our evolutionary history when our antecedents were less talented at logic or planning; and as part of our biological heritage, we have no option but to try to understand and proceed along the same course. This applies as much to our values concerning other species as it does in other value domains. Attitudes at two extremes bracket our perspectives on the mental processes of other species. Our society often finds it suitable to ignore continuity of process and to appeal to the lore of human uniqueness in our dealings with other species. This is understandable in a country that slaughters over 1.3 billion mammals annually for meat and euthanizes over 10 million each year as unwanted pets (Rueter 1984). It is convenient as well to dehumanize slaves and enemies. At the opposite pole is a tradition that pictures nature as a paradise filled with sentient beings that experience only health, happiness, and imagined immortality. Followers of this tradition may view other species as human surrogates and attribute to them all the qualities they wish their fellow humans possessed.

ACTIVISTS AND THE CONSEQUENCES

Some supporters of animal welfare movements wish to eliminate the use of all animals in research. Others want to improve conditions and reduce suffering and waste. A third group represents the vast spectrum of attitudes between the self interests of the investigator and those of the consumer who eventually benefits from, and pays for, research.

The effects of a lobby may not be what was expected. The first group may be pleased at anything that increases the dollar cost of research since more costs means fewer experiments and less use of animals. The second group has reason to be pleased with progress since there have been very clear improvements in laboratory animal management. An excellent example of progress has been the Primate Supply Information Clearinghouse's success in recycling animals to eliminate needless euthanasia and reduce the demand for imports. On the other hand, changes translated through a bureaucratic network can sometimes mean more expensive stainless steel, but not necessarily an improvement in the quality of health or life.

All of the effects on research are not positive - people and animals suffer and die if experiments are slowed and marginal projects (often meaning basic versus applied) do not materialize. For example, a delay in research on AIDS or Lyme's disease due to flaws in our value system produces the same suffering as delays due to logistics, funding, or ignorance. A consequence that is particularly germane to the wellbeing of primates in research is the impact that changing regulations have had, and are likely to have, upon laboratory research in ethology. Ethology is at greater risk than most research areas for two reasons. First, there is less evidence that people's lives are at stake in most ethological experiments. Secondly, there are many scientists sympathetic to Passingham's criticism of ethology: ..."Remember that the gap between man and chimpanzee in the development of the brain is wider than the gap between chimpanzee and shrew. What would we find out about the social life of a chimpanzee by watching a shrew?" (Passingham 1982; page 21). The answer to this question is fundamental to all biological sciences: we learn about processes. Furthermore, the real differences may be small when compared to the similarities. The details of a system do matter, but what transfers from one subject to another are the processes. This is particularly important when studying biological systems that adapt rapidly to different ecologies, a major characteristic of behavioral systems among primates.

It is possible, no, it is probable, that various anthropoids have mental properties that differ from each other in ways that fit each species' eco-niche. Indeed, one would expect anthropoid mentalities to have all the properties of those of humans, but configured in numerous contrasting ways by evolutionary adaptation. Details of the central nervous system of our particular species may make sense only in our human niches. It does not imply that other species do not think, have anxieties, feel pain, experience fear... Such a perspective only implies that we would expect these processes to be cast in a biology that differs for each species. This approach also recognizes the importance to behavior of noncognitive models such as fixed action patterns, innate releasing mechanisms, imprinting, instinct, circadian rhythms, pheromones, stimulus filtering, and hormonal cycles.

It is exciting to study species who share the elements of our own mental processes but yet have species-specific differences in their mental biology. Anthropologists have often proposed that the greatest challenge to human intelligence is to understand our own evolutionary origins (for example, Oliver 1959). It would appear that the study of mental processes in living animals, including our own species, is a much more difficult topic, and certainly no less important. As Goodall (1986, page 16) phrases the problem, "Only by working with immense patience, understanding, insight, and skill with individual chimpanzees in captivity is it possible to investigate questions that relate to the upper limits of chimpanzee intellectual ability." We can not approach problems of perception and intelligence in primates without also becoming more sensitive to pain, stress, and aberrations that are due to cruel and inadequate housing and treatment.

These interpretations are particularly relevant to questions of animal ethics and mental wellbeing and there are scientists who hold strongly differing views. Benjamin (1983) argues that if animals are not conscious, we may have no ethical obligation to protect them or treat them humanely. He furthermore takes the position that we are obligated to treat them like persons only if they use language to bring possibilities to mind, use the word "if" in counterfactual speculation, make arrangements for tomorrow, and regret yesterday. The way one approaches the question of mental wellbeing depends upon the presumptions we make about nonhuman primate mental processes.

HOW CAN AN ETHOLOGIST HELP? (OR, ANECDOTES FROM A GUENON COLONY)

It is important to treat animals as individuals. The following anecdote illustrates the problem faced by Gallup et al (1980) in trying to determine which nonhuman primates show self-awareness. No standard protocol matches personality variation in a colony. The first and last time that we used a chair restraint illustrates several points. Some of our animals were tolerant of medication, sutures, and bandages while others were not. Usually, a bandage could be placed over a wound if it was needed to promote cleanliness in the dirty world of the cage. The animal could then be released to the social environment where hygiene was better than in a small wire holding cage. One adult male was an extreme exception. A serious fight injury required numerous sutures for a major upper leg wound. He removed both bandage and sutures several times. Finally, after he removed even steel sutures held in place with a plaster jacket, we were forced to chair him to prevent him from reaching the injury. The standard macaque chair was adaptable to vervets, but not this vervet to a chair. Though he was in great distress, we felt he would adjust with time, but after several days, the veterinarian made the observation that the monkey was blind! Although his eyes were open, he did not track or respond visually to us or to objects, yet I knew that prior to his injury he was a healthy normal adult male. If I could not demonstrate vision in a vervet stressed by chairing, it is not surprising that restraint, sedation, and paint marking failed to demonstrate self-recognition in Gallup's monkeys. We returned the vervet to his social group where his sight was instantly evident. He removed his sutures but recovered completely.

Since hiding is one of the primary responses of solitary guenons to a terrifying situation, it sometimes can explain some extraordinary actions. Escapees from our colony are likely to be found in the nearest hole in the wall, in a bucket, a window, or even a file cabinet. The side came loose on one of the shipping crates when our Sykes' monkey group was imported from Uganda. The crates were unloaded beside the runway, and my first sight of the adult male was to see him swing open the crate, look around, and then retreat. He stayed inside the box and held the side closed with his hands.

Animals who live in complex social groups may respond in a very different way. The introduction of a vervet group into an indoor cage after their trans-Atlantic flight from Uganda is an example. When released into a room, the animals displayed strong anxiety and dispersed. Upon introduction of the last group member, the adult male, the group's behavior changed. All of the animals, old and young, rushed to him and tried to hide under him or as close to him as they could get. He was lifted off the floor by a mound of monkeys. This "mound" followed him around the room, repeating the lifting process several times. The group was literally hiding under the adult male, and it was clear that the male performed a role which was very important to the rest of the group.

Our project with this newly imported group required that we take PAP smears from each adult female five days a week for a year (Bramblett Pejaver & Drickman 1975). This required frequent animal handling, and the routine did not go well at first. Since we went into the group cage to take samples, it required two people to take each sample: one to catch the female, another to keep the adult male away. He defended the females as we approached them, and our monkey housing had no provision for easy separation of individual animals.

After a few very uncomfortable sample-collecting sessions, the male monkey solved the problem. As we entered the cage, he made a successful dash for the outside. Turning back, he closed the door on us (shutting us inside the cage) and began displaying vigorously toward us from the outside. The monkey demonstrated a solution to us (which he probably learned in a previous laboratory) that became the daily routine. He would leave the cage as we entered, make a violent display from the outside as we took the samples, and calmly re-enter the cage as we left. He performed his role without danger to himself or us and he made no attempt to escape the building. Female responses were varied; most habituated to the routine, some even to the point of presenting calmly for the swab. One female never cooperated, and obtaining the final sample presented the same challenge as the first.

Another series of events illustrates a different phenomenon. We used a capture net to restrain animals at first. Subsequently, we allowed the animal to enter a small carrying cage as an alternative to being netted. Recalcitrant individuals were netted. With time, we used the net less and less, until it became a rare occasion. However, we were surprised to find that the less the net was used, the greater the animals' fear of it became. Indeed, the net was particularly fearsome to juveniles who had never been netted, but who observed the adult's reactions to sight of it. Eventually the net became primarily an object for human display, full of holes that made it difficult to use anyway (Bramblett Pejaver & Drickman, 1982).

We tried to follow a predictable schedule that was worked out with sensitivity to the animals' requirements. When there is a consistent routine, both animals and caretakers can understand each other's signals and the expected sequence of activity. There is minimal fear and maximum ability to recognize and address problems by both parties. For example, if the animals had to be locked indoors during outdoor maintenance, they were given time and opportunity to check the membership of the indoor cages and to move from one to another to find a compatible mix of social companions before we locked the doors. This reduced tension, fighting, and other forms of aggression.

In a subsequent project we planned to use urine samples to study endocrine activity. As a pilot project, six adult males were trained to urinate on demand into a glass beaker (Kelley & Bramblett 1981). This was accomplished through the use of peanut rewards without removing animals from their group cage or separating them from their 29 companions. The ease of training and the cooperation that the animals exhibited lead me to think that one of the tragedies of modern laboratory animal research has been the tendency to use hardware and isolation to solve research problems instead of patience, a well-prepared protocol, and a supply of rewards (peanuts). Rarely have laboratory or zoo managements in the U.S. taken full advantage of available behavior-modification techniques to ease, much less enrich the lives of animal exhibits. The idea is neither new nor untested (Markowitz 1977; Schmidt & Markowitz 1977). (See Turkkan et al., this volume.)

SPECIFIC RECOMMENDATIONS FOR HOUSING AND CARE OF GUENONS

1. Guenons live in social groups. The most stimulating thing you can possibly add to the environment of a guenon is a conspecific.

2. If conspecifics are not available, try the most logical species available as a companion. Guenons are very flexible about polytaxa societies. Any combination has to be carefully watched in order to recognize problems. Although they readily form mixed-taxa groups, addition of animals to an existing group is likely to result in severe aggression and injury.

3. The composition of the group should be appropriate to the species (Bramblett & Coelho 1987). A multi-male multi-female species, should not be housed in one-male groups or male-female pairs.

4. If research design requires isolation or single animals, allow them a familiar hiding place. Visual barriers that allow the animal to hide are the ultimate escape mechanism for guenons.

5. Guenons are inquisitive and respond positively to novel materials in their surroundings. Objects do not have to be complex to be useful. For example, metal water bowls from their transport cages made a popular play item. Juveniles wore them like little tin army helmets during play. Mirrors provide a stimulating addition to the habitat. Guenons spend a great deal of time using the mirrors. They keep track of companions by watching their reflections and interact with self images. Objects for manipulation have high value. Something as simple as a handful of pebbles may provide hours of attention and manipulation. Captive guenons sometimes carry smooth pebbles in their cheek pouches for days, retrieving them for manual play.

They did not try to eat them, but seemed interested in manipulating and rubbing them with their hands. In an indoor cage, cover the floor with hay. The result is less need for cage washing and a tremendous improvement in the novelty and comfort of the animal quarters. (See Chamove & Anderson, this volume.)

6. Organize management routines around the animals' social roles. Even if your manipulation conflicts with the animal's role obligations (such as maternal defense of an infant), patience and compromise can produce a protocol satisfactory to both experimenter and subject in most cases.

7. Use positive reward techniquesto modify behavior. In this way one can avoid intrusive technology or isolation of the animal, at the same time enriching the life of caged primates.

8. Incorporate as many organic materials and structures as possible into the animals' housing. A real tree is far more interesting than one made of steel. It can be bent, broken, eaten, and enjoyed! In a few days, the remnants of the real tree must be replaced, and this has the happy effect of producing a new stimulus.

WHAT NOW?

The first recomendation is to appoint an ethologist to the staff of each major primate facility or laboratory. Make the ethologist responsible for monitoring each animal on a regular, systematic, and frequent schedule. In most cases ethologists are already at hand, but they should be explicitly integrated into all research with nonhuman primates; ethology should be just as fundamental a part of research design as nutrition or microbiology. This puts a professional in the role of advocate who can speak for the animals when necessary. With a proper background and given appropriate responsibility, the ethologist should be able to improve conditions within the limits of research design and resources. The primary benefit of such a resource person to the institution is improved quality control. Experiments will be less confounded by laboratory stress and behavioral pathologies. Any research program can benefit from regular observation of each animal, and if the observations lead to the correction of problems, both animals and scientists benefit.

A second recommendation is that we house anthropoids in social groups. The most stimulating, diverse, and biologically important addition that we can usually make to a captive animal is a companion.

A third recommendation is that we recognize the importance of mental wellbeing among standards of good laboratory care. For example, if having objects for manipulation means slightly greater fecal contamination of the cage, we must be willing to compromise hygiene. Natural materials are permeable to urine, but unless there is a clear health threat, we must allow natural elements in primate housing. In most standard laboratory cages today, the animal is forced to spend most of the day on a surface covered with urine and feces. It is ludicrous to feel good about hygiene just because the steel cage can be sterilized in a cage washer once a day! Yet current trends condemn housing in outdoor areas large enough to support grass in favor of "clean" concrete or steel.

ACKNOWLEDGEMENTS

This colony has been supported over the past twenty years primarily by the Department of Anthropology and the College of Liberal Arts of The University of Texas at Austin. Sharon Bramblett assisted with preparation of the manuscript. I would like to thank Susan Maki for stimulating my interest in laboratory enrichment.

— 2 —
What Is Psychological Well-Being? Can We Know if Primates Have It?

Roger K. Thomas and Rosanne B. Lorden

> Animals lack what we have gained, but we lack what they still have...With them we are still in the Garden of Eden; and with ourselves, the Fall is everyday (Fowles 1970, pp. 61-62)

Fowles wrote the above in his discourse on "relativity of recompense," his view that a sentient creature's happiness is relative to its awareness of its existence. In Fowles's opinion, animals are blissfully ignorant.

We will consider whether nonhuman primates are blissfully ignorant or whether they are aware of their existence. We assume that awareness of one's existence is a prerequisite to knowing or having "psychological wellbeing" and that two prerequisites for the "awareness of one's existence" are having adequate "consciousness" and "self-awareness." Before we can consider whether nonhuman primates have these capacities, we must consider the general status of such concepts. Hereafter, we will not use the "shudder quotes" (Putnam's [1981] useful phrase, p. 216) but the discomfort they indicate will remain.

Since previous regulations were limited to considerations of physical wellbeing, the introduction of regulations concerning the psychological wellbeing of nonhuman primates implies that a distinction between physical and psychological wellbeing is reasonable. Whether it is reasonable depends on the context. In the present regulatory context, making rules to provide for psychological wellbeing implies that something exists to be regulated. This needs to be examined, as does the scientific meaningfulness of a distinction between physical and psychological wellbeing.

ONTOLOGICAL AND EPISTEMOLOGICAL CONSIDERATIONS

Questions about the existence of psychological wellbeing as distinct from physical wellbeing call for a more general consideration of "existence" and "reality." These are ontological questions about what *is*. Inseparable from such questions are epistemological questions about what can we know about what there is. Ernst von Glasersfeld (1987) examined some of these issues in connection with Piaget's "radical constructivist epistemology." An important conclusion of radical constructivist epistemology is that we can never know what there is.

Ontological and epistemological beliefs have implications for theory and for choice of methodology. For example, does legitimate methodology include introspection, intuition, empathic introspective discourse, observational, correlational, and experimental studies of nonverbal behavior, neurophysiology, etc? All scientists have ontological and epistemological beliefs, although some may not realize important implications of their beliefs. We want to be clear about our views, because they guide what we write later about consciousness, self-awareness, and psychological wellbeing. For rhetorical purposes, if no other, we submit that our views are the ones that behavioral scientists ought to hold.

Ontologically, we believe there is a real world of physical entities that exists independently of perceivers. Epistemologically, we believe (1) that what one can know about the real world will be incomplete for logical, practical (e.g., economic and temporal) and theoretical reasons (e.g., implications of the uncertainty principle; Heisenberg 1958) but (2) that reasonable agreement can be reached by perceptive and knowledgeable persons about the parameters of the physical entities and their spatial-temporal relationships. "Perceptive and knowledgeable" recognizes that special knowledge may be necessary (e.g., learning electron microscopy to "see" a bacterium or knowing the theory of relativity in order to estimate the composition and dimensions of the universe). We also assume that one's knowledge of the physical world is constructed, in part, from one's prior knowledge and theoretical commitments (Kuhn 1970). Therefore, we combine a realist ontology with a constructivist epistemology (see Manicas & Secord 1983; Leary 1984). According to Manicas and Secord, "...it is precisely the task of science to invent theories that *aim to represent* the world" (p. 401; italics added), and this aim is best accomplished by the "...application of imaginative theory that simultaneously guides observations, analysis, and experiment" (p. 405). We agree with Leary (1984) that "...concepts and theories are best construed as constructive approximations or tentative mappings of the real world as opposed to 'real knowledge' in some simple objectivist sense" (p. 918).

With our beliefs stated, we now address the status of the concept of psychological wellbeing and its proper place within a scientific framework, for these are scientific questions and not matters of personal values or anthropomorphic projections.

PSYCHOLOGICAL-PHYSICAL WELLBEING AND MIND-BODY DUALISM

The distinction between psychological and physical wellbeing is an aspect of mind-body dualism, a topic with a considerable literature to call upon. The three contrasting beliefs about mind-body are: (1) only the mind exists: idealistic monism; (2) only the body exists: materialistic monism; or (3) both mind and body exist: dualism. We emphasize *belief*, because there is no way to establish the "truth" or logical priority of any of these views.

The range of human beliefs can be exemplified by those who believe in faith healing versus those who believe in administering rabies antitoxin in case of a bite by a rabid animal. Beliefs are influenced by one's knowledge as well as other factors, perhaps chief among them what one is most "confident" or "comfortable" with, whatever that means! Confidence and comfort as criteria of belief are not stated lightly, but we must leave this issue and move on.

Idealistic monism and dualism posit mind as a nonphysical substance. While there are many versions of these two views (see Churchland 1984), all agree that consciousness is nonphysical and does not reduce to the concepts or explanations of physical science; consciousness requires a new science and a new methodology. Positive evidence of a nonphysical mental substance is lacking, however, and there is no clearly articulated theory of mind-stuff, no statement regarding its elements, no laws of its operations or of its structural connections (Churchland, 1984).

We prefer the materialist view that mind reduces to body and the psychological reduces to the physical, in part, because:

> . . . there are many clear cases of physical causation but not one clear case of nonphysical causation... Most philosophers now agree that no argument has successfully demonstrated why mind-body causation should not be regarded as a species of physical causation (Fodor 1981, p. 132).

Further, we believe that a materialist view is more compatible with the scientific approach to evidence and reasoning and is the appropriate one for considering the psychological wellbeing of nonhuman primates.

There are several versions of materialism with different implications for concepts like consciousness, self-awareness, and psychological wellbeing. We will consider briefly four theories, drawing heavily on Churchland's (1984) and Fodor's (1981) accounts.

Radical behaviorists hold that all reference to mental causes can be eliminated from the language of psychology in favor of reference to an organism's responses to environmental stimuli. The radical behaviorist, according to Fodor, "...is prepared to dismiss references to mental causes, however plausible they may seem, as the residue of outworn creeds" (p. 250).

Logical or philosophical behaviorists acknowledge the existence of mental states but translate them into behavioral hypotheticals expressed in the language of observable stimuli and responses. Note that there might be an infinite number of behavioral hypotheticals pertaining to a particular mental state. For example, "The monkey has psychological wellbeing" (mental state) might be translated as: (1) If the monkey eats (sleeps, copulates, etc.), then it will do so in a way that is normal (statistically) for its species; (2) If cage space is sufficient, then the monkey's movements will be normal (e.g., not showing stereotypies); (3) Etc.

Central state identity theorists hold that mental states are identical with states of the brain. An identity theorist might ascribe psychological wellbeing to those nonhuman primates whose brain neurophysiology is sufficiently similar to a human's...on the assumption (questionable; see Concluding Remarks) that psychological wellbeing can be defined adequately for humans. A difficulty for central state identity theory is its specificity, that is, it casts its lot with the carbon-based nervous systems of Earth's fauna and fails to allow for other systems that might be capable of mental states, such as computers or a hypothetical alien's silicon-based nervous system (Churchland 1984).

Functional theorists avoid the problems of identity theory's specificity by emphasizing the "causal role" of mental states as opposed to structural commonalities among them. The analogy of computer software versus hardware is applicable here. Functional theorists define mental states in terms of "...their unique causal roles in a complex economy of internal states mediating sensory inputs and behavioral outputs" (Churchland, 1984, p. 36). Further, in a complete departure from logical behaviorism, "Functionalism construes the concept of a causal role in such a way that a mental state can be defined by its causal relations to other mental states" (Fodor 1981, p. 136). Also unlike logical behaviorism, functionalism is not reductionistic; it does not require the elimination of mentalistic concepts from theoretical explanations. While functionalism avoids the problems of identity theory's specificity, "The functionalists would not be disturbed if brain events turned out to be the only things with the functional properties to define mental states" (Fodor, p. 137).

We endorse the functionalist view of mental concepts. However, it is not clear how psychological wellbeing qualifies as a functional mental state.

Churchland chose "pain" (and Fodor chose "headache") to illustrate functional mental states. Consider Churchland's example of how pain causes both behavioral outputs and other mental states. "Pain...causes distress, annoyance, and practical reasoning aimed at relief; and it causes wincing, blanching, and nursing of the traumatized area" (p. 36).

We observe that there is an unbalanced relationship between pain and the three examples of mental states said to be caused by pain. Consider the following: (1) we can think of pain directly causing annoyance or distress or practical reasoning to gain relief but not vice versa; (2) pain is caused only by localized physical trauma, but annoyance, distress, or practical reasoning may be caused by a number of conditions not related to pain nor physically localizable; (3) although we won't make the case here, we suggest that mental states like annoyance are reducible to mental states, like pain, that are reducible to their physical substrates. Therefore, mental states like annoyance appear to be superordinate to mental states like pain.

Psychological wellbeing also appears to be a superordinate concept. Consideration of psychological wellbeing as a functional mental state awaits an adequate account of the sensory inputs, behavioral outcomes, or other mental states that define it. We believe that psychological wellbeing is too ill defined to serve as a scientific concept. However, in view of its spreading use, it would be silly to ignore the concept, no matter how ill defined it may be. A scientifically-acceptable definition might be developed by identifying a suitable subset of behavioral hypotheticals (see Logical Behaviorism above) but that would require a separate chapter. In order to move on, we will simply assume that the psychological wellbeing that is to be addressed via regulations would first be defined in a scientifically-acceptable way.

CONSCIOUSNESS, LANGUAGE, AND SELF-AWARENESS

We are assuming that consciousness and self-awareness are prerequisites for awareness of one's existence and wellbeing. These are mental states which, as scientific concepts, may be no better defined than psychological wellbeing, but there is a vast literature on consciousness on which we can draw. Natsoulas (1978) began his analysis of consciousness by distinguishing between $consciousness_3$ and $consciousness_4$. The subscripts refer to the third and fourth definitions in the 1933 Edition of the Oxford English Dictionary. Quoting Natsoulas:

> The main definiens of the fourth entry reads in full, "The state or faculty of being conscious as a condition or concomitant of all thought, feeling, and volition; 'the recognition by the thinking subject of its own actions or affections' (Hamilton)." This concept of consciousness amounts to a restriction of the "wide, colorless use" (Dewey, 1906) to which the third entry is addressed.

> Consciousness3 means simply being aware of something, anything, whereas I exemplify consciousness4 by being aware of, or by being in a position to be aware of, my own thought, feeling, volition, or other mental episode (i.e., event, process, or occurrent state).

Natsoulas (1983 1985) states that consciousness$_3$ is simply being aware of something, such as an internal state or environmental event. This subsumes activities such as perceiving red, hearing a noise, holding an event in short-term memory, and perhaps Segal's (1976) category of consciousness as synonymous with stimulus control of behavior. Consciousness$_4$ is being aware of being aware and includes such categories as intentions, abstractions, self-concepts, and thoughts.

Many have expressed views on the relationship between language and consciousness. In Carlson's (1987) recent account, "Consciousness...can be viewed as being synonymous with verbal processes" (p. 407) and "... the *sole* criterion for consciousness is verbal acknowledgement" (p. 403). "Private nonverbal processes are conscious if we can describe them...we are conscious of *external* events only if we can think (and verbalize) about them" (p. 407). Because even the most linguistically accomplished of the apes could barely provide such descriptions, it might appear that this view precludes nonhuman animals from having consciousness. Carlson also said,

> Although consciousness may be synonymous with activity of verbal mechanisms of the brain, we need not conclude that only humans are conscious. The evolutionary process is incremental: new traits and abilities build upon ones that already exist...Certainly your dog can learn to communicate with you. The fact that it can learn to tell you when it wants to eat, go for a walk, or play probably means that it is conscious, also, (p. 406).

However, Carlson did not equate consciousness with *communication per se,* noting that "Species-typical communications...probably do not fall in the domain of consciousness" (p. 404).

We maintain that consciousness$_4$ is a prerequisite for an organism's appreciation of its psychological wellbeing as distinct from its physical wellbeing. We will now consider some of the evidence for consciousness$_4$ in nonhuman animals.

Self-Awareness in Nonhuman Primates

The first reference should be Donald Griffin's book, *The question of animal awareness* (1976; 1981), because it has made him the best recognized authority on the subject. However, Mason's (1976) incisive review of the first edition found Griffin's examination of the title question to be unsatisfactory. Because Griffin's book has influenced, perhaps, too many uncritical readers, it is worth quoting quote extensively from Mason's review:

> What bothers me most about this book is not the cautious tone of these rather inoffensive conclusions, but the fact that so much time is spent building up to them - defending propositions that do not require defense...By now we should need only the briefest reminders that animals have knowledge of their environments, that such information is processed, stored, and retrieved, that animals are sensitive to the consequences of their actions, that they are capable of intentions and expectancies...one might hope that some useful distinctions could be drawn between "having a mental experience" (that is, being aware of one's internal state or of the external environment) and "being aware of having such an experience"...he has not explored such questions systematically in relation to the problem of animal awareness at either the substantive or the theoretical level (p. 930)."

Natsoulas (1978) also questioned whether Griffin had addressed awareness in animals at the level of $consciousness_3$ or $consciousness_4$. Natsoulas suggested that the closest Griffin came to considering the distinction was when he discussed "intention" and "It is not clear whether Griffin was suggesting that the function of an animal's intentional behavior involves $consciousness_4$ or merely $consciousness_3$" (p. 141).

In the second edition of his book (1981, p. 27) Griffin acknowledged Mason's review but only to make a minor point, not to answer the principal objections Mason had raised. Griffin also cited Natsoulas but, again, in a secondary way. Based on Griffin's "Summary and Conclusions," we think he would agree that the animal literature still did not permit one to assert clearly whether any animal performances reflect $consciousness_4$.

Self-recognition: The most direct tests of self-awareness that have been reported involve recognizing oneself in a mirror. The procedure was introduced by Gallup (1970) who subsequently asserted that it was a test of self-awareness and consciousness (Gallup 1977 1982). Animals given access to mirrors typically respond initially as they do toward *other* members of their species, but after a few days of exposure to the mirrors, chimpanzees (*Pan troglodytes*) and orangutans may respond (some don't) with self-directed behavior (Gallup, 1982). Gallup also introduced a test in which dye was placed on an eyebrow ridge or ear while the animal was anesthetized. The dye marks could be seen by their bearers only in the mirror, and self-directed responses to the dye marks, aided by the mirror, were taken as evidence of self-recognition. In less formal testing, the Rumbaughs have obtained videotaped evidence of a chimpanzee examining the inside of his mouth with the aid of a flashlight and a real-time video image. The chimpanzee gave every sign of being aware that it was his own mouth he was viewing in the video monitor. (See photograph taken from the videotape in Rumbaugh Washburn & Savage-Rumbaugh: this volume. Ed.)

Gallup (1982) also reported *failures* of self-recognition by 11 species of monkeys, two species of gibbons, and gorillas. The report of failure by gorillas was supported by three experimental studies. William McGrew, during formal discussion of a conference presentation by Gallup (1986; attended by RKT), objected that the gorillas tested were in zoos and were, as a result, abnormal. Gallup responded that he had no bias in the matter, that he hoped a successful experiment with gorillas could be conducted, and that he was merely reporting the available data. In any event, the evidence for self-recognition is limited to chimpanzees and orangutans. (But see Bramblett: this volume, for descriptions of sophisticated mirror use in vervet monkeys. Ed.)

Robert (1986) investigated the development of mirror responses in an infant chimpanzee and an infant orangutan. She concluded that the chimpanzee's responses were similar to those of children of the same age, while the orangutans's were similar to those of younger children. It should be noted that the orangutan was 2 years old at the time of the study, whereas the chimpanzee was 7 months. While there was some indication of self-directed behavior, the results were negative for the dye-mark tests of self-recognition. Nevertheless, given that the ape infants behaved toward the mirror as did human children of roughly comparable age, and that adult chimpanzees and orangutans show self-recognition, it is reasonable to suggest that the development of self-awareness in the apes might have progressed beyond that revealed by the mirror tests. Nonetheless, there is as yet no compelling evidence that a chimpanzee or an orangutan might contemplate its psychological wellbeing.

"BEING IN A POSITION TO BE AWARE..."

Natsoulas (1978) suggested two criteria for $consciousness_4$, (1) "...being aware of, or...(2)...being in a position to be aware of, my own thought, feeling, volition, or other mental episode (i.e., event process, or occurrent state)" (p. 140). The second criterion provides a reason to ask whether nonhuman primates show the kinds of cognitive processes that would be essential to $consciousness_4$ and self-awareness. Owing to time and space constraints, we will address only a few of the many topics on cognitive processes in animals that might be addressed. We will begin with Carlson's (1987) view that consciousness is synonymous with verbal processes, because it is related to the Premacks' (1983; Premack & Premack, 1983) findings that language training is essential to successful performance on some tests of abstraction. The Premacks cited several examples of tasks used with language-trained and nonlanguage-trained chimpanzees. We will take issue with their principal example, abstract "same/different" judgments.

"Same" and "Different" Judgments

The importance of such judgments was indicated by Henry Nissen (1958) who wrote:

> ...all reasoning reduces to three processes: responsiveness to identity and to difference, and thirdly the balance or relative weight given to each of these...The term reasoning is usually reserved for higher level integrations - that is, for response to similarities and differences among units...All class concepts require simultaneous responsiveness to identities among the members of the class, and to differences between them and members of other classes. The balance between the two we may call sagacity; "judgment" might be an even better term (p. 194).

Premack (1983) distinguished between *same* and *different* judgments based on appearance and those based on abstraction. Using a matching-to-sample paradigm and letters to symbolize objects, one may illustrate the distinction as follows. An appearance-based task might involve presenting AA and BC and reinforcing responses to AA when "same" is correct or reinforcing responses to BC when "different" is correct (changing objects frequently to preclude memorizing the objects; e.g., CC vs. DE, FF vs. GH, etc.). An abstraction-based judgment, following Premack's example, might involve "XX as sample with YY and CD as alternatives or XY as sample with CD and BB as alternatives" (p. 128).

The Premacks (1983) argued forcefully that the successful performance of abstract same and different judgments demands a form of analogical reasoning that requires language. However, they either overlooked or discounted the report by Smith King Witt and Rickel (1975) of the performance of the chimpanzee, Casey, on a similar abstract problem. Smith et al. concluded that Casey "...learned a generalized form of the sameness-difference matching from sample concept" (p. 471) despite never having been "...previously trained on any type of learning task" (p. 469).

Burdyn and Thomas (1984) gave squirrel monkeys programmed training that led to conceptual judgments of same and different. The cues to choose exemplars of same were exemplars of the concept "triangularity" and the cues to choose exemplars of different were exemplars of "heptagonality." Because the triangle and heptagon cues in no way resembled same and different, the Premacks presumably would agree that the judgments involved abstraction. The squirrel monkeys also performed successfully when an interval as long as 16 seconds (for the best monkey) was introduced between the withdrawal of the triangle or heptagon and the presentation of exemplars of same and different. Successful performances under such circumstances meant that the monkeys were able to use conceptual symbols as mediational cues; that is, despite their physical removal, the cues served a time-bridging function. Whether the programmed training with these symbols in the same-and-different tasks may be construed as language training is unresolved, but Burdyn and Thomas's (1984) and Smith et al.'s (1975) studies raise questions about the Premacks' (1983; Premack & Premack, 1983) contention that language training is necessary for abstract representation.

Nevertheless, it is impressive that the Premacks' most experienced chimpanzee, Sarah, often performed new problems posed to her immediately and without special training. Burdyn and Thomas's squirrel monkeys had hundreds of trials of programmed training before reaching criterion on the final task, and Smith et al.'s chimpanzee also had extensive programmed training. Undoubtedly, Sarah's extensive experience with many kinds of verbal and conceptual problems facilitated her performances on new tasks. Her "education" may be compared to that of humans. Opposed to the Premacks' view that language training is essential, we suggest that what's essential is adequate "instruction" on the contingencies involved in new tasks. One may ask whether the Premacks' chimpanzees that failed on the same-different tasks had sufficient training on the relevant reinforcement contingencies.

Piaget's Theory Applied to Nonhuman Primates

Piaget's theory addresses cognitive development, the extent of which is relevant to the capacity to be aware of one's psychological wellbeing. Although Piaget's research was done with children, the theory was meant to apply to nonhuman animals as well (e.g., Piaget 1971). Piaget's theory posits four major periods in cognitive development. First is the *Sensorimotor Period,* which extends from birth to about 24 months in humans and involves the initial development of the child's concepts of space, time, causality, object permanence and so on (cf Phillips 1975, Piaget & Inhelder 1969). Second is the *Preoperational Period,* from about 2 to 7 years. According to Phillips (1975):

> The essential difference between a child in the Sensorimotor Period and one in the Preoperational Period is that the former is relatively restricted to direct interactions with the environment, whereas the latter is capable of manipulating symbols that represent the environment (p. 62).

Except for transitivity, which will be considered below, relatively little research explicitly on the Preoperational Period has been done using nonhuman animals. However, there is a large and growing body of evidence that animals use concepts that represent environmental entities or events (e.g., Thomas 1980).

The Preoperational Period is followed by the *Concrete Operational Period.* Again, according to Phillips:

> Since birth, the dominant mental activities of the child have changed from overt actions (in the Sensorimotor Period) to perceptions (in the Preoperational Period)...the Concrete Operational child conserves quantity and number, constructs the time and space that he will live with as an adult, and establishes foundations for the kind of thinking that is the identifying feature of the next and final period of his intellectual development, Formal Operations (p. 117).

Conservation of quantity and number is representative of the Concrete Operational Period. Several studies have investigated conservation by nonhuman primates. We will consider these below.

The Period of Formal Operations was summarized by Piaget and Inhelder (1969) as follows:

> The great novelty of this stage is that by means of differentiation of form and content the subject becomes capable of reasoning correctly about propositions he does not believe, or at least not yet; that is, propositions that he considers pure hypotheses. He becomes capable of drawing the necessary conclusions from truths which are merely possible, which constitutes the beginning of hypothetico-deductive or formal thought (p. 132).

To the best of our knowledge, there has been no research to substantiate the occurrence of formal operations in nonhuman primates.

Now we will examine briefly some studies with nonhuman primates concerning object permanence, transitivity, and conservation concepts representative of the Sensorimotor, Preoperational, and Concrete Operation Periods, respectively.

Object permanence: The most studied Piagetian concept in animal research (see Thomas & Walden's review, 1985) is "object permanence." According to Piaget (e.g., 1954), the child does not realize initially that objects have independent existence. The development of that knowledge occurs in six stages extending to about 24 months. Because all the animal studies report success in the first three stages, we will begin our description with stage 4. A typical stage 4 task would have the subject observe an object being hidden in location A and then moved to location B. The subject that has not yet developed the concept of object permanence will search for the object in location A. In stage 5, the subject performs stage-4 tests successfully but fails to search appropriately when the object is hidden from view (e.g., covered with a cloth) while being moved, especially if more than one move is involved (e.g., from A to B to C). In stage 6, the subject performs successfully even when the object is hidden from view and moved more than once.

At the time of Thomas and Walden's review (1985) investigators had reported the partial development of object permanence for birds, hamsters, woolly monkeys, and a stumptail macaque. Complete development of object permanence had been reported for cats, dogs, squirrel monkeys, cebus monkeys, rhesus monkeys, a gibbon, chimpanzees, and gorillas.

Redshaw (1978) compared the development of gorillas and humans on 14 developmental tasks including tests of object permanence. She concluded that the gorillas were consistently ahead of the humans, completing the tasks at a mean age of 43.5 weeks versus the humans' mean age of 54 weeks. Wood Moriarity Gardner and Gardner (1980) compared two chimpanzees, ages 18 and 30 months, with humans, ages 8, 18, and 24 months, on tests of object permanence. The 18-month human and chimpanzee performed comparably, but the 24-month human performed slightly better than the 30-month chimpanzee.

Cornell (1978) raised methodological and theoretical questions about the test procedures typically used with humans (procedures that have been adapted relatively directly for use with nonhumans) and Thomas and Walden raised further methodological objections to the animal studies. A rigorous interpretation is that none of the studies with animals or humans is entirely free from methodological objections.

Most recently, Natale Antinucci Spinozzi and Poti (1986) , who also questioned much of the earlier work, devised new methods to avoid the flaws as they saw them and tested a Japanese macaque and a gorilla, both 22 months old. They concluded that the gorilla showed the fully-developed concept of object permanence but that the Japanese macaque showed development only through stage 5. The gorilla's development of the concept was said to be comparable to that reported for humans.

Transitivity: Transitivity, an example of cognitive functioning in the Preoperational Period, involves giving the subject the information that a certain relation holds between A and B and between B and C and testing whether the subject can infer that the relation holds between A and C (e.g., $A > B$ and $B > C$ implies that $A > C$). Menzel (1969) trained and tested seven adult chimpanzees in ways that permitted him to distinguish between a "logical" and a "statistical" solution. Briefly (but see Menzel for his thorough examination of the issues), Menzel tested the chimpanzees on tasks with logically complete information (permitting transitive inference) and logically incomplete information (where the apes could only learn the reinforcement contingencies associated with particular pairings, AB, BC, etc.). The statistical solution was applicable in both tasks and the data indicated that the chimpanzees' performance in both cases was consistent with a statistical solution. Menzel concluded, "The evidence for an inferential rather than learned basis of ordering must be judged inconclusive...(and)...it would appear unnecessary to invoke logical transitivity to account for their performance" (p. 488).

McGonigle and Chalmers (1977) tested squirrel monkeys using a nonverbal test of transitivity based on one used with children by Bryant and Trabasso (1971) and obtained results similar to those reported by Bryant and Trabasso. However, McGonigle and Chalmers described a "binary (statistical) decision model" (somewhat comparable to Menzel's "statistical" solution) that might account for the results as opposed to Bryant and Trabasso's "coordination model" (somewhat comparable to Menzel's "logical" solution). Gillan (1981) studied transitive inference in chimpanzees using procedures comparable to McGonigle and Chalmer's (1977) and concluded that his findings supported an "integration" interpretation (somewhat comparable to Menzel's "logical" solution or Bryant and Trabasso's "coordination model") as opposed to a "nonintegration" one (somewhat comparable to Menzel's and McGonigle and Chalmers's "statistical" interpretations). Although Gillan included a number of control measures and concluded that McGonigle and Chalmers's binary decision model could not account for the results (Gillan did not cite Menzel), he misrepresented McGonigle and Chalmers and did not use some of the tests they did which help distinguish between the logical and statistical solutions.

Conservation: Important in the Concrete Operations Period are conservation tests which assess whether the subject knows that quantity has not changed despite changes in appearance. A typical example is to show the subject two identical vessels containing equal amounts of liquid: the pretransformational display. While the subject observes, the liquid is then poured into two new vessels, one tall and narrow and the other short and wide: the posttransformational display. A nonconserving subject usually indicates that the tall, narrow container has more liquid. Alternatively, two identical rows of objects (e.g., M & M candies) are shown as the pretransformational display. Then, without removing any candies, one row is shortened: the posttransformational display. A nonconserving subject judges the shorter row to have less candy.

Czerny and Thomas (1975) using squirrel monkeys investigated the ability of animals to make conservation-related judgments, and Thomas and Peay (1976) studied conservation of length by squirrel monkeys. Pasnak (1978) studied prerequisites to conservation using macaques (*Macaca mulatta*). Woodruff Premack and Kennel (1978) investigated the chimpanzee Sarah's ability to conserve liquid and solid quantity. Muncer (1983) studied two chimpanzees' abilities to conserve liquid quantity and number of M & Ms. While at least some of the subjects in each experiment appeared to perform successfully, Thomas and Peay (1976) noted the principal problem that makes all the studies inconclusive. Briefly, the problem is whether animals that correctly judge the stimuli in the postransformational condition to be "equal" or the "same" in quantity do so because they are good "perceptual estimators" (see Gelman 1972) of equivalence or because they use the concept of conservation. Despite the careful controls used by Muncer (1983), Thomas and Peay (1976), and Woodruff et al. (1978), these two possible bases of correct responding cannot be distinguished absolutely. Thomas and Walden (1985) concluded that it may not be possible to determine whether a perceptual or conservational solution was used in the absence of the subject's verbal explanation. They noted that none of the language projects with nonhuman animals has developed the linguistic skills an animal would need to explain its responses so that we could decide whether they were governed by perception or conservation.

Summary

The evidence whether nonhuman primates are "...in a position to be aware..." that is, show evidence of self-awareness and higher cognitive functions such as abstraction is, at best, inconclusive. Self-recognition (which is not equivalent to being aware of one's psychological wellbeing) has only been demonstrated for a few chimpanzees and orangutans, not all of those tested. Abstract representation (measured by sameness-difference judgments) has been demonstrated for chimpanzees and monkeys but such evidence does not demonstrate a capacity to be aware of one's psychological wellbeing. The assessment of cognitive development within a Piagetian framework is fraught with methodological questions which when rigorously applied preclude any conclusion about the state of a nonhuman primate's cognitive development.

CONCLUDING REMARKS

Psychological wellbeing is too ill defined to be used as a basis for regulatory action. We believe that there is no compelling reason to distinguish between psychological and physical wellbeing. If psychological wellbeing were defined adequately, regulations pertaining to it would likely be regulations pertaining to physical conditions. Whether a list of meaningful physical conditions, other than those already specified in existing regulations, can be constructed remains to be determined. We do not deny the possible relevance of social conditions, simulations of natural habitat, and so forth, but there should be scientifically valid evidence that such things matter to the animal before they are included in the regulations.

Psychological wellbeing can be compared to concepts like "freedom." Legislation can be drawn to protect freedom, but what does it address? It addresses physical activities such as speaking, writing ("freedom of expression") and attending the church of one's choice ("freedom of religion"), etc. In each case, there are behavioral (physical) correspondents.

Relatively well-educated and suitably motivated humans have the only chance of understanding what psychological wellbeing might mean to a human, much less a nonhuman animal. Other than certain physical assurances (e.g., adequate food and water and habitats that provide safety, sanitation, and thermoregulation), how could legislation be drafted to insure the psychological wellbeing of humans? Even within the relatively narrow culture of, say, Caucasian Americans, what regulations could assure the psychological wellbeing of one who prefers life in the country versus one who prefers life in a big city? Humans have hardly determined what psychological wellbeing means to us. The pitfalls of anthropomorphic projection by a human of what psychological wellbeing might mean to a nonhuman primate should make such projection prohibitive.

Despite the cognitive achievements of nonhuman primates, there is no compelling reason to believe that they have the capacity to be aware of their psychological wellbeing. Yet, we do not overlook the possibility of such capacity, especially in the great apes. In view of the uncertainty that remains, we close with the sixth stanza from Judson Mitcham's (1985) "Notes for a prayer in June"...

> Before our eyes, that heavy old coin disappears,
> while it stays where it is. We're aware
> sundown is a lie now, though we see it
> the same way. Chimpanzees, I have read,
> might pause from their foraging or play,
> sit quietly and gaze into the west
> till moved by darkness.
>   Perhaps, in their eyes,
> nothing seems magical, or it all does. From us
> comes a forced final nod toward the sleight
> of relentless method, how it turns
> pure mystery to laughter in the end.
>   Still,
> we know there's a magic we begin with, tricked
> by love's act into this ruled world.

— 3 —
Resolving Issues of Psychological Well-Being and Management of Laboratory Nonhuman Primates

Kathryn Bayne, D.V.M.

PROBLEMS IN DEFINING PSYCHOLOGICAL WELL-BEING

A recent Congressional amendment to the Animal Welfare Act (1985) mandates a physical environment to enhance the nonhuman primate's "psychological well-being." Psychological well-being, however, is an ambiguous expression. *Webster's Ninth New Collegiate Dictionary* (1987) defines "well-being" as "the state of being happy, healthy or prosperous." To prosper, Webster's says, is to flourish or thrive, which, in the case of animals, must refer to good health. Since any assessment on our part of the happiness of a nonhuman primate could only be subjective, I will therefore focus in this paper on the part of the definition concerning health, specifically mental health.

Many ways to assess well-being in animals have been proposed, including reproductive success, body weight, physical condition, physiological indices of stress like cortisol and ACTH, and behavioral assessments. Whether such measures in fact show how to optimize the environments of nonhuman primates is unclear, however. For example, stereotypic behavior is considered undesirable in any primate management program, but that does not mean that the absence of stereotypies is *prima facie* evidence of mental or psychological health. No single measure of well-being is likely to apply to all nonhuman primate species in all circumstances. Therein lies the difficulty in implementing the amended Animal Welfare Act.

EXISTING USDA AND USPHS GUIDELINES

Existing guidelines specify minimum standards to assure the physical well-being of laboratory animals, including nonhuman primates. The Animal Welfare Act regulates the use of animals at public and private research and educational facilities. *Public Health Service policy on humane care and use of laboratory animals* governs the use of animals at facilities funded by the USPHS and relies extensively on the recommendations in the *Guide for the care and use of laboratory animals* (*Guide*: NIH 1985).

Biomedical scientists and veterinarians tend to equate the parameters required for physical health with those required for mental health. Thus, it is assumed that if the recommendations in these publications succeed in safeguarding the animals' physical health, a positive contribution is also made to the animals' mental health.

APPROACHES TO ENVIRONMENTAL ENRICHMENT

Species-Typical Activities

In general, investigators take two approaches to improving the well-being of nonhuman primates through environmental enrichment. The first approach attempts to increase the time an animal spends in species-typical activities (activities recorded in free-ranging populations; for example, see Anderson & Chamove 1984; Reinhardt Houser Cowley & Champoux 1987; and Westergaard & Fragaszy 1985). This approach assumes that any enrichment strategy that increases normal activities will concurrently reduce behavioral pathology. The form environmental improvement takes may include opportunities to engage in such activities as social interaction, exercise, and manipulation of objects and devices introduced into the home cage. Techniques being employed to bring about beneficial changes in the activity budget include social housing (in pairs or groups), increasing the variety of staple foods offered daily, providing small between-meal treats, introducing cage manipulanda, or adding novel forms of stimulation (e.g., radio music) to the colony room.

Quality of Human-Animal Interaction

The second approach to environmental enrichment is to improve the quality of interaction between animals and people. The animal care technicians who are trained to care for the animals, the laboratory technicians who perform experimental procedures, the investigators who design the research protocols, all have the potential to exert a profound impact on the quality of life of the nonhuman primates in their charge. A similar benefit might also be realized by employing individuals expressly to feed treats to the animals, interact with them, or play games with them, such as taking turns manipulating a cage object.

SCIENTIFIC NEEDS VERSUS THE NEEDS OF THE ANIMALS

A problem to be faced in implementing any enrichment strategies is the variety of needs and restrictions placed on animal housing by the people who work with the animals. There are three main groups to consider: (1) scientists, (2) veterinarians, and (3) animal care technicians. Each of these groups interacts with the animals differently and their reasons for those interactions also differ (e.g., tuberculin testing versus transfer to a testing room), as do the frequency and amount of time they spend with the animals. Some people see only anesthetized animals while others work with fully conscious animals. Thus, the need for hands-on accessibility to monkeys differs.

DEVELOPMENT OF THE NIH MANAGEMENT PLAN FOR NONHUMAN PRIMATES

The National Institutes of Health (NIH) have initiated a program of research to develop ways to measure and to improve the psychological well-being of nonhuman primates. The goal is to develop a management plan with both short-term and long-term options for environmental enrichment, tailored to subsets of animals according to their housing condition (e.g., isolator units, single cage in a colony room, pair caging, group caging). An important component of the research program is the statistical documentation of any benefits derived from a particular enrichment strategy, including how long benefits continue, and the category of animals benefitted (e.g., by age, sex, weight, previous rearing history). The current practice of simply noting whether a particular enrichment strategy is or is not being used by the animals will be expanded to routinely include detailed observations of the types of behavior that increase or decrease when, say, a particular enrichment device is placed in an animal's cage.

INTERVIEW SURVEY

The initial step in the development of the NIH Plan was to assess the flexibility of nonhuman primate housing requirements by surveying the scientific and support personnel working with these animals. The survey was conducted by interview, each interview lasting about one hour. Table 1 presents a sample of the survey questions, which were intended to assess the diversity of research requirements at NIH. The interviews took place in the fall of 1987. In all, 56 investigators and five attending veterinarians, representing ten Institutes, were interviewed. The survey issues were discussed fully in each interview.

REPRESENTATION BY INSTITUTE

Table 2 summarizes the proportional representation of each Institute in the survey sample. The differences accurately reflect proportional differences in the number of individuals working with nonhuman primates in the different Institutes.

REPRESENTATION BY PRIMATE SPECIES

The survey generated information about the percentage of each species among the nonhuman primates at NIH. Table 3 shows that the rhesus monkey (*Macaca mulatta*) is the most frequently used species; squirrel monkeys (*Saimiri sciureus*) are second and cynomolgus monkeys (*Macaca fascicularis*) third. These three species make up about 80% of all nonhuman primates used at NIH.

Whereas Table 3 shows the proportional numbers of animals of each species at NIH, Table 4 shows the percentage of scientists using each species. Approximately 78% of the investigators at NIH who use nonhuman primates study just two species, *M. mulatta* and *M. fascicularis*. Many neuroscientists and behaviorists stated during the interviews that they preferred the rhesus monkey as a research subject because of its intelligence and aggressiveness. They found these qualities helpful both in training the monkeys and in maintaining their interest in experimental tasks.

The survey results showed that a wide variety of housing for nonhuman primates exists at NIH. Rhesus monkeys are maintained in single cages that house either single adults, or pairs or groups of infants; in juxtaposed single cages, each housing a single animal but with a removable partition between them to permit controlled visits; in isolator cages housing single adults or paired juveniles; and in a variety of group enclosures – corn cribs, modified dog runs, and a two-acre corral.

Table 1

Sample Survey Questions for NIH Investigators

- o What kinds of tasks do the animals engage in during the experiment? Briefly characterize the use of the animals.
- o What age animals do you use?
- o What species do you use?
- o What long-term planning could be implemented at the source of your animals and at the current facility that would benefit the animals and your research?
- o How are your subjects currently housed?
- o Do you foresee problems with pair/group housing of current stock in the future?
- o How do you measure the psychological well-being of your subjects?
- o Do you consider your animals' environment suitable?
- o Do you consider their environment enriched?
- o What suggestions do you have for cage improvements?

Table 2

Percentage of Interviewed Investigators from Each Institute

National Institute of Mental Health	42.9%
National Institute of Child Health and Human Development	20.6
National Cancer Institute	6.3
National Eye Institute	6.3
National Institute of Allergy & Infectious Disease	6.3
National Instituteof Neurological & Communicable Disorders and Stroke	4.8
National Institute of Aging	4.8
National Health Lung & Blood Institute	3.2
National Instituteof Dental Research	3.2
Clinical Center	1.6

Table 3

Percentage of Each Species of Nonhuman Primate Used at NIH

Macaca mulatta	61.5%
Macaca fascicularis	12.1
Saimiri sp.	11.3
Callithrix sp./*Saguinus* sp.	6.8
Aotus trivirgatus	4.5
Pan troglodytes	2.2
Macaca arctoides	1.3
Cebus apella	0.3

Table 4

Percentage of NIH Investigators Surveyed Who Use Each Species of Nonhuman Primate

Macaca mulatta	58.1
Macaca fascicularis	20.3
Saimiri sp.	6.8
Callithrix sp/*Saguinus* sp.	5.4
Macaca arctoides	2.7
Cebus apella	2.7
Aotus trivirgatus	2.7
Pan troglodytes	1.3

Each of these housing arrangements presents its own advantages and disadvantages with regard to well-being. Certainly, animals housed in groups have the benefit of dynamic social partners, but they also face the hazards and stresses of dominance hierarchies. The stresses of social living may be greater for captive monkeys than for their wild conspecifics because confinement prevents them from moving a safe distance away from the center of the group. (But see other chapters in this volume for suggestions on redesigning group caging to provide visual barriers for privacy and escape from conflicts.) Furthermore, dominance hierarchies may make it difficult to assure equal access to enrichment devices and furnishings.

Animals housed alone, on the other hand, while they have limited auditory or visual contact with other animals, can be given frequent access to people and enrichment devices. It is these animals that may benefit the most from environmental enrichment.

RESULTS FROM INTERVIEWS WITH NIH SCIENTISTS

Singly-housed animals are of particular concern at NIH because the survey indicated that approximately 83% of adult captive primates at NIH are caged alone. However, the survey revealed that in 95% of cases, singly-housed animals are removed from their cage at some time during the study. Fifty-four per cent of those studies require regular removals, one to two times each week; and 36% of removals involve animals who are awake to experience the novel stimuli along the route to the test room. Many scientists pointed out that their animals receive an optimal diet and medical attention, indicating that their physical needs are met per *Guide* standards.

Based on the standards for animal care and housing set forth in the *Guide* (1985) and the most recently promulgated USDA regulations (1978), 62% of the investigators believed their animals' current housing was suitable. However, 52% indicated their willingness to house their primates in pairs or groups, whether or not they were doing so at the time.

How long a captive primate is active on a research protocol that requires individual housing could influence the choice of enrichment methods for the animal. The survey indicated that the average nonhuman primate participates in a research study for 9.6 months, indicating that there is a place for relatively short-acting enrichment methods. Thus, even though it would be less useful for long-term studies, a strategy that was demonstrated to be beneficial for only a few months before the animal lost interest in it could be helpful for animals used in short-term studies.

Twenty-nine per cent of the studies surveyed require training animals. All of these studies used positive reinforcement in training. Only about 9% of the investigators surveyed require their animals to be fed a restricted diet while on study or in training. There is, therefore, ample opportunity for dietary enrichment without compromising research.

Of the 56 scientists surveyed, 54 believed their primates' research environments were already enriched by the standard of the day, often through participation in the research itself, which typically had a behavioral component (manipulating devices, viewing targets on computer screens, etc.). Many scientists indicated that the frequent opportunity to leave the home cage added substantial variety and interest to the animals' lives.

More than half (57%) of the scientists interviewed were already making some effort to assess the psychological well-being of their nonhuman primate subjects, although they were not monitoring behavior in a formal or objective way. Informal assessments relied on body weight, quality of pelage, alertness, and overall attitudes. Table 5 summarizes these informal methods of assessment. A few scientists said they did not feel qualified to make such assessments, suggesting there is a need for more systematic analysis by personnel with behavioral expertise. In response to this identified need for behavioral assessment of individual animals, the NIH Plan will call for trained individuals whose responsibility will be implementation and assessment of environmental enrichment programs.

An important survey question was what the investigators thought could be done to the home cage environment to improve their animals' psychological well-being. The scientists work closely with their animals and rely on having normal animals as experimental models. Their ideas may have particular utility because they know well the constraints under which their subjects live and work.

Their suggestions appear in Table 6. The most frequent recommendations were for larger and more complex cages and for more varied cage furnishings. Next came recommendations for companionship, and variety in diet and ambient stimulation. Taken as a whole, the list encompasses many of the suggestions proposed by the experts represented in this volume.

Table 5

Methods Used by NIH Investigators to Assess Psychological Well-being

- o Physiological measures
- o Performance during experiment
- o Mood changes
- o Behavioral abnormalities (stereotypies, self-abuse, masturbation, coprophagy, urine-drinking)
- o Level of aggression
- o Facial expressions
- o Quality of hair coat
- o Response to people
- o Dominance over cage neighbors
- o General behavior normal for species
- o Appetite and body weight
- o Quality of vocalizations
- o Attention to surroundings
- o Immunologic status
- o Consistency in behavior

Table 6
Enrichment Suggestions by NIH Investigators

Increase size of cage/more vertical space	20.2
Apparatus, toys, rope, swings	17.0
Perches, shelves (including use of wood)	10.8
House in pairs or groups	7.7
Vary diet	7.0
Natural lighting	3.9
Provide television, music	3.9
Use non-metallic materials for cage	2.3
Decrease number of animals in a room	2.3
Increase size of rooms	2.3
Provide central exercise area	2.3
Improve quality of caretaker interaction with animals	2.3
Paint walls in colors	1.6
Wire mesh walls to increase visibility of cage neighbors	1.6
Central social area	1.5
Control over environment	1.5
Provide animals with hiding places	1.5
Improve cage washing system	1.5
Improve watering system	1.5
Provide nesting material for pregnant females	0.8
Flexible barriers between animals	0.8
Padded cages for recovery from anesthesia	0.8
Mirrors, house across from each other	0.8
Provide time outdoors	0.8
Feed throughout day	0.8
Improve room lighting	0.8
Increase predictability of daily routine	0.8

Implementing these suggestions presents difficulties, however. First, if a facility is currently accredited or is seeking accreditation from the American Association for the Accreditation of Laboratory Animal Care (AAALAC), any alteration to the cage environment must be acceptable to AAALAC. Any object to be introduced into an animal's living cage or into the colony room must meet *Guide* and USDA standards.

Second, enrichment strategies must not interfere with experimental design. For example, food treats cannot be given to animals on restricted diets; subjects must remain accessible to the scientist and veterinarian; and certain enrichment strategies (e.g., social housing) should not be combined with certain experimental procedures (e.g., tethering, used by 16% of the scientists). Therefore, the selection of an enrichment strategy must be made with due consideration to the living and working environment of the animal.

Finally, no object can be introduced into an individual cage that would obstruct its squeeze panel (but note that many non-obstructing devices are available). As only 9% of scientists interviewed currently train their animals to go into transport cages, the majority use squeeze panels to transfer animals or to restrain them at the front of the cage (e.g., for administration of daily medication). When more animals have been trained to freely enter transport cages, it will be possible to introduce a greater variety of toys, manipulanda and other furnishings into their cages without concern for obstructing the squeeze mechanism.

RESULTS FROM INTERVIEWS WITH VETERINARIANS

The veterinarians had several suggestions for improving psychological well-being. Four indicated that social housing would be an acceptable means of enhancing the well-being of compatible animals. Two recommended perches and manipulanda as cage enhancements. A complete list of their suggestions appears in Table 7.

Current attempts to enhance the well-being of primates at NIH include providing cages with nylon balls (Nylaball: Nylabone Corp., P.O. Box 127, Neptune, NJ 07753), a food delivery apparatus, manipulanda, and cage furniture (e.g., nest boxes and perches for New World monkeys), as well as (where they are not ruled out by other priorities) social housing, varied diet, and placement of animals in a room where workers in the building can visit them, thereby increasing their contact with people and the overall stimulation in their lives.

It became apparent from the survey that the veterinarian of a large facility cannot closely monitor the success of enrichment strategies and the mental status of all the nonhuman primates in his or her care when the number is very large -- up to a thousand animals in one case -- and the responsibilities for care of the animals' physical health are demanding.

Four of the veterinarians surveyed try to assess the psychological well-being of the nonhuman primates in their care. The signs they attend to are listed in Table 8, with self-mutilation and stereotypies being the most common veterinary concerns.

Table 7

Recommendations by NIH Veterinarians for Housing Changes

Social housing
Perches in the cage
Manipulanda in the cage
Puzzle feeders on the cage
Train animals to enter transport cage freely
Exercise area
Visual privacy barriers for group-housed animals
Computerized watering system for isolator cages

Table 8

Signs of Psychological Well-being Utilized by NIH Veterinarians

Self-mutilation
Stereotypies
Over-reaction to a stimulus
Distress calls
Physical appearance
Appetite

The veterinarians interviewed had some concerns about potential housing changes. Animals in their charge must often be handled (e.g., to be anesthetized), and for this the veterinary staff needs ready access to individual animals. This question can be resolved by proper training of personnel in capture and restraint techniques or by redesigning caging to allow individuals to be isolated from the group. Several chapters in this volume provide examples of training animals to accept temporary isolation from the group for experimental purposes, and even, in one case (Turkkan Ator Brady & Craven: this volume), tolerating a cuff for blood pressure readings. Requiring personnel to enter group cages to capture individual animals carries with it, however, the risk that animal care technicians will be bitten or scratched. These types of wounds are of particular concern where disease transmission is a possibility, including the potentially fatal Herpesvirus simiae (Herpes B virus). Veterinary, as well as scientific, input into novel cage design is important for heading off hazards to both animals and personnel.

Schmidt and Markowitz (1977) pointed out, in connection with enrichment of zoo habitats, that adding behavioral engineering devices to group housing may make it easier to detect illnesses in the animals. Such devices can increase the amount and variety of activity that cage occupants display, making it easier to detect an animal whose activity has fallen off abnormally– a change that is more difficult to notice in less active animals.

Two of the veterinary respondents voiced concern over the cost of re-caging an entire building. One veterinarian believed the animals' needs for exercise and manipulative opportunities could be met with inexpensive alterations to existing cages.

SELECTING ENRICHMENT STRATEGIES

Because nonhuman primates live in a variety of housing conditions and take part in studies with very different requirements, there is no one solution to the question of enrichment; rather, a spectrum of enrichment strategies will be needed. Many researchers are testing a variety of ways to improve the well-being of nonhuman primates (e.g. Anderson & Chamove 1984; Clarke Juno & Maple 1982; Hearn & Dixson 1984; Westergaard & Fragaszy 1985). These investigators come from differing fields and publish their results in many different specialty journals, making it difficult for people with day-to-day responsibility for the conduct of research and animal care to stay abreast of all the work they might find useful. The University of Washington Primate Information Center publishes a monthly newsletter of current primate references as well as bibliographies on this and related topics. Dr. Scott Line of the California Primate Research Center and Dr. Melinda Novak of the University of Massachusetts have also compiled bibliographies on psychological well-being of captive primates.

Even so, an informal forum for exchange of up-to-date information on primate well-being is needed. An electronic conference system or computer bulletin board, readily accessible to a large number of institutions across the country, could be used to exchange the latest results – positive or negative – of research on environmental enrichment; and practical suggestions for enrichment would be immediately available to interested individuals.

CONCLUDING REMARKS

It is an accepted working premise in biomedical research that the research animals used as experimental models must be healthy, yet the means of producing and maintaining healthy animals are less clearly defined. Guidelines for physical well-being of laboratory animals are established, but the impact of these guidelines on psychological well-being is unknown.

NIH research represents the typical diversity of areas in which nonhuman primates are used as subjects. The animal care technicians, laboratory technicians, and veterinarians who support the scientists must understand the scientific requirements of each study and the special needs of the nonhuman primates used in that research. The individuals working with nonhuman primates are acutely aware of the potential difficulties in resolving scientific requirements and the psychological needs of the animals.

The National Institutes of Health are committed to achieving a better understanding of the physical and psychological needs of captive primates and how to meet them. The NIH intramural plan for managing nonhuman primates will be based on sound research, a continuing dialogue among NIH personnel, and exchange of information with extramural sources.

II. SUBORDER: PROSIMII
Family: Lemuridae (Lemurs) Indriidae (Indris)

— 4 —
Maintenance of Primates in Captivity for Research: The Need for Naturalistic Environments

Michael E. Pereira, Joseph M. Macedonia
David M. Haring and Elwyn L. Simons

ABSTRACT

The physiology and behavior of nonhuman primates comprise biological systems that are coadapted to particular physical and social environments that are normally present throughout development. Maintenance of wild-caught and laboratory-born primates in small, individual cages precludes social interaction and other experiences likely to be essential to regulation of physiology and behavior in these animals. Just as inbreeding can destroy much genetically-controlled developmental homeostasis, so maintaining intrinsically gregarious animals like primates in isolate caging is likely to engender idiosyncratic physiologico-behavioral response patterns. The validity of biomedical or behavioral research conducted with such subjects remains indeterminate without appropriate control procedures, viz., comparisons of individually-caged subjects with animals living in species-appropriate social groups and environments. Because the Duke University Primate Center is dedicated to the breeding and conservation of prosimian primates, all colony members have always been housed in social groups. For some lemur species, however, even housing animals socially in large indoor rooms outfitted with naturalistic substrata has proven inadequate to promote normative behavior. For sifakas (*Propithecus verreauxi*), providing them with outdoor living space produced dramatic improvements in activity patterns and reproductive effort. The endpoint of our husbandry program for lemurs is placement of social groups in large natural habitat enclosures to prepare them for the possibility of reintroduction to protected natural ranges. With the completion of our newest enclosures, the Duke Primate Center has provided nearly 100 lemurs, members of seven endangered species, more than 11 hectares (28 acres) of forest in which to acquire skills in climbing and traveling in trees, foraging, and avoiding predators. Simultaneously, the welfare of individual colony members remains our top priority while the most advanced research on the behavior and reproductive biology of understudied, endangered primates is facilitated. Such research should ultimately promote future reintroduction efforts. The husbandry and research done at the Duke Primate Center, therefore, are humane and valuable for the primates involved, for both the individuals and their populations.

PRIMATE NATURAL HISTORY AND BIOMEDICAL RESEARCH

In designing facilities for maintenance of nonhuman primates in captivity, it is essential to realize that primates are social animals. Naturally-occuring primate social groups frequently number between 50 and 100 members (Altmann & Altmann 1970; Kummer 1968; Rowell 1972), and some number well over 100 (Baldwin & Baldwin 1971; Poirier & Hu 1983). Thirty-nine of the 54 primate genera (Richard 1985), or 70% of extant forms, coalesce into social groups that include two or more unrelated adults and their infant and juvenile offspring. Even primates formerly thought to lead "solitary" lives, like orang-utans (*Pongo pygmaeus*) and bushbabies (*Galago* and *Euoticus* spp.), are now known to be genuinely social (see, e.g., Clark 1978; Rijksen 1978; Harcourt & Nash 1986). Much remains to be learned about the structure of primate societies and the nature of primate social interaction through study of these species.

The primary adaptations of primates have evolved in environments that include complex social interactions within groups and populations of broadly stable composition. In virtually every species studied, maturing individuals develop strong, long-term affinitive relations with their mothers, siblings, and unrelated conspecifics between birth and adulthood. In most Old World monkeys, for example, infants are nurtured intimately by their mothers for about a year. In the matrix of the social group, infants explore and play; their mothers retrieve them from occasional "mistakes" they make in interaction with both social and physical features of their environment. Hours of sleeping and nursing are accompanied daily by maternal grooming. Through observation during infancy, youngsters learn their family's social status (Berman 1980 1982ab; Altmann 1980; Pereira 1988b), what and how to eat (Watts 1985; Whitehead 1986; Pereira 1988a), and how to detect and respond to predators. For two to four years after weaning and before puberty, juveniles further develop relations with their mothers, siblings, and peers, and initiate important relationships with unrelated group members (Pereira & Altmann 1985). Affinitive interactions, including greetings, grooming, social play, and sleeping associations, balance the agonistic interactions and competition that accompany group living (Judge 1982;

Smuts 1985; de Waal 1986). For primates, normative behavioral development is a life-long process that depends on continuous input from myriad ecological and social stimuli (Pereira & Altmann 1985).

Consider now that the vast majority of biomedical research using primates evaluates data gathered from individuals confined in tiny, barren, isolated cages. Subjects are taken from their social groups in the wild as immatures or adults while others are born into laboratory cages. Laboratory-born primates usually begin life with their mothers in small isolation cages, and are separated from them before the normal age of weaning (cf Altmann 1980; Berman 1982ab). They spend their early juvenile years either alone or with a small group of peers in other barren cages, and then spend the rest of their lives in a cage less than one meter tall that offers about three-quarters of a square meter of floor space. These are the current space recommendations in the NIH *Guide for the Care and Use of Laboratory Animals* (USDHHS 1985) for proper management of baboons in laboratories.

Under such conditions, nonhuman primates cannot express normal patterns of physiology or behavior. Indeed, behavioral biologists would predict that subjecting primates to these conditions must produce multifarious aberrancies in both systems. This prediction derives from the axiomatic premise that normal primate physiology and behavior are systems coadapted to particular physical habitats and social environments invariably present throughout natural development (Figure 1). A primate deprived of regulatory environmental stimuli should thus be expected to exhibit atypical features in its baseline and responsive physiology and behavior. A critical conclusion emerges from this argument that must be addressed both by the scientific community and by society: valid biomedical research cannot be conducted on primates or other social species in existing laboratory environments. Because we must suspect the normalcy of physiologico-behavioral responses in any social organism maintained in restrictive, abnormal environments, the validity, and thus value, of any research based on such subjects is questionable. Fox (1986a) cogently expands this argument, reviewing generalities and heuristic examples for many species commonly used in animal laboratories.

The 1985 NIH *Guide* states that good biomedical research must consider the effects of environmental factors on the wellbeing and biologic responses of research animals. The *Guide* also warns, however, that proper management includes genetic and environmental control to minimize variations in animals' responses to experimental regimens. Unfortunately, this principle has traditionally been misinterpreted to mean that extensive inbreeding and minimal, sterile, isolate housing will minimize any baseline variance in physiology and behavior among research subjects. The presumption concerning inbreeding and variability is, in many cases, demonstrably false: phenotypic variance in laboratory rodents, for example, often increases with degree of inbreeding (Russell & Burch 1959; Lane-Petter & Pearson 1971; Festing 1979). Variability in the physiology and behavior of social mammals probably also increases with the degree of isolation and restriction imposed by their captive environments. Just as inbreeding destroys much genetically-controlled developmental homeostasis, so must maintenance of intrinsically gregarious animals in small, isolate cages generate a multitude of idiosyncratic physiologico-behavioral response patterns by precluding essential regulatory experience.

A vast literature exists describing effects of housing style and group composition on the physiological functioning of laboratory animals. Over the past 30 to 40 years, research has related differences in captive social environments to differential development of brains, musculoskeletal anatomy, neuroendocrine systems, immunological systems, behavior, and reproductive performance (Fox 1986a reviews much important literature; see also the pertinent discussion by Fagen [1981] of related research). At the 1983 public hearing in Chicago preceding revision of the NIH *Guide*, Pereira and Altmann (1983) provided the Institute of Laboratory Animal Resources' (ILAR) Committee with a bibliography of 150 articles selected from this literature. Each article documented changes effected by social or other environmental factors on aggression, reproduction, infanticide, pituitary-adrenal or pituitary-gonadal function, resistance to chronic physical ailments or infectious disease or response to drugs or toxins in laboratory animals. The most surprising feature of the available literature reviews is the revelation that relationships between captive environments and physiologico-behavioral functioning have been virtually ignored for primates (see review in Nash & Chilton 1986). This has occurred despite primates' status as the most valuable animal models for biomedical research (but see Newell-Morris & Fahrenbruch 1985, for primatologists' criticism of Cercopithecinae and other primates as models for human biology).

In response to present concerns, the ILAR committee recommended that "when appropriate, group housing should be considered for communal animals" (USDHHS 1985, p. 12), but concluded that "there is little objective evidence for defining adequate care in relation to social environment. The data are limited and contradictory and lack sufficient substance to establish absolute recommendations" (p. 12). Other readers of the literature believe a more extensive response is needed to the current ratio of knowledge to ignorance regarding these issues (e.g., Pereira & Altmann 1983; Fox 1986a). Lack of uniformity in effects of housing on physiology and behavior detracts in no way from the import of that vast body of research. Features of biomedical laboratories and research methodology that destroy or thwart homeostatic mechanisms should be expected to *generate* variability, within and across species. Additional variability is to be expected in results emerging from different laboratory environments and from species that have evolved vastly different types of sociality and ranging. The degenerate responses often produced by chronic isolate caging even in animals that are less social than primates (see literature cited in Fagen 1981; Pereira & Altmann 1983; Fox 1986a) foretell likely severe effects on species that naturally exhibit pronounced gregariousness and complex forms of social organization.

In the remainder of this chapter, we review recent research conducted at the Duke University Primate Center that illustrates the impact that housing style can have on primate behavior and physiology, and thus on the validity of research using captive primates. We highlight a study that revealed effects of indoor versus outdoor housing on the activity budgets and reproduction of our sifakas (Indriidae: *Propithecus verreauxi*). We conclude by discussing critical implications for researchers studying primates in more restrictive settings, and suggested guidelines for improved housing for primates maintained for biomedical and behavioral research.

The Duke University Primate Center

The Duke University Primate Center is dedicated to the breeding and conservation of prosimians and to training conservationists and researchers for work with these species. Prosimians comprise seven of the 14 Primate families: Lemuridae and Lepilemuridae (true lemurs), Cheirogaleidae (mouse and dwarf lemurs), Indriidae (indris, avahis, and sifakas), Daubentoniidae (aye-ayes), Lorisidae (lorises and galagos), and Tarsiidae (tarsiers). Extant lemuriform species (first five families mentioned above) are found only in Madagascar and all are considered endangered. Many are now seen only occasionally in their homeland (Jolly 1986; Pollock 1986a), and several are not in captivity anywhere other than at the Duke Primate Center (Tattersall 1982; Pollock 1986b). The Duke Primate Center has an unrivaled record of captive survival and breeding with several of the species most difficult to rear in captivity, and we now maintain over thirty prosimian taxa, including about 720 individuals. The real threat of extinction for several prosimian species over the next 10 to 15 years has led to international recognition of the Duke collection as a critical second line of defense against the prospect of extinction for prosimian primates (Tattersall 1982; Jolly 1986; Pollock 1986ab; Richard 1982).

Since the 1960's, the Duke Center has also been a focus of research into the behavior and reproductive biology of prosimians. Because Madagascar was virtually closed to outside researchers between 1972 and 1982, research on lemurs at Duke took on additional importance. Information on olfactory behavior, learning, mother-infant relations, reproductive biology, vocalizations, metabolic rates, development, and social behavior has come from research with the Duke colony (for recent work, see Foerg 1982; Hoffman & Foerg 1983; Glander & Rabin 1983; Izard & Rasmussen 1985; Macedonia & Taylor 1985; Rasmussen 1985 1986; Taylor & Sussman 1985; Ganzhorn 1986 1987; Izard & Simons 1986; Macedonia 1986 1987; Wright Izard & Simons 1986; Wright Toyama & Simons 1986; Pereira & Simons 1986; Cherry Izard & Simons 1987; Gebo 1987; Kappeler

1987ab; McNab & Wright 1987; Pollock & Mullin 1987; Pereira Klepper & Simons 1987; Rasmussen & Izard 1988; Haring in press; Pereira Seeligson & Macedonia in press).

At the Duke Primate Center, however, the survival and reproductive success of every colony member have always been of first concern. Consequently, all colony members have been housed under naturalistic conditions and no research procedures more invasive than withdrawal of blood samples and occasional biopsies have been allowed. Surrounding the main building are more than 50 outdoor runs or cages made of chain-link or lighter-weight fencing that enclose between 25 and 250 cubic meters of space and numerous branch substrata. Most runs have a top fence in lieu of a roof to allow the family group of lemurs to experience natural patterns of sunlight, wind, and rain. Each run also has one or more elevated boxes containing shielded heat lamps to provide refuge from the most severe North Carolina weather.

The lemurs in our outdoor runs provide a unique resource for comparative research on prosimian behavior and reproductive biology. We maintain several different social groups of each species, providing diverse opportunity for research within and across taxa that are together in one place only at the Duke Center. Social groups of many lemur species are also maintained in multi-hectare forest enclosures (see below; also Taylor & Sussman 1985; Ganzhorn 1986; Macedonia 1986; Pereira et al. 1987; Simons 1986; Glander in press; Pereira Seeligson & Macedonia in press). While most large-bodied lemurs are maintained outdoors, our small-bodied mouse and dwarf lemurs, tarsiers, lorises, and galagos are held in social groups inhabiting indoor rooms outfitted with naturalistic substrata a(Izard & Pereira 1987).

DUKE PROGRAM FOR CAPTIVE CONSERVATION OF SIFAKAS

Since the inception of the Duke University Primate Center (1966, and before, when the collection was maintained at Yale University), the Center has been the only institution outside Madagascar to maintain and breed sifakas, obtaining our first sifaka from Madagascar in 1960. Sifakas are large (4-6 kg), diurnal lemurs of which two species are recognized: *Propithecus diadema*, which inhabits Madagascar's eastern rain forests, and *P. verreauxi* (Figure 2), which survives in the island's remaining western deciduous forests and southern spiny forests (Tattersall 1982; Pollock 1986a; Jolly 1986). Like other indriid lemurs (*Indri indri* and *Avahi laniger*), the diet of wild sifaka consists primarily of leaves, although considerable quantities of fruit, flowers, and tree bark are also ingested (Richard, 1978).

Also like the other indriids, sifakas are endangered primates that have proven difficult to maintain in captivity. Until relatively recently, all attempts to sustain and breed indriids in captive settings have had poor results. Because these lemurs exhibit specializations of the digestive tract for processing fibrous plant materials, it is possible that inappropriate diets have been responsible for the premature deaths of many captive individuals. A characteristic "wasting" syndrome, marked by progressive disinclination to eat, lethargy, and eventual death, has led to the abandonment of attempts to maintain three of the four species of Indriidae in captivity.

Beginning in the mid-1970s, mortality among Duke sifakas increased sharply, leaving the captive-born male as the colony's sole survivor. Because the sifakas population continues to dwindle in Madagascar, one of us (Simons), having taken over as Director of the Primate Center in 1978, established unique agreements with the Malagasy government, allowing the Center to obtain new sifakas in 1982 and again in 1984 and 1986. An intensive, stepwise program of husbandry was developed, designed to maintain high survivorship while gradually increasing the degree to which the colony's living conditions approximated natural environments. These measures were expected to improve activity patterns and breeding success.

At first, the sifakas were housed in large rooms (about 85 cu m). The amount of naturalistic substrata in these rooms, including tree trunks and branches, was increased to simulate more closely natural environments. Next, the sifaka diet was improved. An increased supply of mango leaves was obtained through a volunteer network in southwestern Florida and Miami. Later, we discovered that related lemurs (Lemuridae) readily consume leaves from a number of local plants, while rejecting plants that contain high levels of alkaloids (Glander & Rabin 1983; Ganzhorn 1986). Consequently, leaves from several local trees were added to the sifaka diet (see below). In 1985, additional changes in housing and diet led to radical improvements in the health, vigor, and reproduction of these animals.

Modifications of the Sifaka Diet

Because wild sifakas are known to eat mango leaves (E. Simons, personal observation), the Duke sifakas have been provisioned with mango leaves for a number of years. Preliminary nutritional analyses of mango leaves failed to reveal an unusual nutrient profile that might have allowed us rapidly to improve the composition of the sifaka diet (K. Glander, personal communication). In 1985, however, we began to introduce local North Carolina foliage to supplement the regular sifaka diet of mango leaves, cut fruit and vegetables, and monkey chow (Table 1; Haring, in press). Plants seen to be eaten by several lemurid species inhabiting two forest enclosures (Ganzhorn 1986) comprised the list of potential food species. Initially, only plant parts consumed by all lemurid species were offered to the sifakas. In addition to species preferences, seasonal preferences in the plant parts selected by our semi-free-ranging Lemuridae were followed in choosing local foliage for the sifakas.

At the onset of the program, two to three local plant foods were introduced each week. All plant material was first rinsed with water. Then terminal branch parts with their leaves were inserted into plastic water containers secured to tree trunks in the sifaka rooms. If a food was accepted, it became a regular part of the daily sifaka diet. Plant species initially rejected were offered again periodically. Some plants were only sniffed or nibbled at first, but later eaten in sizable quantities. Foods repeatedly rejected were dropped from the diet. During the summer and fall of 1985, four or five of the accepted plant species were presented fresh daily along with mango leaves, fresh cut fruit, and chow. Beginning in the spring of 1986, the sifakas were fed three to five of the previously accepted plant parts plus two to four new species each week.

Fig. 1: Social environment to which behavior & physiology of ringtailed lemurs (*Lemur catta*) are coadapted. Young adult female in center carries her 3-mo-old son while being groomed by her juvenile sister. In front of her are two offspring of her adult sisters; to the left are two adolescent males: her cousin & nephew. Ringtailed lemurs are remarkably socially cohesive in comparison to ruffed lemurs (*Varecia variegata*) and chimpanzees (*Pan troglodytes*). (Photo by MEP.)

Table 1

Diet of Duke Sifakas Prior to 1985

Fruits	Vegetables	Other Foods
apple	broccoli	Science Chow
avocado	cabbage	hard-boiled egg
banana	carrot	cottage cheese
grape	cauliflower	mango leaf
mango	celery	sweetgum leaf
melon	corn	
orange	cucumber	
peach	garbanzo bean	
pear	lettuce	
pineapple	sweet potato	

Table 2 shows that five North Carolina plant species have become dietary staples for the Duke sifakas, with an additional five species eaten on a seasonal basis (Haring in press). We found that naive individuals more readily accepted novel plant species when in the company of colony members previously exposed to those species (also Whitehead 1986). Captive-born infants and newly-acquired wild-caught adults sniffed at length those plant species that were unfamiliar to them but consumed by others in their group. The naive animals also persistently smelled the hands and mouths of animals that had consumed the new foliage.

In addition to providing local foliage during its growing season, the DUPC has begun to freeze leaves during the season of their acceptance for winter provisioning. At present, it appears that leaves of most of the tree species used undergo freezing and thawing with little adverse effect on palatability (Haring in press).

Modification of Sifaka Housing Environments

Early in the summer of 1985, the first attempt began to house a sifaka group outdoors. Previously, only one individual was housed outdoors, due to concern over exposure of these rare animals to potentially dangerous pathogens. That project terminated when the subject in fact became ill. Continued low activity levels and sub-optimal reproduction in the Duke sifakas, however, demanded that outdoor environments be established. In June 1985, a group of sifakas (Group 3) was moved to a large outdoor cage directly adjacent to the main building. Behavioral data gathered from this

Figure 2. Adult sifaka (***Propithecus verreauxi coquereli***) feeding in a forest enclosure at the Duke University Primate Center. (Photograph by DMH.)

Table 2
Local North Carolina Plants Offered As Food to Duke Sifakas in 1985/1986

Species Accepted	Part/Level
American hornbeam (*Carpinus caroliniana*)	lv / *
mimosa (*Albizia julibrissin*)	fr,fl,lv/*
Eastern redbud (*Cercis canadensis*)	sd,fl,lv / *
winged sumac (*Rhus copallina*)	fl,lv/*
sweetgum (*Liquidambar styraciflua*)	bd,lv / *
beech (*Fagus grandifolia*)	lv / +
blackberry (*Rubus* sp.)	lv / +
crabapple (*Malus angustifolia*)	fr / +
red maple (*Acer rubrum*)	fr,fl,lv / +
rose (*Rosa multiflora*)	fl,lv / +
boxelder (*Acer negundo*)	lv / o
blackgum (*Nyssa sylvatica*)	fl,lv / o
kudzu (*Pueraria lobata*)	lv / o
loblolly pine (*Pinus taeda*)	lv / o
tulip poplar (*Liriodendron tulipifera*)	lv / o
Species Rejected	**Part/Level**
alder (*Alnus serrulata*)	fl,lv
white ash (*Fraxinus americana*)	lv
catalpa (*Catalpa speciosa*)	lv
dogwood (*Cornus florida*)	fl,lv
elm (*Ulnus americana*)	sd,fl,lv
grape (*Vitis rotundifolia*)	fr,lv
greenbriar (*Smilax* sp.)	lv
hickory (*Carya* sp.)	lv
holly (*Ilex opaca*)	lv
honeysuckle (*Lonicera japonica*)	fl,lv
mulberry (*Morus rubra*)	lv
persimmon (*Diospyros virginiana*)	fr,lv
river birch (*Betula nigra*)	lv
sourwood (*Oxydendrum arboreum*)	fl,lv
sycamore (*Platanus occidentalis*)	lv
trumpet vine (*Campsis radicans*)	lv
black willow (*Salix nigra*)	fl,lv

lv=leaves; fl=flowers; fr=fruits; sd=seeds; bd=buds
* eaten throughout growing season
\+ eaten only seasonally
o eaten by only a few individuals

group (Group 3) were compared to data previously gathered from the same group while it was housed indoors and to data taken concurrently from other indoor sifaka groups. We predicted that the experimental setting would stimulate activity, including feeding, and lead to improved health and reproductive effort. Table 3 lists the activities studied.

Analyses of activity budgets revealed significant increases for Group 3 in time spent locomoting, playing, and feeding (all $p < 0.05$; Fig. 3), and a significant decrease in time spent inactive ($p < 0.05$; Fig. 4), in comparison to the concurrent activity budgets of four indoor groups (Macedonia 1987 gives details of the statistical analyses). Examination of correlations between frequencies of outdoor activities and ambient temperature, time of day, and degree of cloud cover revealed no effects of these variables.

One variable that could have influenced activity in Group 3 was enclosure size. The outdoor cage this group lived in measured about 234 cu-m, roughly three times the volume of their former indoor rooms (82 cu-m). Therefore, an investigation was undertaken in 1986 that focused upon possible effects of enclosure size in stimulating activity in captive sifakas. One group (Group 4) was provided access to an outdoor extension of their quarters that was considerably smaller (about 60 cu-m) than their indoor room. Members of this group could move freely between indoor and outdoor chambers.

Comparison of Group 4's activity budgets in the novel indoor/outdoor environment with their activity budgets the previous year (indoors only) revealed significant increases in time spent locomoting ($p < 0.005$) and playing ($p < 0.05$; Fig. 5) and a significant decrease in time spent resting ($p < 0.005$; Fig. 4). Increases in time spent feeding approached significance ($p = 0.10$). Most important, we found that well over 90% of time was spent in the outdoor compartment, despite its smaller size. That the larger, indoor chamber was not preferred suggests that, in the previous studies, enclosure volume *per se* played little, if any, role in the observed changes in activity budgets. The results suggest either that sifakas prefer outdoor over indoor housing or new over old cage space. Continuing predominant use of outdoor space during warmer months by sifakas in indoor/outdoor quarters fail to support the novelty hypothesis. Members of Group 4 spent more than 75% of their time indoors feeding upon food placed indoors or resting near heat lamps. Also, during the winter and early spring of 1987, a female sifaka with a young infant appeared to prefer her indoor room. Pereira et al. (1987) discuss the importance of behavioral thermoregulation for prosimian infants.

Table 3

Activities Quantified During Study of Indoor *Vs* Outdoor Activity Budgets in *Propithecus verrauxi*

Activity	Definition	Predict:Inside>Outside
resting	inactive, motionless	DECREASE
scanning*	visual explore/no locomotion	INCREASE
comfort move*	shift position,stretch,scratch	NO PREDICTION
hanging*	suspend motionless	NO PREDICTION
inactivity	pooling of all above	DECREASE
locomotion	vertical cling & leap bipedal hop or walk quadrupedal walk, climb, brachiate	INCREASE FOR ALL
feeding	ingest food or water	INCREASE
autogrooming	use tooth-comb, tongue, or grooming claw on own pelage	NO PREDICTION
allogrooming	groom groupmate as above	NO PREDICTION
playing	swing from rope, rapid ricocheting locomotion, alone or in chase, wrestle	INCREASE
aggression	lunge, cuff, bite toward other	NO PREDICTION
marking	throat-mark (male), anogenital mark "endorsing" (male marks over female's)	NO PREDICTION

* not subjected individually to statistical testing.

The Impact of Naturalistic Conditions

The value of fresh foliage and outdoor housing to captive sifakas is evident at several levels. First, the psychological wellbeing of these endangered primates clearly has been improved by the changes in husbandry described above. Evidence includes the significant reduction of lethargy, accompanied by significant increases in feeding, body weight, and diverse activity, notably including social play. Other studies have implicated play as an important behavioral index of physical and psychological wellbeing (Mason 1965; Harlow 1974; Baldwin & Baldwin 1976; Oliver & Lee 1978). In particular, rejection of play solicitations and failure to initiate play are early signs of distress or illness (observations by MEP on weanling savanna baboons, *Papio c. cynocephalus*, and juvenile ringtailed lemurs, *Lemur catta*; also Fagen 1981). Personal observations during our sifaka program have been as compelling as our limited graphs and statistics.

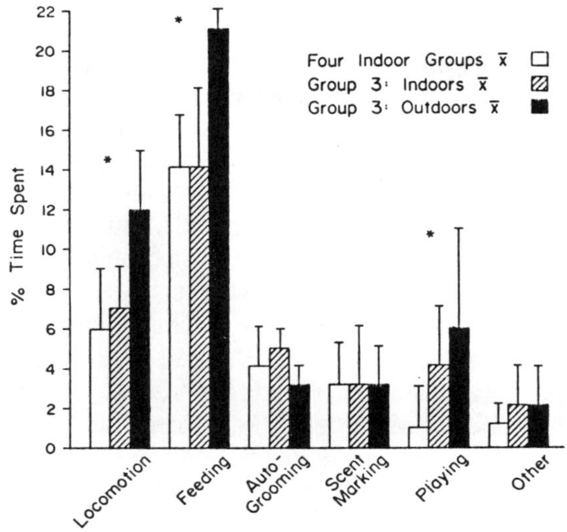

Fig 3: Time spent in 6 activity categories by members of 4 indoor groups and Group 3 (alternately held indoors and outdoors). Group 3 was not one of "4 indoor groups." Bars: group means; brackets: 1 s.d. Asterisks: significant differences.

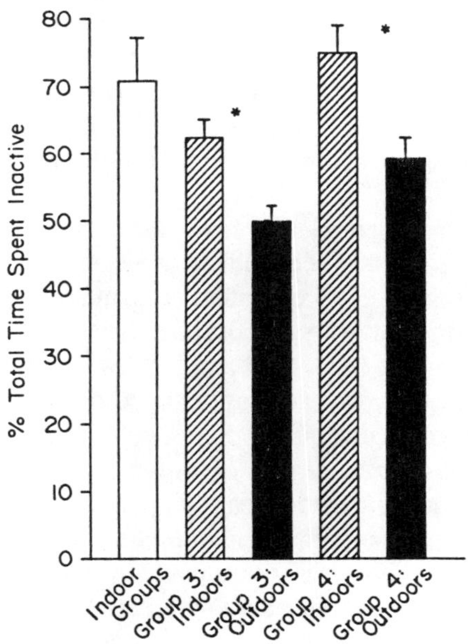

Fig 4: Percentage of time inactive: indoor & outdoor groups. Bars: group means; brackets: 1 s.d. Asterisks: significant differences.

Beyond the added weight and acrobatic play, each sifaka's generally diverse and alert behavior has left no doubt that its psychological wellbeing was improved dramatically by the naturalistic outdoor environment

Second, the long-term impact of the new sifaka management program rests in its ability to maintain good health in these animals while simultaneously boosting the success of their reproductive efforts. Sustained weight gains and a lack of gastrointestinal disorders and other ailments in the sifakas throughout 1986 and 1987 indicate improved health. Remarkably, five conceptions occurred in the sifaka colony between summer, 1985 and summer, 1987. These were the first large steps made toward establishing successful breeding in captive sifakas. Our results stand in sharp contrast to the three conceptions that occurred during the five years preceding the outdoor program (1975-1977 and mid-1982-mid-1985; the colony comprised just one individual between 1977 and 1982).

The ameliorated activity budgets generated by outdoor caging suggested that environments offering still fuller access to natural(istic) sunlight, weather and forage would further improve sifaka welfare and breeding. Therefore, we moved one pair of sifakas into our smallest forest enclosure (0.5 ha) for several months in the late summer and early fall of 1986. The female conceived during that period and gave birth to an infant that survived for four months, the longest period of survival for a sifaka infant conceived in captivity over the previous five breeding periods. Notably, this female and another maintained indoors the previous year produced stillborn infants that year (Table 4). In the spring of 1987, a pair was again placed in our small enclosure and another was introduced to a new 2.25-ha forest enclosure built especially for sifakas. In December 1987 and January 1988, the females of these pairs gave birth to new infants both of which still survive (Table 4). A fourth sifaka born recently at Duke (July, 1986) is now a young adult. Conceived in Madagascar, Sabina joined our colony *in utero*, and she and her mother have occupied one of our large indoor/outdoor complexes since their arrival.

Finally, maintaining our lemurs in forest enclosures allows us to prepare individuals for future efforts to reintroduce these endangered primates to natural ranges in Madagascar. The Duke Center began maintaining *Lemur* social groups in forest enclosures in 1981, when the 0.5-ha enclosure was constructed (Simons 1986; Glander in press). In 1983, 1986, and 1987, we added a 3.5-ha, a 2.25-ha, and a 5.5-ha enclosure, respectively, each containing thousands of trees. Because the Duke Forest is a mix of hardwood and pine species (e.g., *Pinus taeda, P. echinata, Quercus stellata, Cercis canadensis, Liquidambar styraciflua, Liriodendron tulipifera*), our enclosures provide areas comparable in both size and structure to the natural home ranges of lemurs. Conflict due to partial dependence on concentrated provisions is virtually eliminated by delivery of prepared foods to several randomly-selected feeding stations located 75 to 250 meters apart in each enclosure. All the lemurs travel about their "island" habitat with their social groups, supplementing visits to feeding stations with extensive foraging on endemic flora.

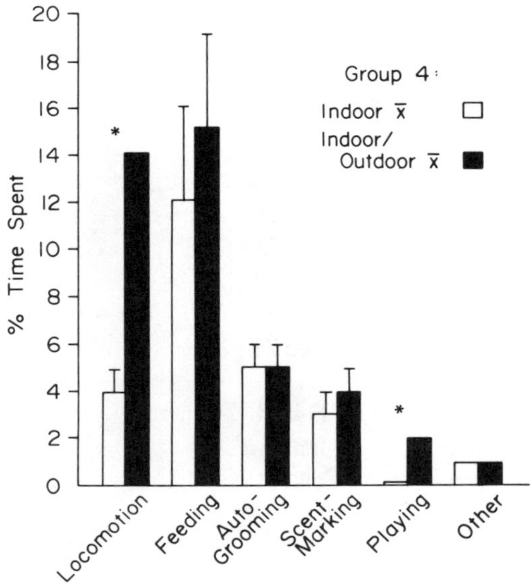

Fig 5: Comparison between indoor & indoor/outdoor study periods for Group 4. Means without brackets: s.d.'s < 1.0. Asterisks: significant differences.

In establishing lemur groups in the Duke Forest, we exchange the great constraints but total security of a cage for the much greater freedom but associated dangers of a semi-wild environment (e.g., raptorial predation, falling from trees). The Duke enclosure projects, however, are an unrivaled success story in captive conservation research with endangered and little-known primates (our four enclosures currently hold eleven groups comprising seven species; Simons 1986). Their naturalistic habitats have promoted mating and effective infant care in every species introduced (Glander in press). After six years, for example, the first group placed in the forest has grown from eight to 31 members. Simultaneously, unsurpassed opportunities have been made available for advanced research into prosimian behavior and reproduction (e.g. Macedonia & Taylor 1985; Taylor & Sussman 1985; Ganzhorn 1986 1987; Macedonia 1986; Pereira & Simons 1986; Taylor 1986; Pereira et al. 1987; Pereira Seeligson & Macedonia in press).

Table 4
History of Sifaka Reproduction at the Duke University Primate Center

Breeding Pair	Infant	Birthdate	Conceived: Captivity or Wild	Housed: Indoors or Out	Lifespan
1970-1977					
Charlotte/Reginald	Charlotte-A	4 May 70	C	NR*	stillborn
	Catherine	31 Dec 70	C	NR	6.3 yrs
	Nigel	10 Feb 72	C	NR	16+ yrs
	Charlotte-B	25 Dec 72	C	NR	stillborn
	Charlotte-C	8 Mar 74	C	NR	stillborn
	Amber	18 Dec 74	C	NR	1.9 yrs
Edith/Sacheverell	Mango	21 Feb 74	C	NR	1.6 yrs
Edith/Nigel	Larke	26 Jan 76	C	NR	0.9 yrs
	Erme	6 Feb 77	C	NR	0.5 yrs
1982-1987					
Theodora/WM	Theodora-A	26 Jun 84	W	I	0.1 yrs
Calpurnia/WM	Calpurnia-A	27 Jul 84	W	I	stillborn
Julia/Augustus	Julia-A	19 Jan 85	C	I	stillborn
	Julia-B	30 Dec 85	C	I	stillborn
Cornelia/Trajan	Cornelia-A	18 Dec 85	C	I	stillborn
	Claudius	5 Dec 86	C	O	0.5 yrs
	Aurelia	8 Dec 87	C	O	LIVING
Flavia/WM	Sabina	7 Jul 86	W	O	LIVING
Octavia/WM	Octavia-A	30 Jun 86	W	O	stillborn
Octavia/Hadrian	Tiberius	20 Jan 88	C	O	LIVING

*: No records are available describing housing regimes for sifakas prior to tenure of ELS as Director of the Primate Center.
W: Conceived in wild C: Conceived in captivity WM: Wild male

Now about 90% of our enclosure animals were born in the Duke Forest. To sample many of their preferred (natural) foods they climb readily to the tops of the highest trees about 30 meters above the forest floor. In fact our forest-living lemurs exhibit skills in foraging and responding to predators as well as in climbing and traveling arboreally that rival those of lemurs in Madagascar. Acquisition of such skills is, of course, essential to success in future efforts to reintroduce lemurs to natural ranges in Madagascar. We are currently developing new agreements with Malagasy officials that will pave the way for reintroduction of captive-born lemurs to protected ranges, perhaps within the next decade.

CONCLUSIONS

Moving social groups of sifakas from generous and well-structured indoor rooms to comparable or even smaller outdoor cages radically changed their behavior. There is little room for doubt that the outdoor housing led to substantial improvements in their psychological as well as physical wellbeing; significant physiological adjustments must have accompanied the observed behavioral changes. Sifakas comprise an important case study of research on captive primates. They (and other rare mammals) have taught us that even relatively good conditions of captivity can impair physiological and behavioral functioning. Housed socially with generous and well-structured indoor space, they become lethargic, fail to reproduce, and sometimes die prematurely. In some way, outdoor living space critically enhances the performance of sifaka neurologic, endocrinologic, and/or gastrointestinal processes. (For effects of housing on brain structure and activity, see Welch 1964; Geller Yuwiler & Zolman 1965; Welch & Welch 1966ab 1968; Bell Miller Ordy & Ralston 1971; also Fox 1986a.)

Biological Dysfunction Under Conditions of Captivity

As yet, systematic efforts have not been made to reveal the biological dysfunctions caused in hardier primates (e.g., macaques) by standard laboratory conditions, conditions characterized by, but not limited to, life-long solitary occupation of caging not much larger than subjects' bodies. For many behavioral biologists, this realization comes as a shock. Scientists aim to reveal facts (or probable facts) by ruling out as many conceivable alternatives as possible through the use of control procedures or observations. Because it is typically impossible for one researcher to corroborate a behavioral study of captive primates with a concurrent or subsequent study of wild representatives of the species, most published research on social behavior is now done either in the wild or under the most naturalistic captive conditions possible. Nevertheless, results from even remarkably naturalistic captive conditions must be considered preliminary and should be accompanied by the caveat that true field work is needed further to evaluate the significance of results obtained under captive conditions (Pereira et al. 1987; Pereira et al. in press).

Validity of Biomedical Research With Captive Primates

An uncompromising demand from the international research community that appropriate efforts be undertaken to validate the results of biomedical research on primates is long overdue. In primates, or any other animal lineage, physiology and behavior have been co-evolving for millions of years, resulting in myriad bidirectional relationships between these two systems. Because primates are born with an evolutionary legacy of profound and complex gregariousness, it is an inescapable, primary hypothesis for all research on captive primates that documented results substantially reflect idiomorphic responses to housing conditions. It is impossible for scientists working with individually-caged animals to interpret any experimental paradigm before identifying the underlying effects of such housing regimes on the systems that they study. Only by establishing and studying species-typical social groups as control subjects can the extent to which any type of biomedical research is compromised by restrictive, isolate caging be known.

Naturalistic Habitat for Laboratory Primates: General Principles

Alternatively, scientists could take the initiative in improving their research by establishing housing conditions that resemble subjects' natural physical and social environments. A presentation of detailed specifications for particular species is beyond the scope of this chapter; the *type* of change needed is at issue here. We need to begin attending primarily to the needs of subjects rather than only to the needs of researchers. For virtually all primates, physical environments should include branches and branch-like substrata, rather than only shelves and stainless-steel grating. For the most olfactorily-oriented primates (viz. prosimians and some Callitrichidae), substrates must *not* be washed too frequently lest behavioral pathology and skin lesions develop from the over-stimulated "need" to scent-mark. While reasonable rates of air exchange should be achieved, laboratories must not become acoustically uninhabitable. Members of most species exchange low amplitude "contact calls" that communicate individuals' identities and levels of arousal (Macedonia 1986; Pereira 1986); the effectiveness of antiphonal calling can be expected to interact with levels of behavioral disturbance, particularly for immatures. Social environments must include the conspecifics needed for subjects to experience a species-typical social group. Most species naturally exhibit an appreciable range in group composition;

researchers can exploit this natural flexibility for different laboratories and applications. In any group-living situation, however, primates must have opportunities to move periodically into temporary visual seclusion.

The resulting captive environments will have "texture," and regular rearrangement of their physical features is essential to help maintain freshness in that texture. Without such rearrangement (or periodic tranfer of groups among cages), even relatively good cage environments are likely to generate undesirable behavioral stereotypies (e.g., pacing and wheeling; Fox, 1986b) in some individuals. Environmental texture provides colony members opportunities to make choices and exert some degree of control over the quality of their moment-to-moment experience; these are options now known to be essential to psychological wellbeing in primates and other captive mammals (Markowitz 1982; Fox 1986a).

Advantages of Group Housing

Group-housing laboratory primates promises to improve, not jeopardize, the quality of biomedical research in many important ways. Not only will animals' physiological systems be much more likely to perform characteristically, but also researchers will be given opportunities to investigate the natural range of performance of their subject systems, rather than only mean responses of indeterminate validity. Because group-living subjects have opportunities to interact (and to choose not to), their researchers are able to gather otherwise inaccessible behavioral data to identify context-dependent changes and other features of natural variability in physiology. Normative variance must be known before deep understanding of any biological system can be achieved. Moreover, juxtaposition of genealogical records with long-term physiological and behavioral data enhances our understanding of the evolution of physiology, behavior, and the interaction of component systems (Sackett Gunderson & Baldwin 1982).

Housing primate subjects in social groups is not only likely to be necessary for good biomedical research; it is also practical. Outdoor housing for social groups produces demonstrably healthier, less expensive, and much less idiosyncratic subjects (Banerjee & Woodward 1970; Hoffman & Stowell 1973; Bernstein & Gordon 1977). Also, there are methods of housing primates economically in modest indoor social groups that would entail feasible transformation of almost any of the laboratory rooms whose walls are currently lined with rows of isolation cages (Hansen 1976 gives methods used formerly at Rutgers' Institute of Animal Behavior). The many scientists who have worked with group-housed primates at research centers in the U.S. and abroad have the expertise needed to advise laboratory personnel on housing animals more naturalistically while maintaining access to individual subjects (see O'Neill this volume). While some procedural changes will be dramatic, such changes are not always synonymous with loss of time or increased work. In fact, many methods long used in progressive primate laboratories reduce standard difficulties with physiological research -- difficulties that threaten animal health, technician safety, and data validity. Primate subjects, for example, can typically be trained to present appropriate body parts for vaginal swabbing, rectal exams, and even blood draws.

Whereas our primary objective in writing this chapter has been to illustrate how unlikely it is that either behavioral or physiological systems function normally in primates maintained in isolated, restrictive cages, we hope that we have also been able to provide a new perspective for those who have begun to question whether wild animals, endangered or otherwise, should ever be kept in captivity (Fox 1986b). We believe that work such as that done at the Duke University Primate Center is humane, both daily and over the long term, for the individual primates involved. Moreover, such efforts are likely to prove invaluable for the future of their species.

ACKNOWLEDGMENTS

We thank M. Philemon Randrianarijaona, Director of the Departement des Eaux et Forets in Madagascar, M. Joelina Ratsirarison, Division of Protection of Nature (DEF), Mme. Rakotosamimanana Berthe, Ministry of Higher Education, and M. Voara Randrianasolo, Parc Tsimbazaza, for helping to establish and maintain the lemur colonies at the Duke University Primate Center. Without their support, effective captive conservation of Malagasy primates would be impossible. We also extend heartfelt gratitude to the Wildlife Preservation Trust International, Ms. Margo Marsh, and Mrs. Jocelyn Alexander for sponsoring the creation of Natural Habitat Enclosure-3 for *Propithecus verreauxi* and *Hapalemur griseus* at the DUPC and to the Geraldine R. Dodge Foundation, Inc. for sponsorship of the special technician position to oversee our husbandry of these two endangered species. Taking this captive conservation program into its most important phase is an inspiring step toward preventing the disappearance of these beautiful primates. We express deep and sincere thanks also to the Mango Connection network of conservationist volunteers in Miami and Fort Myers, who have for years supplied our sifakas with the mango leaves they so relish. Andrea S. Katz, Supervisor of the DUPC lemur colonies, kindly criticized the manuscript as it developed. Finally, the expertise and endless commitment of the entire technical staff at the DUPC must be recognized for the vast amount of research that could never have occurred over the past many years without its support.

Family: Lorisidae (Lorises)
Tarsiidae (Tarsiers)
Cebidae (New World Monkeys)—*Aotus*

— 5 —
Psychological Well-Being of Nocturnal Primates in Captivity

Patricia Chapple Wright, David Marsland Haring,
Mary Kathryn Izard, and Elwyn LaVerne Simons

INTRODUCTION

Nocturnal primates including *Aotus* and *Galago* are being increasingly used as laboratory animals. There are approximately 3,000 owl monkeys and 500 *Galagos* in the US that are used primarily in biomedical research. Because of their large eyes relative to body size these night-active animals are often preferred for studies of vision. The importation of wildcaught animals for research has become increasingly difficult and inadvisable from a conservation point of view, and it is preferable to obtain captive-bred animals from breeding colonies for research needs. Any understanding of the social behavior and psychological needs of these night-active primates will produce better breeding results and healthier, better-adjusted subjects. There are over 30 species of nocturnal primates, many of which have never been kept in captivity. For this chapter we have selected three genera as case studies. Two (*Aotus* and *Galago*) are frequently used in biomedical research, and the third (*Tarsius*) has only recently been shown in zoo exhibits. We hope to explain how a knowledge of the behavior, diet and habitat preferences of these primates in the wild can give us useful insights into keeping these nocturnal primates healthy and breeding in captivity.

AOTUS

Natural History

Aotus, the world's only nocturnal monkey, is found in most forested areas of South America from Panama south to northern Argentina. Unlike most primates, owl monkeys live in monogamous pairs and the father is the primary caretaker of infants (Dixson & Fleming 1981; Wright 1984). One offspring is born each year, gestation is 133 days (Hunter Martin Dixson & Rudder 1979), lactation is 7 months, and the infant is carried on the father's back from the first week until four months of age (Wright 1984). In the wild, offspring remain with the family until three years old (Wright 1985), when they disperse. Establishment of territory and formation of the offspring's own family occurs in a neighboring area. *Aotus* family groups (usually an adult pair, an infant, a yearling and a subadult) defend exclusive territory of about 10 ha (22 acres). Males fight males from other groups, and females fight females, in territorial disputes at bordering trees (Wright 1985; Robinson Wright & Kinzey 1987). Owl monkeys are primarily fruit eaters, but they supplement their diet with both insects and leaves (Wright 1986).

Social Grouping and Reproductive Behavior

Much of the information gained from observations in the wild can be used to improve the breeding and psychological well-being of *Aotus* in captivity. Owl monkeys should be maintained in male-female pairs. Solitary individuals are stressed, but neither should adults be housed with other adults of the same sex. Neighboring groups will fight through cages, so to prevent injury, more than 6 inches should separate cages. Groups do best when not in visual contact with one another.

Fathers should never be removed from cages after birth, because mothers have been known to reject infants if fathers are not present to carry them. Enough room must be provided in cages so that the father with an infant on his back can easily maneuver from perch to perch without endangering the infant. Closely-spaced thin branches should be placed in the cage when the infants are a month old so they can learn to locomote on this "jungle gym". Owl monkeys in the wild are particularly active at dawn and dusk, a time when they enter and exit sleep trees. These twilight bouts of leaping and running also occur in captivity. The cage must have perches at several levels to accommodate this activity.

Aotus monkeys have been known to live over 20 years in captivity (Dennis Merritt: personal observation) and healthy females can produce offspring consistently every 9-12 months. Because females cycle every 16 days (Bonney Dixson & Fleming 1980), laboratories that use owl monkeys should be supplied with enough offspring to obviate imports after the first

Figure 1. *Aotus trivirgatus*, father and 3-month-old infant. (Photo by PCW.)

Figure 2. *Tarsius syrichta*, one-month-old female infant. (Photo by DMH.)

three years. Any infants that are rejected can be hand-raised on human infant formula. In one instance when a mother died after birth, the father continued to carry and raise the infant, with technicians removing it at three-hour intervals for feeding (Evelyn McGuinness: personal communication 1986). At three months of age the infant began to eat infant cereal from a bowl left in the cage. This infant became a well-adjusted adult who bred successfully.

Caging

In the wild, owl monkeys sleep in enclosed vine tangles or tree holes (Aquino & Encarnacion 1984; Wright 1985). Infants are *not* left in the nests like some prosimians, but the family sleeps in contact with one another in these dark areas during the day. In the wild these sleeping sites are scent-marked and the same site is used for years. Nest boxes large enough to accommodate the whole family should be provided as sleeping sites in cages. Wood is the preferred medium. Because of the *Aotus* habit of scent-marking nest boxes from the gland at the base of the tail, personnel might prefer to exchange wooden nest boxes every 2 to 6 months. Plastic nest boxes cause fungus sores, and constant cleaning of nest boxes stimulates excessive scent-marking.

In the wild, mothers maintain a distance of five or more meters from fathers that are carrying infants, except for short periods of nursing every 2-4 hours (depending on the age of the infant) (Wright 1984). If there is not enough room in the cage for her to get away, the mother may become stressed and aggressive toward both infant and father. After birth, two cages can be joined to provide more adequate space.

When infants are three to four months old and weigh about half as much as the father, father-offspring conflicts (also observed in the wild) can occur; the father bats at the offspring to get it off his back. More space will help alleviate such conflicts. However, the offspring should *not* be removed from its family for placement with a member of the opposite sex before age two, when the second younger sibling is born. When the offspring is 2-3 years old, parent-offspring fights, resulting in severe injury, can occur without warning.

In the wild, owl monkeys live in areas where the day varies from 12 hours to 10 hours, and therefore 12-14 hours of darkness is optimal in captive setting. *Aotus* in the wild is more active (playing, traveling farther, fighting, and giving loud calls) at high moonlight levels (Wright 1985). In a captivity *Aotus* prefers red lights or small wattage blue lights rather than total darkness.

Diet

Aotus can be raised successfully on New World monkey chow soaked in fruit juice, or small-pellet chow. Chow should be soaked to prevent extremes of tooth wear in these small monkeys. Fresh fruits such as apples, bananas, avocados, and grapes should be offered in rotation to add interest to the diet. Different individuals prefer different combinations of fruits, but variety is important. Purina Special Chow, Ralston-Purina Co., St. Louis, MO is the preferred chow. Although owl monkeys are nocturnal, like other New World monkeys they need Vitamin D3 (provided in New World Chow) to prevent rickets. Treats of crickets or large insects are relished.

TARSIER

Tarsiers are primates with a geographic distribution limited to the Philippine Islands, Borneo, Sulawesi, and Sumatra (Nowak & Paradiso 1983). They are never found farther than 10 degrees from the equator. Their diet is the most specialized of all primates, because they eat only live animal prey. The characteristics that strictly limit their geographic range provide special problems in captivity and if their special needs are not met, a colony will not survive.

The Duke University Primate Center (DUPC) maintains a colony of two species of tarsier, *Tarsius syrichta* (the Philippine tarsier) and *Tarsius bancanus* (the Bornean or Western tarsier). Tarsiers have proven to be extremely difficult to maintain in captivity, due in part to their requirements for live prey (captive diets consist mainly of crickets, mealworms, and anolis lizards with an occasional grasshopper, mantid, katydid, and cicada during the summer), and their need for a warm, humid environment. They perish at temperatures below 65°F.

Attempts to breed tarsiers in captivity have met with little success. Only the National Zoo has succeeded in having several captive-born *Tarsius bancanus* reach adulthood (Miles Roberts, personal communication). *Tarsius syrichta* have often bred and given birth but the infants usually die within a few days, with the exception of one Philippine tarsier born at the Frankfurt Zoo that survived for 1 year. A hand-raised tarsier at the DUPC born February 1987 is thriving (Haring & Wright: in press).

Diet and Caging

At the DUPC, tarsiers are housed in rooms of about 2 x 2 x 2.75 m containing a wide range of bamboo and vine substrates of varying dimensions and orientations. This allows ample room for animals to maintain a comfortable distance from caretakers and provides a suitable variety of substrates for foraging, resting, and social (grooming, reproductive) needs. Animals are kept either in male-female pairs or in groups of up to three females per male. Each room is maintained at a temperature of 28° to 31°C with relative humidity levels of 80% being optimal, though not always obtainable. Low humidity can cause the animals' tails to become rough and sore looking, with hands and feet also becoming chapped. Each room contains at least one water bowl. Rooms are sprayed with a misting hose 3-4 times a day. The animals lick water droplets from poles after the rooms are sprayed and delight in the moist substrates - often rubbing their heads or even entire bodies on a moist branch. The spraying also benefits the tarsier's main prey items - crickets and anolis lizard.

The DUPC feeds 10,000 crickets per week to its tarsier colony of 17 individuals, and each month receives a shipment of 200 anolis lizards (*Anolis carolinensis*) from Louisiana. The lizards are relished primarily by *T. syrichta*. In addition, bowls containing mealworms are periodically placed in the rooms, although only a few individuals eat them. During summer months, staff catch insects such as grasshoppers, katydids, cicadas, and mantids for the animals, which greatly improve their appetite, as boredom from the same cricket/lizard diet is inevitable. When insects caught by staff are placed in the tarsier's rooms, there is intense hunting and the insects are eaten with great enthusiasm. Except for such highly desirable prey items, tarsiers will not descend to the ground to forage (unless an animal is sick). Because crickets are terrestial and hide, they provide an additional challenge to tarsier husbandry, as the crickets must be encouraged to climb branches instead of congregating on the ground. This is accomplished by keeping the surface of branches and other substrates moist which attracts the crickets. Another problem is that anolis lizards are diurnal and inactive under the red lights of "darkness" when tarsiers hunt. Tarsiers are only attracted to moving prey and a perfectly motionless lizard remains safe.

Due to their fragility, tarsiers respond poorly to regular handling, so handling is kept to an absolute minimum. Reproductive cycles in *T. bancanus* can be monitored visually, obviating catching the animals (Wright Toyoma & Simons 1986a). At the DUPC, tarsiers are on a 12:12 light-dark cycle with lights going out at 11:30 A.M. This allows us time to clean rooms in the morning and to feed and observe in the afternoon. While "daytime" light is provided by Duro-lite Vita-lite flourescent bulbs, incandescent red light bulbs provide sufficient illumination for nighttime behavioral observations.

Because of their low metabolic rate, low body temperature (McNab & Wright 1987), and habit of eating large prey items, the animals spend the majority even of "active" periods sleeping and sitting motionless. The greatest level of activity occurs immediately after the white lights go out and night begins. Because the animals are most active at this time, it is essential that caretakers closely observe the tarsiers then to make sure all animals appear healthy and are hunting successfully. An animal in poor health might be unable to capture prey but could not be singled out from the others later in the afternoon when all the animals are resting. Unless the tarsier's weakened condition is noticed immediately and steps taken to correct the problem, rapid decline is inevitable. Captive tarsiers must remain healthy or they will be unable to hunt and survive even when surrounded by food. More than any other primate, tarsiers rely on the same foraging skills to survive in captivity as in the wild.

Natural History

Effective husbandry of tarsiers requires a certain amount of knowledge of the animal's natural history. The three species of tarsiers are found in the understory regions of rainforests. Edge habitats and secondary growth where insects are abundant are preferred to primary rainforest (Niemitz 1979; Crompton & Andau 1987). Although they are approximately the same size, the three species differ in social behavior and habitat preference. For instance, major differences occur in captive *T. bancanus* and *T. syrichta* in social interactions, activity patterns and use of substrates, all of which may be explained by their different social structure and ecology in the wild. These differences must be recognized in cage design and day-to-day management and husbandry. In general, *T. syrichta* is far more active than *T. bancanus* (Haring Wright & Simons 1985). Upon waking, *T. bancanus* is slow to begin hunting, staring at the observer without movement. *T. syrichta*, however, begins foraging for crickets immediately after the lights go out. However, an individual *T. syrichta* that has been stressed by regular handling will remain motionless until the observer retreats.

T. syrichta and *T. bancanus* in captivity also display different sleeping and resting site preferences. *T. bancanus* choose the top of nearly vertical poles for sleeping. *T. syrichta*, on the other hand, sleep on horizontal poles, stumps, shelves and even the ground. A *T. bancanus* discovered sleeping on the ground would be cause for great concern because it would indicate the animal was seriously ill and unable to cling to a vine. In the wild, *T. bancanus* sleep on low trees in vine tangles (Niemitz 1979). However, it is common to see healthy captive *T. syrichta* sleeping on the ground (generally hiding behind something), and in their natural habitat they have been seen on rocky mountainous slopes sleeping among the roots of trees (P.C. Wright, personal observation). During resting periods *T. bancanus* choose vertical poles and use their tails as props, while *T. syrichta* will sit on shelves or horizontal vines, tails dangling beneath them.

Social behavior is also markedly different in *T. bancanus* and *T. syrichta*. Although both species have estrus cycles of 24 - 3 days and females remain in estrus for 2-3 days (Wright et al. 1986b), in *Tarsius bancanus*, males court females with urine marking and mating calls during the hour before copulation but *T. syrichta* males chase females and copulate with them without extensive courtship. *T. bancanus* females, when pregnant, become very intolerant of adult males and will chase them at every occasion. One incident has been recorded where adult *T. bancanus* males attacked a newborn infant (P.C. Wright, personal observation). It may be that in the wild, Bornean tarsier females give birth and care for their infants in isolation from adult males. In captivity it is essential that pregnant females be housed separately or with enough space to avoid males. Pregnant *T. syrichta* females, on the other hand, show no aggression toward adult males, and in a number of instances *T. syrichta* babies have been born into groups containing unrelated adult males and females with no aggression being directed toward the infant.

It is especially important that pregnant tarsiers have access to more food items. A tarsier gestates for 6 months and in the last 2 months can double intake of food (Izard Wright & Simons 1985). Infant tarsiers weigh 20-30 grams or 20% of the mother's weight (Wright Haring Simons & Andau 1987). Tarsiers nearing parturition should be watched carefully because labor can be difficult and aid might be needed in delivering an infant. If labor is not monitored, both mother and infant may die. The duration of labor in the five births observed at the DUPC has varied from 20 minutes to 7 hours. Female tarsiers transport infants up to seven weeks of age by carrying them in their mouth, much like cats. When mothers are foraging they "park" the babies on a suitable branch, returning to them every half hour to nurse and keep them warm. (Although no nests have been described, Philippine tarsiers might care for their young in some sort of "nest" which might take the form of a hollow tree trunk or vine thicket.)

Effective tarsier husbandry requires the resolution of a wide variety of problems. Cage environment must be optimum for both the animals and their prey. Cage furniture must be designed so that animals are able to effectively forage, rest, socialize, reproduce, sleep and escape/hide from both caretakers and conspecifics. Observant and patient caretakers must be available to monitor each animal's health individually and recognize and take action to correct potential problems. Tarsiers in captivity cannot survive without effective foraging skills, and weak or sick animals very quickly die. An understanding of the animals' ecological niche is essential in designing adequate caging, particularly if behavioral research is to be carried out. The closer the captive habitat matches the animals' natural habitat in terms of climatic conditions and provisions of choices for social interactions and foraging the more content the animals will be in their captive environment.

GALAGO

Galagos or "bushbabies" are small, nocturnal prosimian primates that are found throughout subsaharan Africa (Napier & Napier 1967). The genus includes species ranging in size from the tiny *Galago demidovii* (60 g) to the large *G. crassicaudatus* (1,300 - 2,000 g). A recent revision of the taxonomy of this group has placed the larger forms in the genus *Otolemur* (Olson 1979), but for the purposes of this chapter, we will use the traditional nomenclature with one exception: a medium brown form from coastal Kenya (Nash & Harcourt 1987), formerly referred to as *G. crassicaudatus crassicaudatus*, is now called *G. garnetti* (Olson 1979).

Natural History

Within the genus *Galago*, great variability exists among species in social systems, diets, and behavior (Bearder 1974; Charles-Dominique 1977; Clark 1985; Harcourt & Nash 1986; Nash & Harcourt 1986). Contrary to earlier notions of the "solitary" nature of bushbabies, recent studies have demonstrated a complexity of social systems not previously recognized. In general, the range of a reproductively-active male overlaps the range(s) of one or more females. Subadult offspring share ranges with their mothers. As young males mature, they disperse from their natal ranges, while female offspring may continue to share ranges with their mothers, or disperse at maturity.

Members of the genus *Galago* sleep alone or in family groups of 2-6 animals, including an adult male, an adult female, and their offspring. Leaf nests, dense foliage, or tree holes are used as sleeping sites. In the wild, pregnant galagos leave their sleeping groups and sequester themselves in a separate nest for parturition, rejoining the social group when the infants are a few weeks old (Sauer & Sauer 1963; Vincent 1969). This may be a mechanism for avoiding neonatal death from interference by other group members (Rosenson 1972). The phenomenon has bearing on the management of captive animals at the time of parturition.

Foraging at night is generally solitary, although a female and her offspring may forage together. Wild galagos consume a diet of gums, insects, and fruits. Proportions of these items vary with species, but in general the smaller species (e.g. *G. senegalensis*, *G. demidovii*, *G. zanzibaricus*) are more insectivorous, and the larger species (e.g. *G. garnettii*, *G. crassicaudatus*) are more frugivorous.

Social communication occurs through a wide vocal repertoire, mutual grooming, and scent-marking. Scent-marking can take the form of urine-washing, chest rubbing, head-and-mouth rubbing, and foot rubbing.

Figure 3. *Tarsius syrichta* male with anolis lizard in mouth. (Photo by DMH.)

Figure 4. *Galago crassicaudatus*, the greater thick-tailed bushbaby. (Photo by MKI.)

The DUPC's *Galago* colony was established in 1966, and has since grown to include four species and almost 200 individuals. Over the years, much practical information has accumulated on social groups, housing, and diet that maximize colony health and reproductive output. We consider these factors to be reflections of the psychological wellbeing of these prosimians.

Caging

With so many animals of a particular genus to accommodate, standard *Galago* caging at the DUPC is small by Duke Primate Center standards, but large by the standards of most laboratories. Caging for *G. garnetti* and *G. crassicaudatus* measures 0.6m x 0.9m x 1.5m high, and is constructed of 1 x 1 in. vinyl-coated wire. This wire, although more expensive than other types, holds up well to *Galago* scent marking and frequent cage washing.

Two horizontal partitions with passage holes cut in them create three vertical areas. This triples the usable floor space without enlarging the cage, and allows animals to escape from each other. Wire shelves (12 x 12 in.) are put in one corner of each area, further enlarging the floor space. Each cage is equipped with a polyurethaned nestbox (approximately 12 x 12 in., with a 1-2 in. lip across the front to keep infants from falling out). Holes (1/4 in. in diameter) are drilled in the bottom of the nestbox to provide drainage for urine. Plastic or other nonporous material should not be used for nestboxes because fungal infections result.

Caging for *G. demidovii* and *G. s. moholi* measures 0.5m x 1.7m high and is constructed of 1 x 1/2 in. vinyl-coated wire. One horizontal partition with a passage hole creates two compartments. Several 6 x 6 in. shelves in the lower compartment, along with ladders made of strips of vinyl-coated wire, provide substrates for locomotion.

Diet

At the DUPC the large species (*G. garnettii* and *G. crassicaudatus*) are fed a mixture of Purina High Protein Monkey Chow (Ralston-Purina Co., St. Louis, MO) and Purina Cat Chow for additional protein. In addition, a different fruit (chopped into small pieces) is fed daily. Favored fruits are bananas, apples, canteloupes, grapes, watermelon, raisins and to a lesser extent, oranges (not liked by *G. garnettii*). Cucumbers, carrots, and cooked or raw sweet potatoes are also eaten but not favored. The small species (*G. demidovii* and *G. s. moholi*) are fed Purina High Protein Monkey Chow mixed with mashed bananas to increase palability. Cubed apples are also fed daily. Crickets (two per animal) are fed three times weekly to the larger species, and daily to *G. senegalensis*. *G. demidovii* are fed mealworms daily.

A weekly treat of hard-boiled egg is given to each animal. A vitamin-enriched banana bread with extra oil is given weekly to prevent hairballs. Lactating mothers are given yogurt daily through the first month of lactation to provide extra calcium. Soft foods such as yogurt with honey, applesauce, and baby food are fed to debilitated animals. Water is provided *ad libitum.*

Weights are taken weekly from infants until they reach puberty, in order to monitor proper growth and development. Adults are weighed quarterly to insure that they are maintaining proper weight and that the diet is adequate. The condition of the pelage, which should be dry, fluffy, and clean, is another measure of the animals' health and the adequacy of the diet.

Social Groups

The basic social group for all species of *Galago* at the DUPC is a male-female pair and their unweaned offspring. New breeding pairs are established with weanlings. This results in less fighting and fewer wounds than if unfamiliar adults are put together. Unfamiliar adult animals may need a period of habituation side by side with a wire partition between them or they may be allowed contact during the inactive diurnal period for several days, while being separated during the nocturnal activity period. Provision of two nest boxes and two food bowls can reduce friction in cases of mild aggression. Two adult males should never be housed together. We have had success with housing trios consisting of twin sisters and a male together, but after separation occasioned by the birth of a litter to either of the females, we have not been able to get the females back together. This is true also for mother-daughter trios.

Reproductive Management

The *Galago* colony at the DUPC is housed under a non-fluctuating photoperiod of 12 hours of light and 12 hours of dark. Under these conditions, galagos breed year-round. Females are monitored weekly for reproductive status, and more frequently, sometimes even daily, if evidence of cyclic activity is noted. Vaginal smears are done to monitor reproductive activity (see Izard & Rasmussen 1985 for a description of vaginal cytology). Parturition dates are predicted based on gestation lengths measured from the first day of estrus or the first day spermatozoa are observed in a vaginal smear. Five days prior to the predicted date of parturition, females are physically separated from their mates in one section of the home cage by inserting a wire mesh barrier. Visual, auditory, and olfactory contact is maintained in order to maintain habituation status. Small gauge mesh between the male and female prevents the male from reaching through and harming the infant. It is important to separate a female from her mate prior to parturition (Buettner-Janusch 1964). This practice has significantly decreased neonatal mortality in the DUPC *Galago* colony (Izard & Simons 1986a). Litters range from one to three offspring, depending on the species (Izard & Simons 1986b).

When the offspring reach one month of age, the male is reintroduced to the female and offspring. At weaning (about 100, 140, or 136 days for *G. s. moholi*, *G. garnettii*, and *G. crassicaudatus*, respectively; Izard 1987), the offspring are removed to their own social groups. Isolation of females prior to parturition, reintroduction of the male when the offspring is one month old, and removal of the offspring at weaning results in fewer problems with fights and bite wounds.

Medical Problems

The most common medical problem in the DUPC *Galago* colony is bite wounds. Most of these wounds occur at breeding, or when two unfamiliar animals are introduced to each other. The incidence of wounding upon introduction can be reduced by prior habituation of the animals through a wire partition. Short of missing a breeding opportunity because animals have been separated to prevent fighting, not much can be done to reduce the incidence of bite wounds in breeding pairs. Breeding pairs should be checked closely throughout proestrus, estrus, and metestrus to catch incipient infections from bite wounds.

Most bite wounds occur on the tail, fingers, and head. It is sometimes necessary to amputate part of the tail, especially if it becomes a target of self-traumatization. Most bite wounds in breeding pairs occur during proestrus and metestrus when the male attempts to copulate and the female is not receptive. This conflict can be avoided by separating the animals at this time.

Another major medical problem is excessive tartar, gingivitis, and periodontal disease, seen mainly in *G. s. moholi*. We feel this is a dietary problem and have approached it by adding powdered Vitamin C to the diet, increasing coarse foods, and periodically cleaning teeth with Dentsply Cavitron.

SUMMARY

The three species of primates described in this chapter have on thing in common, nocturnality, but their requirements in captivity differ in many respects. The social structures of each species differ and cage groupings must take this into consideration. Owl monkeys have a strictly monogamous breeding system in which infants are cared for by fathers. Removing fathers prevents the offspring's survival. In contrast, galagos have a solitary social system and males must be removed if infants are to survive. Even within such closely related primates as the two species of tarsiers, infants may be attacked in one species but not the other. In one species, *T. syrichta*, females can easily live with other adult females and a male, even raising offspring is such groupings. *T. bancanus* does not tolerate other adults of the same sex.

The diets of each of these night dwellers also are very different. Owl monkeys are frugivores, tarsiers are insectivores, and galagos combine fruits, insects and gum in their natural diet. Captive diets must take these differences into consideration.

Locomotor patterns differ with each species and caging must accommodate these patterns. Tarsiers are vertical clingers and leapers and require vertical support. Owl monkeys are springers and quadrupedal walkers and galago locomotion varies according to species. Most nocturnal primates prefer nestboxes for sleeping and scent-marking must be considered a normal behavior pattern.

The ultimate measure of a healthy breeding colony of nocturnal primates is a good rate of reproduction, low neonatal mortality, and low mortality rates in general. We feel we are approaching the optimum conditions to provide for the psychological wellbeing of these nocturnal primates.

III. SUBORDER: ANTHROPOIDEA
Family: Callitrichidae
(Marmosets and Tamarins)

— 6 —

Psychological Well-Being of Captive Primates: General Considerations and Examples from Callitrichids

Charles T. Snowdon and Anne Savage

INTRODUCTION

Concern for the psychological wellbeing of captive primates is not new; the history of such concerns began well before recent pressures from animal welfare activists. It has long been recognized that successful captive breeding of primates requires equal attention to physical and psychological needs. Robert Yerkes, writing in 1925, summarized the essentials for successful breeding of great apes as follows:

> "We should emphasize the following points: freedom or reasonably spacious quarters; fresh air and sunshine, preferably coupled with marked variations in temperature; cleanliness of surroundings as well as the body; clean and carefully prepared food in proper variety and quantity; a sufficient and regular supply of pure water; congenial species companionship and intelligent and sympathetic human companionship and, finally, adequate resources both in company and in isolation for work and play." (Yerkes, 1925, pp. 241- 242)

In recent years the NIH *Guide for the Care and Use of Laboratory Animals*, the United States Department of Agriculture (USDA), the accreditation program of the American Association of Laboratory Animal Care (AAALAC) and the evaluation of animal facilities by veterinarians, all have focused on the first half of Yerkes' prescription; concern with physical conditions of sanitation, heat, lighting, fresh air, food preparation and water sources. However, the most recent edition of the *Guide* (National Institutes of Health 1985) raises for the first time concerns about social and psychological conditions and suggests that social housing is preferable to isolation, where appropriate to the species.

In this chapter we raise some general considerations and criteria for developing and evaluating humane environments for captive primates, emphasizing some specific solutions that we have developed to deal with both physical and psychological needs in primates, primarily in Callitrichids.

WHAT ARE THE CRITERIA FOR A GOOD CAPTIVE ENVIRONMENT?

Veterinary-medical criterion. There are several potential criteria for evaluating captive environments for primates. The veterinary-medical criterion is defined as good physical health and freedom from disease. All captive environments that meet current requirements of institutional animal care review committees and the USDA probably satisfy the veterinary-medical criterion. While the veterinary-medical criterion does not require a social or psychological component, veterinary opinion is likely to change based on current research demonstrating profound effects of social environments on immune system functioning (e.g. Coe, Rosenberg, Fischer & Levine, 1987). Animals that are housed in good social environments are better able to defend against disease.

Biological criterion. While the veterinary-medical criterion is necessary for a good captive program, it is not sufficient. A second criterion is biological: how successfully do captive animals breed? The criterion of reproductive success has two levels. We can ask simply whether or not reproduction has occurred. Many breeding facilities report successful reproduction. However, the infants have been hand-reared in nurseries. We submit that this is not reproductive success. Hand-rearing techniques frequently produce offspring who are themselves incapable of reproducing and rearing their young without extensive human intervention.

An environment conducive to reproductive success is one where animals both breed successfully and rear most of their offspring successfully, and where the offspring in turn breed and rear most of their offspring successfully with minimal human intervention. To meet this end, animals need an early social environment where they can acquire the skills of parental care. This means housing animals in social groups approximately normal for the species, keeping infants with primary caretakers for extended periods, and allowing enough space for species-typical behavior to occur. Although the initial investment in adequate group caging may be high, the costs of parental care provided by the monkeys themselves will be considerably less than the costs of human care for infant primates.

Behavioral-ecological criterion. While the biological criterion, like the veterinary-medical criterion, is necessary, we do not think that even these two are sufficient. The final criterion is behavioral-ecological. A good captive environment is one where animals can acquire and retain the behavioral skills they would need to cope successfully with their natural environment were they ever to be released to the wild. While we will rarely want to release captive animals to the wild (Konstant and Mittermeier, 1982; Brambell, 1977, Senner, 1980), reintroduction has been tried with some success (Kleiman, Beck, Dietz, Dietz, Ballou & Coimbra-Filho, 1986). The point we wish to make is not that captive animals must be released to the wild, but an ideal captive environment would allow them to develop and display the skills they would need for survival (Snowdon, 1989).

The ideal captive environment. In our opinion an ideal captive environment must meet all three criteria, the veterinary-medical, the biological, and the behavioral-ecological. The conceptualization of this ideal captive environment provides us with a standard against which we can evaluate both the physical and the psychological wellbeing of captive primates. The starting point for developing an ideal captive environment is to reproduce some of the salient features of the wild environment, and to create opportunities for captive animals to develop skills that they might need in the wild.

PHYSICAL WELLBEING

Animals' physical needs cannot be separated easily from their psychological needs. An animal whose physical needs are not met is unlikely to have its psychological needs met. It is also true that in many cases the physical needs are provided in a psychologically inappropriate way. We will discuss two aspects of physical wellbeing that have psychological implications: diet and locomotion.

Dietary Variety

It is possible to determine the basic nutritional needs of an organism and to provide a purified homogeneous diet that meets these needs, but as Markowitz and others in this volume have documented, the presentation of food in time and space is very important to psychological wellbeing.

We have been maintaining successful captive colonies of cotton-top tamarins (*Saguinus oedipus*) and pygmy marmosets (*Cebuella pygmaea*). The marmosets and tamarins (Callitrichids) are small, arboreal neotropical primates. They typically give birth to twins, and parental care is shared by fathers and other group members (Cleveland & Snowdon, 1984). The twins weigh approximately 25% of their mother's weight at birth and females frequently conceive on the first postpartum ovulation. Because of the energetic load for reproductive females, they are rarely involved in parental care other than nursing.

We have made several changes in dietary and feeding regimes over the years. We have found that variety is very important. First, there are striking individual preferences for different foods and failure to provide variety means some animals have access to preferred foods while others do not. The protein needs of our animals appear to be greater than those provided in commercial chow, so we provide three feedings a day with early morning and late afternoon feedings providing supplementary protein. The contents of the afternoon feedings vary over days so that animals have access to a wide variety of attractive foods, such as yogurt, cottage cheese, hamburger, chicken, and mealworms. With the major feeding of chow at midday we feed fruit and vegetables and these also vary from day to day and seasonally (Snowdon, Savage & McConnell, 1985). Studies of cotton-top tamarins in the wild (Neyman, 1977) show that tamarins feed on a wide variety of fruits and insects, so it is important to provide a wide variety of food in captivity.

Several years ago we had difficulty in getting some parents to care for their twins properly. We noticed that animals carrying infants were hesitant to approach food dishes that had been placed near floor level in the cage. When we moved the dishes to parts of the cages that were at least one meter above the floor, this ambivalence toward food stopped, and animals carrying infants readily approached the food dishes and fed. The success of rearing infants has increased markedly since this simple change in food placement. For arboreal primates like ours it seems only logical in hindsight that food resources should be placed at an elevated level within the cage.

Foraging Analogues

Various foraging analogues for captive primates have been developed to serve three functions: (1) to simulate foraging behavior in the wild, (2) to enrich the environment regardless of whether it simulates nature, and (3) to serve as therapy for disturbed animals.

Animals rarely encounter food in great abundance in the wild, and they often must search for it. Distributing food over the day and making it harder to find approximate the natural conditions. Kleiman et al. (1985) described techniques for developing foraging skills in captive lion tamarins (*Leontopithecus rosalia*) prior to their reintroduction to the wild. Foods were hidden in a variety of places in a very large cage, and the animals learned to find the hiding places. McGrew, Brennan & Russell (1986) designed a simple sap feeder made of wooden doweling drilled lengthwise with holes that were filled with gum arabic. The feeders were presented to common marmosets (*Callithrix jacchus*), a species that extracts sap from trees in the wild. Using this feeder the captive marmosets readily learned to gnaw at the device to extract the gum arabic.

Sheryl Hayes (personal communication) developed a complex, inexpensive, feeder for enrichment in *Cebus* monkeys. The animals had to perform several different actions to obtain food. One of us (A.S., unpublished results) designed a complex feeding box for use with cotton-top tamarins. The animals had to perform a variety of manual actions to locate food. Sheryl Hayes (personal communication) has hidden food in hanging baskets (also see Molzen & French: this volume) and in a box filled with wood shavings and finds that tamarins spend long periods foraging. Chamove, Anderson, Morgan-Jones & Jones (1982) observed six species of primates before and after the floor of their cages was spread with deep woodchip litter and found in all but the most arboreal species an increase in time spent foraging and a decrease in aggression. (See also Chamove & Anderson: this volume.) Thus, there are a variety of simple techniques for enhancing feeding and at the same time developing foraging skills in animals. Studies by Plimpton, Swartz and Rosenblum (1981) and Rosenblum and Smiley (1984) have demonstrated that foraging tasks can serve a therapeutic role for behaviorally-disturbed bonnet macaques (*Macaca radiata*).

Caging for Arboreal, Neotropical Primates

Exercise areas. Marmosets and tamarins are clawed, arboreal primates and many standard small, stainless-steel primate cages are not suitable for the development and maintenance of their locomotor and sensorimotor skills. There are a variety of possible solutions. Hearn, Abbott, Chambers, Hodges and Lunn (1978) held common marmosets in relatively small cages, but in the center of the room was a large exercise cage. Each group of monkeys was given access between their home cage and the exercise cage for one day each week. Not only did the animals benefit from having their own time in the exercise cage, but it was Hearn et al.'s impression that the animals benefitted by watching other monkeys being active in the exercise cage. (See also Blackmore, this volume.)

Airconditioning ducting as tunnels. McGrew and McLuckie (1986) have used flexible, transparent air- conditioning ducting to allow cotton-top tamarins to leave their cages and go to an outdoor exercise area or explore new environments. The technique was a laboratory analogue to dispersal in the wild, and revealed that subadult females were the first to leave the home cage, traveled the greatest distance and stayed away the longest. Subadult males were next in adventurousness and the breeding adults were the least likely to leave the home cage. This technique allows animals to move about and explore new environments on their own volition without any handling. Younger animals play in the ducting, moving rapidly from one place to the next (W. McGrew & E. Price, personal communication).

Cage furnishings. The environments we have designed for our marmosets and tamarins are quite large, ranging 3-12 m^3 for the cotton-top tamarins (Snowdon Savage & McConnell 1985) up to 20 m^3 for the pygmy marmosets. All of these cages are filled with a high density of natural branches as well as ropes to simulate vines. The branches and ropes provide a surface on which the claws can get a good grip, and they are arranged to give the animals a variety of textures, diameters and degrees of firmness. Thus, some branches are quite solid and do not give when an animal lands, other branches as well as the ropes are quite loose, with a lot of give. The animals must learn to accommodate their jumping to these different surfaces. The bulk of the branches and ropes are placed a meter or more above ground level to simulate an arboreal environment, and all food and water containers and nest boxes are placed high in the cage. Every two to three months the branches and ropes are removed and replaced with a new configuration. This not only provides for adequate sanitation, but also gives the animals a varied environment that reduces the likelihood of motor stereotypies. The animals must frequently adjust their routes to get from one place to another within the cage.

We have some evidence for the success of this complex caging environment. Several years ago we conducted a pilot study at Monkey Jungle in Goulds: Florida, where we released four of our captive-born cotton-top tamarins to the five-acre rain forest at Monkey Jungle for a short time. Three of the four monkeys showed excellent locomotor skills; worked their way through the canopy, leapt from tree to tree and climbed and rested on a wide variety of substrates. We think our indoor caging comes close to providing the locomotor skills these monkeys would need in the wild.

It is important to develop housing that goes well beyond the minimum standards. A more complex environment allows members of the group to engage in a wider variety of social activities, and it allows the development and maintenance of locomotor and sensory skills. Thus, caging of appropriate size and complexity contributes to both physical and psychological wellbeing.

PSYCHOLOGICAL NEEDS

There are three aspects of psychological needs we would like to emphasize. First is the need to design captive environments that give animals opportunities to develop behavioral competence. Behavioral competence in primates generally involves social interactions since most critical behavior is acquired through experience, and several recent studies have shown social learning to be of even greater importance than previously thought. Even trial-and-error learning, however, requires a complex and diverse physical environment. Second, the development of a sense of control over the environment has been shown to be a critical component of psychological wellbeing. Animals cannot passively receive environmental events; they must be able to act on the environment and consequences must result from their actions. This is a key aspect to most good environmental enrichment. Third, we must develop alternative research strategies to allow us to study animals while they remain in complex physical and social environments.

Early Social Experience

Early social experience is important in many aspects of primate development. We will discuss three examples in areas where early experience was thought previously to be of little importance.

Parental Care: Even sophisticated biologists often assume a "maternal" or "paternal" instinct, but we have learned from breeding marmosets and tamarins in captivity that parental skills are acquired through direct contact with someone else's infants. This contrasts with the data on the development of parental skills in rhesus macaques, where exposure to a mother or to peers while growing up appears to be sufficient for the development of normal maternal skills (Harlow, Harlow & Hansen, 1963). Data from a variety of species indicates that if marmosets and tamarins are removed from their groups before they have had the opportunity to help their parents care for new infants have a much lower likelihood of rearing their own offspring successfully [saddle-back tamarins (*Saguinus fuscicollis*: Epple, 1978; golden lion tamarins, Hoage, 1977; cotton-top tamarins, Cleveland & Snowdon, 1984, Snowdon et al., 1985, Tardif, Richter & Carson, 1984 a,b]. Cleveland & Snowdon (1984) documented that juvenile tamarins initiated play with infants although they rarely carried them. On the other hand, subadult (or peripubertal) tamarins carried infants as much as adults did. We suggest that tamarins should be housed with their families through at least two additional sets of twins to gain experience of parental care before breeding themselves.

Predator Recognition: Many biologists and psychologists assume an "antipredator" instinct, believing, often in the absence of proof, that predator avoidance should be an innate reaction derived through natural selection. However, rhesus macaques do not have an innate fear of snakes. Mineka, Keir & Price (1980) showed that wild-caught rhesus macaques still feared snakes several years after capture whereas captive-born monkeys showed no fear at all. Subsequently, Mineka, Davidson, Cook & Keir (1984) showed that captive-born monkeys could learn to fear snakes simply by observing a wild-caught monkey displaying fear. The fear of snakes induced by observation extends for at least six months after the initial training session. The fear is also resistant to traditional therapies for phobias (Mineka & Keir, 1983), but if monkeys watch a naive monkey behaving non-fearfully toward a snake, the observer monkeys appear to be "immunized" against acquiring snake fear (Mineka & Cook, 1986). More recent work shows that naive monkeys can learn snake fear through watching videotapes (M. Cook & S. Mineka, personal communication).

We have begun testing our captive born tamarins with live snakes and model snakes. So far not a single tamarin born in our colony has displayed any signs of fear to a snake (S. Hayes, personal communication). Benjamin Beck (personal communication) reports that golden lion tamarins react to aerial predators, but the possibility of prior exposure was not controlled in his captive lion tamarins. Controlled studies need to be done with monkeys known never to have seen aerial predators before experimental testing.

The weight of evidence indicates that detection of predators and the appropriate avoidance behavior must be learned, and that this learning most effectively occurs in a social environment. If one agrees with our contention that we should maintain in captive populations the behavioral skills needed to survive in the wild, then we should include in our captive environments opportunities for monkeys to learn to detect and respond appropriately to predators.

Communication: A final area where social learning is likely to be very important is in the ontogeny of communication. Most of the evidence indicates that nonhuman primates are different from birds and human beings in that early auditory experience seems to have little effect on subsequent vocal development. In the calls that have been studied the genetic component appears to predominate (Newman & Symmes, 1982). However, a recent study by Seyfarth and Cheney (1986) on vocal development in wild vervet monkeys (*Cercopithecus aethiops*) indicates the strong likelihood that observational learning and social reinforcement are important in developing both competent vocal structure and appropriate usage of vocalizations. Thus, social housing of animals should be important for normal vocal development as well.

In the future it would be more conservative to assume that important aspects of behavior depend upon appropriate early social experience than to assume that behavioral competence is a function of innate factors and maturational variables. We should house our animals to maximize the likelihood that they will develop behavioral competence through social interactions.

Development of Sense of Control

We have already discussed way to enrich the captive habitat to induce many of the types of behavior that would be displayed in the wild. Another aspect of environmental enrichment has to do not with naturalistic simulations but with increasing activity by other means. The issue of work, raised by Yerkes (1925) in the quotation that began this chapter, is important even though environmental enrichment is rarely thought of in terms of work. We rarely think of work as being psychologically positive, yet most of us would be quite unhappy if we were deprived of the opportunity to work or to engage in other stimulating activities. Shouldn't we consider the importance of work for primate wellbeing as well?

Markowitz (1982) offers a variety of innovative techniques for getting animals to work for food. (Also see Rumbaugh Washburn & Savage-Rumbaugh on food-reinforced videogames and puzzles, King & Norwood on fishing, Gilbert & Wrenshall on diving for raisins, Chamove & Anderson on foraging through litter, Bloomsmith on simulated termite foraging, Molzen & French on extractive foraging – all in this volume. Ed.) volume). Animals given such opportunities usually show marked improvements in their general behavior.

The theory of the value of work is that organisms that have learned to control their environment in specific ways will be readier to seek new ways to control it when challenged with novel situations. This sense of ability to control has important psychological and physiological implications. Hanson, Larson and Snowdon (1976) presented rhesus monkeys with loud, continuous white noise. One group of monkeys could press a lever to terminate the noise while another group, exposed to exactly the same amount and pattern of noise, had no lever with which to control its termination. When cortisol levels were measured directly after noise exposure, the monkeys that could control noise termination had significantly lower cortisol levels than the monkeys that could not. In fact, the cortisol levels of the animals that could terminate the noise did not differ from baseline values. In social interaction tests carried out a few hours later, the monkeys that controlled noise termination were less aggressive than the animals that had no control over the noise. However, in the next part of the study the control lever was taken away from the animals that had had it, and their cortisol levels following exposure to noise they could no longer control rose to values much higher than those seen in monkeys that had never had control.

Mineka, Gunnar & Champoux (1986) studied the ontogeny of control in monkeys. They reared yoked groups of young monkeys in environments with levers, chains and other manipulanda. In half the groups, operating the manipulanda produced small food rewards. Each experimental group was yoked to a group that received the same food at the same time but with no connection to pressing levers or pulling chains. Thus, the first groups experienced a correspondence between their actions and a positive environmental change (food) while their counterparts experienced no such correspondence. When tested subsequently, animals with experience of control over food were less afraid of threatening stimuli, coped better with brief separations from peers, and showed lower cortisol levels than the animals that did not experience control over food.

These two studies have several implications for behavioral enrichment programs. First, the best enrichment assures animals a direct correspondence between their actions and some important change in the environment. Simply giving animals new objects or exposing them to music or video screens would not be expected to have much effect on their psychological wellbeing. Rather, animals must be able to develop some sense of control over the environment. The animals must be able to operate on the environment with good effect. Second, the study on control of loud noise showed that once an animal has control, to lose it may be worse than never to have had it. Further research is needed in a greater variety of situations to determine how long-lasting and how deleterious are the effects of losing control once it has been established.

Designing Noninvasive Research Techniques

For many nonscientists as well as scientists, research is synonymous with invasiveness: good research is thought to require isolating animals from one another and banning any sort of enrichment as a violation of strict experimental control. It is thought that animals must be prodded and handled, blood samples drawn and surgery performed before research results can be considered significant. While particular research problems may require that animals be isolated, blood drawn or surgery performed, we suggest that researchers think more creatively about possible alternative ways to gather the same or similar data without using invasive techniques or isolation of captive primates.

We have been working extensively with a highly-endangered species, the cotton-top tamarin, where our ethical commitment is to minimize interference with their social and reproductive behavior. However, the colony is supported entirely by research funds, and thus research results must be of good quality. Out of necessity we have developed a variety of noninvasive or minimally-invasive research techniques that can be easily applied to research on other species as well. We are certain that others can devise their own ways of gaining good-quality research data with less invasiveness than at present, once they decide the goal is worth the effort.

A major interest of comparative psychologists is learning and perception. Years ago the common marmoset was tested using a Wisconsin General Testing Apparatus, and the results indicated that marmosets were very slow to acquire a learning set (Miles & Meyer, 1956). We have used the same type of apparatus with cotton-top tamarins, but instead of taking animals out of their social groups and testing them in a sound-attenuated isolation chamber, we have moved the apparatus to the side of the group cage. We have found very rapid rates of learning-set acquisition (Jeff Lande, unpublished results), comparable to the very best obtained by Harlow (1949) with rhesus macaques. Subsequently, the same apparatus has been used to test color discrimination in cotton-top tamarins using Munsell color chips (Savage, Dronzek & Snowdon, 1987). The monkeys can be trained to work at the apparatus one at a time for food treats; often in the middle of testing they will climb to sit with one of their companions for a few minutes and then return to complete the testing. We have been able to achieve much higher levels of performance by testing in the animals' home cages (individual animals will work for 35-45 min a day, some accumulating over sessions more than 3000 color-discrimination trials).

Studies of monkeys' discrimination between different natural vocal signals required nearly 30,000 traditional operant-discrimination trials to establish clear discrimination (Petersen, Beecher, Zoloth, Green, Marler, Moody & Stebbins, 1984). In contrast, Kim Bauers (unpublished results) used a playback technique to present cotton-top tamarins with different calls through speakers placed in the home environment. Using their spontaneous reactions to different call types, she found a significant discrimination between call types after only a single trial with each animal. Testing in the home cage eliminates the need to separate animals from their social groups, and appears to result in higher levels of performance on learning and perceptual tasks.

Information can be gathered on captive animals to complement data from field research using a variety of simulations that do not require moving or handling animals. McConnell and Snowdon (1986) simulated territorial encounters between groups of cotton-top tamarins similar to those seen in the wild (Neyman, 1977; Anne Savage, unpublished observations) by simply opening doors between adjacent colony rooms. This increased the amplitude of calls from strange groups by about 5 dB. After about 30 sec of hearing the intragroup calls of the group in the other room, one or two animals would orient toward the other group and begin to give territorial long calls. They also scent- marked and displayed other territorial behavior. Because close observations could be made, Mc Connell and Snowdon were able to describe different roles for males and females in territorial encounters and found an orderly pattern to the calling throughout the bout. We previously described the study by McGrew and McLuckie (1986) that simulated dispersal in captive groups of tamarins. Both of these studies illustrate how information about the natural behavior of cotton-top tamarins can be obtained in naturalistic simulations on captive animals. Because of the difficulty of making precise observations in the wild, information of comparable quality would be difficult or impossible to obtain on wild animals.

Monitoring the hormonal patterns of captive animals is important for evaluating reproductive status and for some measures of psychological wellbeing. Some believe, though, that all hormonal measurements must be taken from serum samples. With small primates, serum measurements can be quite dangerous since repeated blood drawings quickly produce dangerously low hematocrit levels. An alternative is to assay urinary hormones (Benirschke, Lasley & Ryder, 1980). We have developed techniques for measuring urinary estrogens (French, Abbott, Scheffler, Robinson and Goy, 1983; Ziegler, Savage, Scheffler & Snowdon, 1987), luteinizing hormone and chorionic gonadotrophin (Ziegler, Bridson, Snowdon & Eman 1987). We are currently developing assays for progesterone, prolactin and cortisol.

Urine samples are collected each morning at onset of daylight from unrestrained animals. The animals we wish to sample are given food treats to keep them near a collection container that is held underneath them until they urinate. Most animals will urinate shortly after light onset. We can collect daily samples throughout an entire reproductive cycle, including pregnancy, the day of parturition and the early days of lactation without compromising the pregnancy or the survival of the infants. Since the hormonal collections are noninvasive, there is no effect on subsequent behavioral observations. Furthermore, the technique has allowed us to provide the most complete description of hormonal changes through pregnancy and lactation available for any primate species (Ziegler et al. 1987 a) and to learn about social mechanisms of reproductive suppression in subordinate females (French, Abbott & Snowdon, 1984; Ziegler et al., 1987 b; Savage, Ziegler & Snowdon, 1988).

Each of these techniques has provided information that would have been impossible to gather otherwise or could have been gathered only at considerable risk to the animals' wellbeing. The development of novel, noninvasive research techniques not only serves the wellbeing of the monkeys but can often improve the quality of the research.

EVALUATION OF PSYCHOLOGICAL WELLBEING

It is easy to generate ideas for promoting psychological wellbeing of captive primates, but it is much more difficult to evaluate them. We first need to agree on a theory of psychological wellbeing. The contributors to this volume have different views about what belongs in such a theory. There is an extensive psychological literature on subjective wellbeing in humans, but little evidence of much progress in the millennia since the Greek philosophers first wrote about happiness and wellbeing (Diener, 1984). If it is so difficult to define wellbeing for human beings with whom we can readily communicate, how will it be possible to define wellbeing for animals?

One measure that is often proposed is preference. If we give animals a choice of different environments and let them choose, presumably they will go where they experience greater wellbeing. There are serious problems with preference as an index of animal wellbeing. In the first place preference is relative to the options we have offered the animals and the particular variables that differentiate those options. While we might envision a large-scale research project that systematically varied all the parameters of cage size, cage complexity, diet, social structure, work options, etc., there are neither primates nor time enough to complete such an undertaking. We must find other ways to make efficient decisions about wellbeing.

A second problem with preferences is that what animals or human beings prefer in the short run may harm them in the long run (Line, 1987). For example, blowflies (*Phormia regina*) prefer the sugar fucose over all other sugars, but they cannot metabolize fucose. If given a free choice, they will eat fucose exclusively and starve to death (Dethier, 1976). Our own species exhibits short-term preferences for foods high in cholesterol and sugar or for drugs like nicotine, alcohol and cocaine that have long-term deleterious effects. Thus we cannot simply define wellbeing by the choices an animal makes.

Another approach to wellbeing is to provide an animal with all its physical needs without asking any work of it. We have already argued, however, that work is an important component of wellbeing. Animals should not get all of their resources "for free". Animals that can act on their environment and receive valuable consequences can better withstand subsequent stress and adapt more readily to environmental changes than those that are not required to work. Thus, one component of wellbeing might be an animal's ability to adapt readily to new social and physical circumstances with a minimum amount of stress.

We have proposed to evaluate captive environments by how well the animals in them would be expected to cope were they to be released to the wild. While this may seem an extreme criterion, it does point to explicit criteria for each species. Animals should be able to rear their offspring with minimal human intervention. They should forage and locomote normally.

They should be in social groups approximating in composition those in nature. The animals should display cycles of activity and rest similar to those of wild animals, and they should show species-typical behavior rather than cage-induced stereotypies.

Some physiological measures have been used to evaluate well-being. In the study by Hanson et al. (1976), normal cortisol levels in the face of environmental stress was the index of wellbeing. Cortisol has been used in many other studies as an index of stress. More recently, Coe and Scheffler (1989) have proposed measures of immune system function to evaluate wellbeing. They found that the number and types of leukocytes in the blood were sensitive to even very minor social stress such as separation of juveniles from their peer groups for as little as four hours. If baseline levels of cortisol and leukocytes are known for a species, changes in these values can be sensitive indicators of wellbeing.

We will never know what psychological wellbeing really means to another species, but we cannot let this stop our progress in improving captive environments for primates. The collective intuitions of the contributors in this book provide an excellent set of guidelines for our continuing efforts to improve captive environments for primates.

ACKNOWLEDGMENTS

The research described in this chapter was supported by USPHS grants MH 29,775 and MH 35,215 and an NIMH Research Scientist Award to the first author. We thank Kim Bauers, Margaret Elowson, Sheryl Hayes, Tina Widowski and Toni Ziegler for discussion and criticism of the ideas presented here. We are especially grateful to Evalyn F. Segal for her critique of the paper. We thank the staff at Monkey Junge for their assistance.

— 7 —

The Problem of Foraging in Captive Callitrichid Primates: Behavioral Time Budgets and Foraging Skills

Eileen M. Molzen and Jeffrey A. French

INTRODUCTION

A primary goal of primate conservation is the perpetuation of free-ranging populations in their native habitats. For some species, however, this goal is presently not attainable for a variety of reasons, including severe environmental degradation, extreme pressure from human population growth, and small population sizes. The immediate conservation goals for these species are survival and propagation in captivity. Such is the case for the golden lion tamarin (*Leontopithecus rosalia*), an extremely endangered Neotropical primate from the coastal rainforests of Brasil. The lion tamarin is one of the few species that currently exists in greater numbers in captivity than in the wild (Kleiman, 1983; Kleiman, Ballou, & Evans, 1982). The National Zoo is directing an innovative program for reintroducing captive-born lion tamarins into protected native habitats. Although the program has had some success, there is still much to be learned about preparing animals for successful reintroduction (Kleiman Beck Dietz Dietz Ballou & Coimbra-Filho 1986). For at least the near future, maintaining populations of lion tamarins in captivity represents the only realistic option for this species.

Lion tamarins and other callitrichid primates live in small social groups composed of a single breeding adult female and male, several sets of offspring ranging in age from infancy to 3-4 years of age, and perhaps one or several unrelated adults or subadults of both sexes. They have traditionally been considered monogamous primates, and the majority of evidence from the field and from captivity supports this classification. In the wild, lion tamarins tend to forage in close association with other group members, and consume a variety of food types including invertebrate and vertebrate prey, fruit, leaves, and perhaps tree sap (personal observation; Dietz & Kleiman,

1986; Peres, 1986). **Benjamin B. Beck**

A major concern for lion tamarins (and other captive primates, for that matter) is the promotion of "naturalistic" patterns of foraging in laboratory and zoo environments. Foraging issues for captive primates divide conceptually into two categories: *behavioral capacities* and *foraging problems* (Cant & Temerin, 1984). First, the specific behavioral capacities necessary for acquiring, preparing, and consuming food items may be compromised in captivity. Skills such as tool construction and use, food extraction, predation, and recognition of toxic plants and animals are often neither required nor maintained in captive settings.

The second issue, foraging problems, is concerned with a number of questions, including when to forage, how long to forage, and what to do instead of foraging. Behavioral time budgets appear to be profoundly influenced by captive husbandry, but in ways that are as yet poorly understood or appreciated. The unrestricted provisioning of laboratory and zoo primates reduces the time needed to obtain food, rendering captive primates extremely efficient foragers, as measured by success per unit foraging effort or time. But this leaves the captive forager much more time for alternative activities than its wild counterpart, and lacking other behavioral alternatives, captive primates often fill the time engaging in pathological patterns of social and individual behavior (Erwin & Deni, 1979; Spinelli & Markowitz, 1985).

Foraging Skills of Wild Callitrichid Primates

Both types of foraging concerns apply to tamarins and marmosets (Primates: Family Callitrichidae). In the wild, callitrichid primates display a variety of specialized foraging skills. Tamarin diets typically consist of fruits, plant parts, tree sap, and both invertebrate and vertebrate prey (Garber, 1984; Hershkovitz, 1977; Lacher, da Fonseca, Alves, & Magalhaes-Castro, 1981; Terborgh, 1985). Each of these foods requires specialized foraging adaptations. Free-ranging lion tamarins show, for example, conspicuous morphological and behavioral adaptations for "extractive foraging." *Leontopithecus* have slender, elongated forelimbs, and a partially syndactylous middle finger that is used to locate, capture, and handle hidden and embedded grubs, insects, and other food stuffs (Hershkovitz, 1977). Extractive foraging is considered one of the most important skills that a captive lion tamarin must acquire prior to successful reintroduction to the wild (Benjamin Beck, personal communication; Kleiman et al., 1986).

As for time budgets, behavioral profiles of captive callitrichids appear to differ markedly from free-ranging individuals. In all Neotropical primates, including callitrichids, foraging occurs frequently throughout most of the day, and can occupy 10% to 50-60% of the time available (Garber, 1984; Pook & Pook, 1981; Yoneda, 1984). In a study of behavioral time budgets of capuchin monkeys in Peru, Janson (1985) found that 65% of group activity involved foraging, while feeding itself constituted 24% of the budget. Clearly, foraging and other food-related feeding activities comprise a major component of behavioral budgets in nonhuman primates (also see other chapters in this volume, such as Bloomstrand; Chamove & Anderson; Pereira Macedonia Haring & Simons; and Rumbaugh Washburn & Savage-Rumbaugh. Ed.)

Standard husbandry for captive tamarins in the majority of facilities is not unlike that for other captive primates. The animals are provisioned with a large food supply, readily accessible in one central location, once or twice daily. The time required to forage is thus greatly reduced (e.g., Vogt, 1978).

Foraging is one of the areas that has received the greatest attention from writers on psychological wellbeing in nonhuman primates. However, as is the case in many controversial and emerging disciplines, there is a great deal of heat (in the form of rhetoric, argumentation, and theoretical speculation) and very little light (good, objective empirical data collected explicitly to address a specific hypothesis). Our preference is to defer to empirical data for *both* the evaluation of hypotheses *and* the generation of legislation regarding the captive housing and husbandry of nonhuman primates. Therefore, in this chapter we present behavioral profiles of captive lion tamarins both before and after the introduction of a novel foraging device that was designed to induce two changes in behavior.

First, we hoped to lower foraging efficiency by making foraging more difficult so that the tamarins would have to try persistently to find food. Second, we hoped to promote some of the skills necessary for extractive foraging by embedding food out of sight of the tamarins, in places that required blind probing for successful acquisition.

METHODS

Subjects and Housing

The three family groups observed in this study contained a total of 14 golden lion tamarins, *Leontopithecus rosalia*. All were captive-born and housed in the University of Nebraska at Omaha primate facilities. Tamarins were housed in family units, visually isolated from other family groups, in large cages (2.0 x 2.0 x 1.8 m) constructed of wood frames and wire-mesh walls. Nest boxes, branches, ropes, and other accoutrements of the captive callitrichid provided substrates for locomotion play and resting. The visual isolation of cages allowed us to observe one family without disturbing other families, and also assured the novelty of the foraging device when it was introduced to other family groups.

The University of Nebraska at Omaha is a signatory member of the Cooperative Research and Management Agreement for *Leontopithecus rosalia*. All of the tamarins in the our Callitrichid Research Unit are on breeding loan from the National Zoological Park, Washington, D.C. All animals in the colony were born in captivity to legally-acquired animals. The tamarin facilities are fully accredited by AAALAC (American Association for Accreditation of Laboratory Animal Care). All research is predicated on one theme: acquire as much useful information as possible with minimum disturbance to the tamarins. For additional details of research, maintenance, and management, see French & Stribley (1985; 1987).

We divided the tamarins of each family into two categories, Adults and Juveniles. A juvenile was any animal who did not occupy a breeding role in a family group. Adults, therefore, were the breeding pair of a family. Table 1 gives a description of the age-sex compositions of each family.

Standard Husbandry Protocol

Husbandry practices were similar to those at many other callitrichid facilities. The tamarins were fed once or twice daily with a basic diet of commercial marmoset diet ("Zu-Preem," Hills Diets), along with a selection of fruits and vegetables. Two or three times a week a variety of additional food items was provided, such as peanuts, cheeses, yogurt, rolled oats, mealworms, and hardboiled eggs. The marmoset diet and supplemental items were placed in ceramic dog bowls atop a small platform on the cage floor (Figure 1). Although both simple in practice and economical of staff time, this method of provisioning does not promote the expression of foraging skills normally used by free-ranging lion tamarins, nor does it lead to activity budgets resembling those of free-ranging New World primates.

Alternative Feeding Protocol

In designing a more challenging foraging task, we had two goals in view: first, to stimulate extractive foraging skills , and second, to increase the amount of time required to acquire food. We developed a device that required the animal to forage from a suspended feeding station. We started with four-cup transparent food-storage containers. A 3-cm hole was cut in the center of each opaque lid, and a 7.5-cm length of smoothed PVC pipe was inserted and secured with dental resin so that it extended about 5.5 cm above the lid's surface. To ensure that the tamarins could not remove the lid, four small holes were drilled into the rims of both the lid and bowl and then fitted with four 1.5-cm S-hooks. One end of a 33-cm length of wire was wrapped around each S-hook and the other end of the wire was connected to a small dog-chain clip so that we could suspend the bowl approximately 27 cm below the top of the cage. (Figure 2)

Each bowl was filled with broken corn cob (commonly used as absorbent litter in rodent vivaria, and readily available) to about 5 cm below

Table 1

Social Composition and Ages of Families of Golden Lion Tamarins at the University of Nebraska at Omaha

	Adults	Sex	Age	Juveniles	Sex	Age
1	Shimmer	F	10	Darwin	M	3
	Brutus	M	15	Elysium	M	3
				Hans	M	2
2	Emily	F	5	Kuff	F	1
	Rufus	M	5	Link	F	1
				Myriah	F	0.4
				Niko	M	0.4
3	Mom	F	12	Fritz	M	3
	Dad	M	16	Gerry	F	3
				Jurgen	M	2

M = male, F = female; ages in years; infant (0.4-year-old) tamarins in Family 2 not included in observations.

the edge of the PVC pipe. The bowls were then baited by scattering 10 to 12 raisins (a relished tamarin treat) near the top layer of corn cob. This ensured that to obtain the raisins from the bowls the tamarins would have to engage in extractive foraging, that is, obtain food from a restricted location without benefit of visual cues, discriminating by touch between edible raisins and inedible corn cob.

The bowls were so placed that an animal could not reach into a bowl while hanging from the side of the cage or sitting on one of the branches in the cage. The only way the tamarin could support itself while working to obtain a raisin was either by hanging by the hind limbs from the top of the cage, or sitting on the bowl's lid and grasping the PVC tube or one of the device's suspension wires for support (Figure 2).

Once the tamarins discovered that the bowls might contain some raisins, they became an object of focused attention. To introduce uncertainty in food availability, the tamarins were given both a baited and an unbaited bowl during all phases of the experiment. They continued to receive their standard rations in the usual way, in ceramic dog bowls placed on a platform on the cage floor.

Figure 1. Standard feeding protocols in most captive callitrichid colonies involve placing food in the same predictable location at the same predictable times each day.

Figure 2. Golden lion tamarins acquiring raisins from the novel foraging device.

Behavioral Assessment

To establish baseline behavioral profiles, a series of observations was conducted during a period of standard husbandry (*Standard*). Next, the foraging bowls were put in place and a second series of observations was conducted (*Novel*). During each observation session, first one focal subject's activity was recorded on a checksheet every 15 sec for five min, then another focal subject was observed for 5 min, and so on until all family members had been observed ("instantaneous sampling", Altmann, 1974). The order of observing focal individuals in a family varied randomly over sessions. At each sampling instant, the observer recorded which one of eight activities the focal tamarin was displaying. The categories were normal, frequent activities of golden lion tamarins in captivity (Stribley, 1984). (Table 2)

Statistical Analyses

Ten Standard observation sessions were conducted for each family group. Because of husbandry requirements unrelated to this study, the number of Novel observation sessions for each family phase varied from 7 to 10. Thus, the number of individual activity samples for each tamarin ranged from 340 to 400. Our major concerns in the analyses were whether or not the novel feeding device affected activity profiles, and whether and to what extent adults differed from juveniles in their reactions to the foraging task. Therefore, a 2 X 2 Analysis of Variance was conducted for each activity, with phase (Standard vs. Novel) and age (adult vs. juvenile) as factors. Differences in the proportion of time engaged in activities were considered to be statistically significant if chance probability was below 0.05.

RESULTS

Figure 3 shows the activity budgets of adult and juvenile tamarins under Standard feeding protocols. Neither adults nor juveniles spent much time actively exploring the environment (Visual/Explore) or searching for food (Look/Forage). The two categories combined reflect an average of 8.7% of the activity budget for adults and juveniles combined. Eating occurred as seldom, occupying approximately 9% of the daily activity budget of both adults and juveniles. Travel represented approximately 10-15% of the time budget, and resting filled a substantial (20-30%) amount of time. Grooming and other social activities were also prominent, together accounting for approximately 35-40% of the time.

The introduction of the foraging device produced a significant change in behavioral profiles. Figure 4 presents activity budgets for juvenile and adult tamarins in both Standard and Novel feeding conditions. The

Table 2
Definitions of Activity Categories of Behavioral Profiles

1. Look/Forage:	Any action displayed in the search for or acquisition of food.
2. Visual/Explore:	Animal visually, orally or manually investigates any aspect of the physical environment except food sources.
3. Eating:	Consuming food.
4. Drinking:	Consuming water.
5. Travel:	Actively locomoting about the cage: running, climbing and walking
6. Rest:	Inactivity; not doing anything that falls in another category.
7. Grooming:	Sifting fur through fingers or teeth; includes allo- and autogrooming.
8. Other Social:	Any interaction betwen two individuals other than grooming; includes aggression and play.

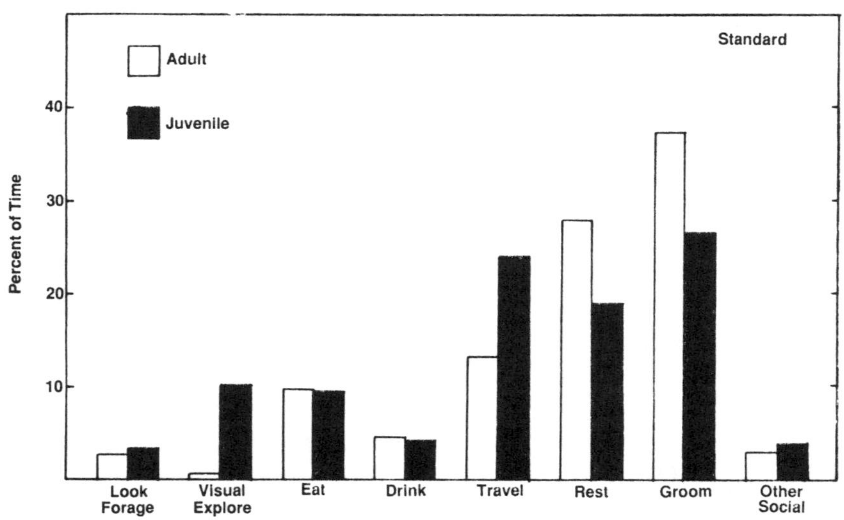

Figure 3. Behavioral profiles of lion tamarins during Standard feeding practices.

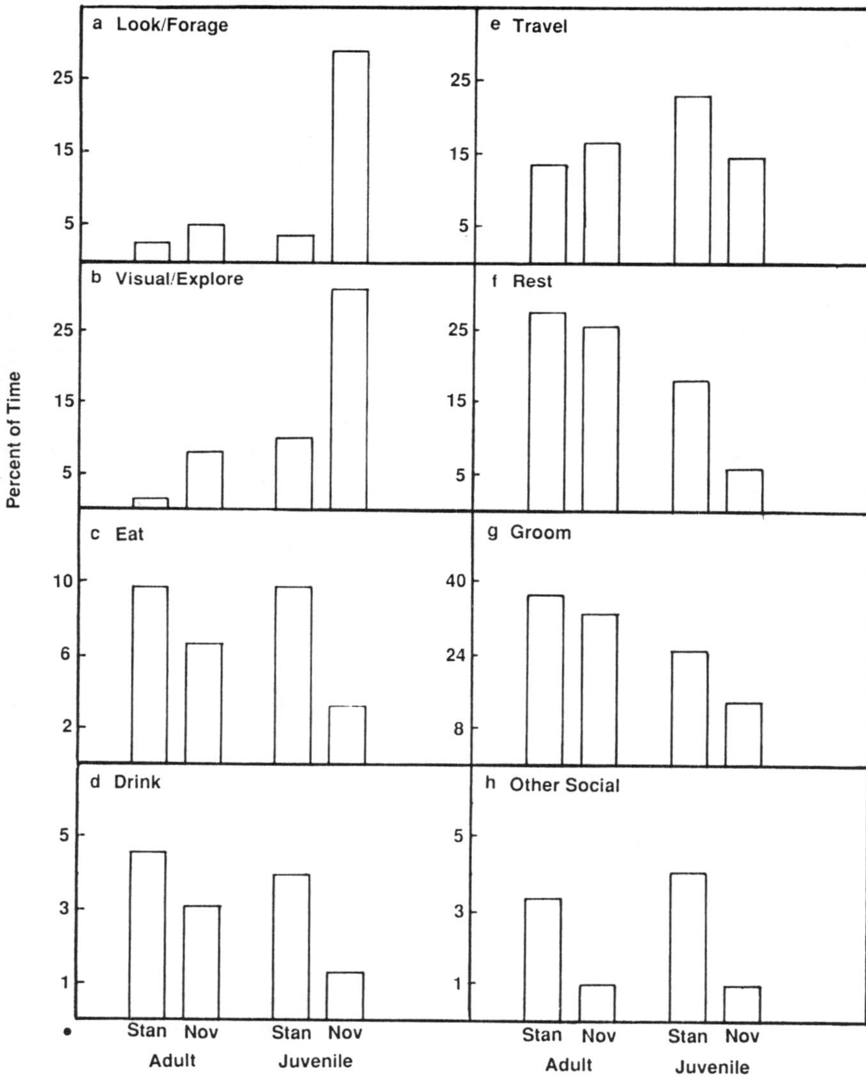

Figure 4. Behavioral profiles of lion tamarins as a function of age and phase. Scale values differ among behavior categories.

Table 3
Mean Foraging Efficiency for Adults and Juveniles During Standard and Novel Feeding Conditions

Age		Standard Feeding	Novel Foraging
Adult	(n=6)	2.51	0.54
	(SEM)	(2.11)	(0.26)
Juvenile	(n=8)	0.68	0.05
	(SEM)	(0.16)	(0.01)

proportion of time tamarins spent investigating their environment, both actively (Fig 4a; Look/Forage) and passively (Fig 4b; Visual/Explore), rose significantly in the Novel phase ($F(1,12) = 7.80, p < 0.016$; $F(1,12) = 13.79, p < 0.003$, respectively). Juveniles spent significantly more time in these activities than did adults in both phases ($F(1,12) = 5.49, p < 0.037$ and $F(1,12) = 20.62, p < 0.001$, respectively). Juveniles did more active investigating (Look/Forage) during the Novel phase than adults did, but the interaction between age and experimental phase just failed to reach significance ($p = 0.055$).

Eating occupied less time while the foraging device was present. Fig 4c indicates that eating decreased in both juveniles and adults ($F(1,12) = 19.20, p < 0.001$). Likewise, drinking took up less time during the Novel phase (Fig 4d; $F(1,12) = 11.95, p < 0.005$).

The foraging device had different effects on locomotion and resting for adults and juveniles (Fig 4e and f). Following introduction of the foraging bowls, Adults were observed travelling significantly more often, but juveniles nonsignificantly less often (Age x Phase interaction, $F(1,12) = 8.03, p < 0.02$).

Resting accounted for about the same amount of adults' activity budgets whether feeding conditions were Standard (28%) or Novel (26%). In contrast, the foraging devices significantly reduced the proportion of time juveniles rested from 19% to 6% (Age x Phase interaction, $F(1,12) = 5.54, p < 0.04$).

The novel feeding arrangement also affected social interactions. Grooming occurred less frequently in the Novel phase than during the Standard phase (Fig 4g; $F(1,12) = 11.40, p < 0.005$). The decline was less for adults (from 37% down to 34%) than for juveniles (from 26% down to 14%), but the Age by Phase interaction failed to reach statistical significance.

Other Social Behavior was also less frequent during the Novel Feeding phase (Fig 4h; $F(1,12) = 14.44, p < 0.003$), but this category was negligible in any case (< 2%).

We attempted to assess the influence of the new foraging device on foraging efficiency. To calculate efficiency, the number of instantaneous samples when a tamarin was observed to be eating was divided by the number of samples when it was observed foraging. The ratio is an estimate of the proportion of time invested in eating, relative to the effort expended to acquire something to eat. A value greater than 1.0 reflects an efficiently foraging tamarin, one who spends more time eating than foraging. Conversely, a value much less than 1.0 suggests greater foraging effort, with animals spending more time foraging than eating. Table 3 presents the foraging efficiency ratios for adults and juveniles when the novel foraging device was and was not present. It is apparent that the device markedly reduced foraging efficiency, especially for juveniles.

DISCUSSION

The activity profiles of lion tamarins under standard provisioning protocols for captive callitrichids differed from reported activity budgets of free-ranging Neotropical primates primarily in time expended in foraging activities. Introducing unfamiliar foraging devices that were explicitly designed to make it harder to secure food treats had a significant influence on the activity budgets of both adult and juvenile tamarins. Time spent actively exploring the environment and foraging increased, while time spent eating and drinking decreased. Adults travelled more, rested less, and engaged in fewer nongrooming social interactions when the device was present. Juveniles showed similar changes, with the exception of travelling, which did not increase but rather decreased. An indirect index of foraging efficiency revealed that, in fact, the device reduced foraging yield and increased foraging effort to levels similar to those observed in free-ranging populations (Hershkovitz, 1977; Pook & Pook, 1981; Terborgh, 1985; Yoneda, 1984). These dramatic changes in behavioral profiles were produced even though the foraging device was supplemental to, rather than a replacement for, standard provisioning.

There were pronounced differences between adult and juvenile tamarins both in behavioral profiles and in skills and attention directed to the foraging task. Although both adult and juvenile profiles were affected by the foraging devices, changes in juvenile profiles were more pronounced. Food-related behavior increased more in juveniles (a nine-fold increase in Look/Forage and a three-fold increase in Visual/Explore) and foraging efficiency fell proportionately more.

Not surprisingly, our observations indicated that juveniles were more likely to be interacting with the devices at any sample instant than adults.

The juveniles' interest in the foraging devices was slow to habituate, and individuals often continued to search for raisins long after the day's supply had all been consumed. In every family, juveniles were the first to "solve" the foraging problem (find raisins and extract them from the bowl). Adults paid attention to the bowls only after a juvenile had successfully extracted raisins from them. Although we did not quantify it in this study, it was clear to us that juveniles ate the majority of raisins from the foraging devices; in all aspects, they were more flexible in adapting to the new foraging demands.

A similar pattern of differential results by age was documented by Kleiman et al. (1986) in their program of training lion tamarins to forage, avoid predators, and locomote prior to their reintroduction to the Poco das Antas reserve in Brasil. Greater flexibility in learning new food-acquisition skills has also been documented in the young of other species of nonhuman primates, most notably in the novel food habits of Japanese macaques (e.g., Kawai, 1965). The propagation of extractive foraging skills in tamarins parallels the pattern of the appearance and spread of novel feeding tactics in Japanese macaques: novel food-processing skills first arise in juveniles and then spread to adults. This may be a general rule for the first appearance and subsequent spread of novel foraging skills in nonhuman primates.

Kleiman et al. (1986) evaluated lion tamarin foraging efficiency on a variety of foraging tasks as part of a rehabilitation program for captive-born tamarins prior to reintroduction to the wild. They provided tamarins with a graded series of foraging tasks, starting from a standard task equivalent to ours, another with standard food bowls placed in novel locations, and one in which food was embedded in "pseudo-naturalistic puzzle-boxes." Visits to the food source per hour (a measure of foraging effort) increased with increasing complexity of the foraging task. Takes per hour (a measure of foraging success) decreased with increasing complexity. Hence, takes per visit, an index of foraging efficiency, decreased to 0.61 in the puzzle-box condition. Although our measure of foraging efficiency is not directly comparable to that of Kleiman et al., the magnitude of the decline in efficiency is equivalent in the two studies.

It is hard to say whether the modifications our foraging device produced in behavioral profiles constitute an improvement in tamarin "psychological wellbeing". The difficulties inherent in defining, estimating and assessing psychological wellbeing are discussed at length elsewhere in this volume (see especially Bramblett; Thomas & Lorden; and Novak & Drewson). In our anthropocentric view, however, we believe the changes we observed were positive. The amount of time the tamarins were inactive decreased, and they took a greater interest in actively exploring their environment. Although affiliative behavior such as grooming also declined, acceptably high levels of grooming persisted in the Novel phase in both adults and juveniles (34% and 14% of the daily activity budget, respectively).

Our observations were conducted over approximately four months.

We do not know whether habituation to the foraging devices would occur with longer exposure. Interest remained high, however, throughout the Novel phase. Hence we feel confident that the devices can be effectively used as a matter of course in day-to-day feeding protocols for tamarins. Studies planned for the future include (i) exploring the feasibility of providing the entire day's ration in the foraging devices; (ii) attempting to simulate natural foraging more closely by installing three to five devices per cage but baiting only a random sample of them each day; and (iii) varying the food items placed in the foraging devices each day to try to sustain the tamarins' interest indefinitely.

ACKNOWLEDGMENTS

We wish to thank the following persons for their assistance in various phases of this project: Ralph Hermann, DDS, provided able assistance in the construction of the foraging devices; Theresa Dethlefs and Betty Inglett kindly provided constructive criticisms of an earlier version of this chapter; Denise Hightower and Bill Oldenhuis provided expertise in the care and maintenance of the tamarin colony; and Scott Lyman contributed time for photographic documentation of foraging in our tamarins. We thank Doug Armstrong and Lee Simmons of the Henry Doorly Zoo: Omaha Nebraska, for excellent veterinary care. Portions of the data discussed in this chapter were presented at the meeting of the American Society of Primatologists in June, 1987. EMM's travel to the meeting was paid from UPFF Fund B of the University of Nebraska at Omaha. This work was supported in part by Public Health Service grant HD 23139-01 to JAF.

Family: Cebidae (Capuchin-Like Monkeys)

— 8 —
Free-Environment Rooms as Alternative Housing for Squirrel Monkeys

James E. King and Vicky R. Norwood

INTRODUCTION

Although squirrel monkeys occasionally appeared in laboratories as early as the 1930s, it was not until the early 1960s that they became popular as research subjects. The popularity of these attractive monkeys is largely attributable to their relatively small cost and small size, important considerations in the economics of establishing and maintaining primate colonies. Several excellent surveys of captive squirrel monkey husbandry that include breeding information have been published (e.g. Abee 1985; Dukow 1983: 1985; Hopf 1981). This literature asserts that squirrel monkeys can be successfully housed in relatively small cages (about 0.15 cubic m) for indefinitely long periods. However, it is argued with increasing frequency that maintaining animals for long periods in relatively small, sparse cages can cause profound behavioral and physiological harm. Conventional caging may be detrimental if it prevents the animal from expressing its natural repertoire (see Erwin & Deni 1979; Fox 1986; Markowitz 1982). In this chapter we will first describe some basic characteristics of squirrel monkeys that may be relevant to optimal housing. Second, we will review conventional advice for housing squirrel monkeys along with some potential problems. Third, we will describe our experiences with an inexpensive and very successful alternative to cages to house squirrel monkeys. Finally, we will describe the introduction of fishing as one source of psychological enrichment for group-housed monkeys.

SALIENT CHARACTERISTICS OF SQUIRREL MONKEYS

Squirrel monkeys are small, South American monkeys whose natural habitat includes large areas of South America, particularly the Amazon basin and portions of Central America. Squirrel monkeys are divided into numerous subspecies whose classification is still subject to vigorous discussion (Hershkovitz 1984; Thorington 1985). Subspecies differ somewhat in weight, coloration, and behavior.

Squirrel monkeys are omnivorous with most of their diet composed of fruit and insects foraged from their arboreal environment (see Baldwin 1985). They are attracted to flying insects and are adept at capturing them in flight (Fleagle Mittermeir & Skopec 1981). Their diet also includes supplemental amounts of small mollusks and vertebrates. Boinski and Timm (1985) reported that squirrel monkeys (*Saimiri oerstedii*) in Costa Rica prey on tent-making bats.

Squirrel monkeys are intensely saltatory. Leaping is a favored mode of locomotion and over 40% of their extensive traveling is accomplished by leaping. This percentage is higher than for other South American monkeys occupying similar habitats. Smaller but still significant amounts of leaping occur during foraging and feeding (Fleagle et al. 1981). Their day range is variable depending upon the food supply and extends from 1 to 4 km (Izawa 1976; Baldwin & Baldwin 1972). This compares with an estimated day range of 0.4 km for howler monkeys (Milton 1980). In large captive enclosures, squirrel monkeys display extensive use of available space. Fragaszy (1979) observed that individual squirrel monkeys ranged over about twice as much space as titi monkeys within an approximately one-hectare field and that males traversed as much as one km during a one-hour observation session. They are extremely active, stay only briefly in one location and frequently change location (Fragaszy 1985).

A striking recent discovery is that levels of adrenal and gonadal hormones in squirrel monkeys may be as much as ten times higher than in Old World monkeys and humans. Furthermore, levels of these hormones, particularly cortisol, vary considerably with psychological variables such as stress (see Coe Smith & Levine 1985). These anomalies may make squirrel monkeys particularly susceptible to cardiovascular and renal deterioration (Strickland & Clarkson 1985). A clear possibility is that these disorders will be exacerbated by stressful housing.

Another characteristic weakness of squirrel monkeys is their susceptibility to dehydration because of a relatively short intestine. The short intestine may be an adaptation to their substantially insectivorous diet (Fooden 1964; Hill 1960). Stress-induced dehydration in squirrel monkeys should therefore be avoided or promptly treated if it does occur.

In brief squirrel monkeys are highly active, adapted to frequent leaping, and possess the skills and propensity to capture flying insects and small vertebrates. They also have high and variable levels of cortisol and other hormones that may predispose them to cardiovascular and renal deterioration as well as other stress-related disorders. Ideal laboratory housing for squirrel monkeys should be considerate of these characteristics.

CONVENTIONAL HOUSING OF SQUIRREL MONKEYS

Several excellent discussions of advice for caging, feeding, and breeding of squirrel monkeys are currently available (e.g. Abee 1985; Ausman Gallina & Nicolosi 1985; Dukelow 1983: 1985). Most authors agree that individual caging is not optimal although it is sometimes necessary because of space constraints or experimental requirements. Large gang cages are sometimes used as an alternative. For individual cages, a floor area of 0.15 m^2 with 51 cm of vertical height is a common standard. Usual standards of cage cleanliness and sanitization are emphasized. Perches and individual water bottles with sipper tubes or an automated watering system are recommended. Temperature should be maintained between 21 and 24° C with relatively high humidity (about 50%) to avoid drying respiratory surfaces. The air circulation system should provide at least 8-12 air changes per hour.

We believe that adherence to the standard husbandry guidelines for individual caging over long periods of time can have deleterious effects on squirrel monkeys. Discomfort and psychological stress may be occasioned by the following sources.

Lack of Exercise. Although squirrel monkeys remain quite active in small cages, the opportunity for leaping and sustained quadrupedal locomotion is virtually eliminated. Turnquist (1983) noted that joint mobility was more variable in cage-housed than in free-ranging patas monkeys and that the increased variability was especially high in monkeys that had spent their entire lives in captivity. In patas monkeys, joint mobility decreases with age until about 18 months at which time it stabilizes. This decrease in mobility is caused by the increasing size and strength of opposed muscle sets. We have observed that when individually-caged monkeys are released into larger enclosures, an initial period of awkward activity ensues characterized by inaccurate leaping from one structure to another and a tendency to fall from elevated locations.

High Noise Levels. Sound levels in colony rooms of individually-housed monkeys can be quite high. Abnormal behavioral and physiological effects of loud noise have been frequently reported (see Gamble 1982; Peterson 1980; for reviews). For example, Barrett and Stockham (1963) found that ten minutes of intensified colony sound such as cage rattling and talking caused corticosterone increases of at least 100% in group-housed rats. An undergraduate research project at the University of Wisconsin ("Monkeys Agree" 1973) showed that 100-db noise levels from recorded jackhammers and rock music increased cortisol levels in rhesus monkeys during the first hour of exposure followed by a decrease to near baseline levels.

Hosing as a Source of Stress. It is usual in colony rooms to hose down drop pans under the cages every day. Squirrel monkeys usually move as near the tops of their cages, away from the water, as they can. Hosing remains aversive even after years of caging and continues to evoke a profusion of loud shrieking.

Low Light Levels in Cages. Although illumination level of the colony room may be within standard ranges, levels within cages may be much lower. Within layered rows of cages, illumination is diminished at the bottom levels. Long-term exposure to low illumination may have deleterious consequences for diurnal species.

Human Proximity. Caretaking activities in colony rooms include daily cage washing and feeding that bring the caretakers near to each monkey's cage. Squirrel monkeys respond variably to this proximity depending, among other things, on how long the monkey has been caged. The inability to withdraw from the human is a potential source of discomfort and stress for the animal.

Tail Sores. Squirrel monkeys do not have ischial callosities as an adaptation for sitting. Among New World monkeys, squirrel monkeys are especially susceptible to sores on the dorsal surfaces at the bases of their tails (Clewe & Duvall 1966) resulting from sitting with the tail tucked under the body and brought forward so that much of the monkey's weight is supported at the base of the tail. Tail sores are exacerbated when monkeys sit for extended times on narrow perches like the small-diameter dowels or pipes that are commonly used in individual cages. Tail sores can be lessened by use of wire-mesh platforms instead of narrow pipes (Clewe & Duvall 1966).

Imposed Proximity to Urine and Feces. Individual cages with conventional drop pans force the monkey into close and often direct contact with urine and feces. Squirrel monkeys are capricious in their food handling, repeatedly taking an item briefly, then discarding it only to retrieve it soon after. This cycle of discarding and retrieving food inevitably results in their eating food from the drop pan that is contaminated with urine and feces. Use of larger food items or smaller mesh on cage floors simply leaves the food inside the cage where it is even more accessible to the monkey.

FREE-ENVIRONMENT ROOMS AS AN INEXPENSIVE ALTERNATIVE

Most primatologists would agree that the welfare of monkeys is better served by housing them in large groups rather than in individual cages. An alternative is the outdoor enclosure, but adverse climate, lack of suitable land, and high construction costs often prohibit outdoor housing. Another possibility is a large indoor gang cage. These are usually quite expensive and despite their size present a severe, psychologically unstimulating environment.

In January 1983 we decided to establish two free-environment rooms. Individual and gang cages were removed from two conventional colony rooms and the monkeys were simply released into the rooms. Each room was approximately 7.4 x 4.3 x 2.4 m with sealed tile floors and walls. One room had a center floor drain, the other a gutter with drain along one wall. Each room was illuminated by at least four fixtures each containing four Vita-lite fluorescent lights. Other room fixtures were hot and cold water taps, covered electrical outlets, and temperature and humidity controls with protective shields.

Various climbing devices were introduced including a 6-ft rack of 12 rodent gang cages (approximately 18 x 24 x 12 in) with fronts removed; racks with no cages; small live trees; step ladders; ropes; and plastic chains. Tree branches were also dispersed through the rooms. These provided the monkeys with a variety of surfaces and areas for exercise, especially leaping, as well as private and secure resting areas. A double-door system was installed in one free-environment room to prevent escapes. This expedient was quickly shown to be unnecessary as the monkeys in the room without the double door rarely attempted to escape when the door was opened. Thus, with the exception of the expense of installing the double door, no significant costs were incurred in converting from one conventional caging to the free-environment rooms. Figure 1 shows our conventional colony room with individual cages and Figure 2 shows the free-environment room with the double-door system.

Individual monkeys were identified by collars bearing small letter beads that spelled out the monkey's name and colored beads that identified the animal by a two-color code. The beads were obtained from a hobby store. Beads were threaded onto plastic-coated jewelry wire, individually fitted to each monkey, and secured by crimping beads from a jewelry store.

Our first free-environment room housed 11 females ranging in age from 1 to over 18 years and 13 males with the same age range. We emphasize that only five of the males were sexually mature; the remaining eight were immature. One of the sexually mature males had only three limbs. During the three months that we kept him in the room he was highly mobile although he did not climb as much as the other monkeys. He was removed after three months when aggression between him and other males increased to unacceptable levels.

Figure 1. Individual cages in conventional colony room.

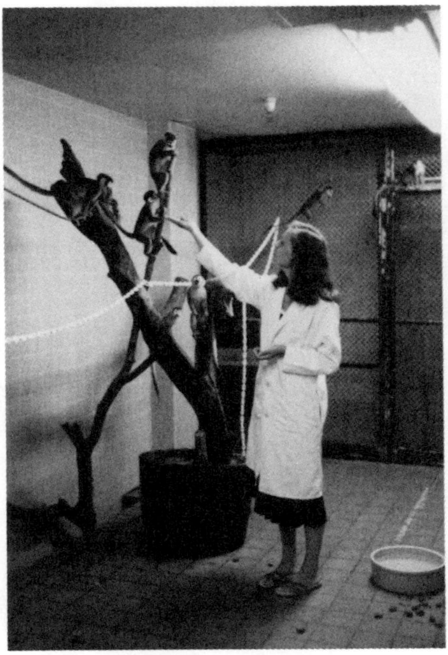

Figure 2. Free-environment room with double door in the background.

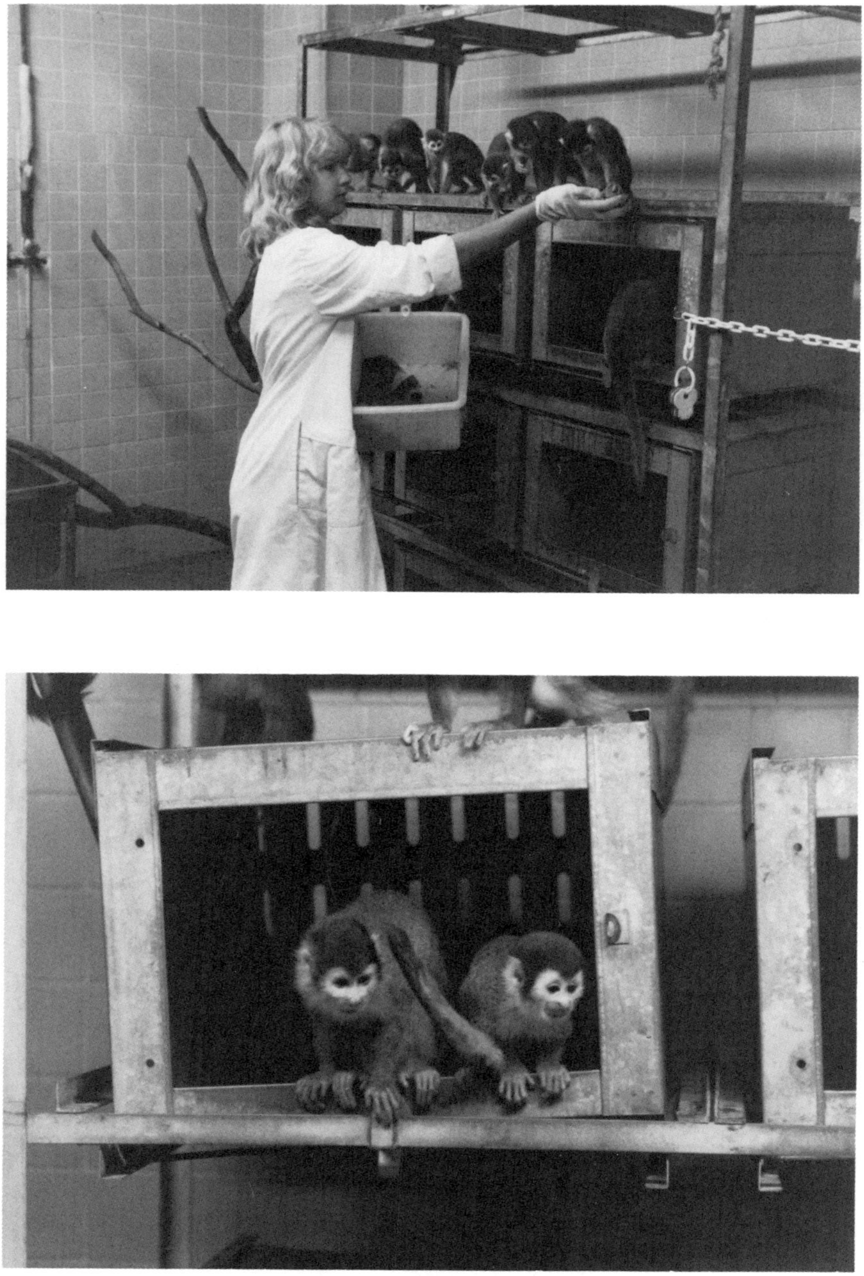

Figure 3. (Top) Caretaker feeding monkeys in free environment room. (Bottom) Two monkeys in nesting box made by removing front door from a small cage.

Two of the other sexually mature males were totally blind. One was a young adult and the other was aged (estimated age over 15 years). The aged male remained in the room until his death 1-1/2 years later. He eventually learned the layout of the room but spent more time on the floor than the other monkeys. He adapted well to the other monkeys. The other blind monkey was a young adult male about six years old. This animal was repeatedly attacked by other monkeys and was removed after one week.

Our second, more experimental, room housed 14 males ranging from 1 to over 18 years old. Six were sexually mature. Occasional fighting broke out among the two young adult males and the aged males. The small number of young adult males and the absence of females are necessary to maintain social stability in the room. The population is adjusted as the immature males reach sexual maturity.

Immediately following the establishment of these two free-environment rooms, a few monkeys incurred sprains and broken teeth, probably resulting from falls attributable to their rustiness in leaping and climbing. Indeed, many monkeys were reluctant to climb at all during their first few days in the rooms.

Two deaths resulted from attacks by other monkeys. One adult male was killed during a period when five adult males were in the mixed-sex colony. An adult female was also killed. She was a surrogate-reared monkey whose social behavior was abnormal. It is clearly important to adjust the sex and age composition of group-housed monkeys to avoid destructive aggression.

When necessary, monkeys can be captured in a net. More experienced monkeys usually consent to being carried out by a caretaker.

The free-environment rooms are cleaned every morning. Disinfectant is applied to the walls and floors twice a week. Cleaning requires about half an hour of caretaker time each day. In contrast, a colony room with 30 individually-housed monkeys requires about 1-1/2 hours of cleaning time per day in addition to the time required for periodic cage sanitizing.

Monkeys are fed chow twice a day from ceramic crocks distributed about the floor. In the late afternoon an additional feeding of fruit, vegetables, or small amounts of hard-boiled eggs is provided in the crocks. The monkeys quickly adjust to the presence of the caretakers and researchers. The top of Figure 3 shows a caretaker feeding the monkeys and the bottom shows two monkeys in a nesting box.

The free-environment rooms were markedly quieter than the rooms with individual cages, mostly because there was no cage- or perch-rattling in these rooms.

FISHING AS A SOURCE OF ENVIRONMENTAL ENRICHMENT

In their natural habitat squirrel monkeys often capture small vertebrates which supplement their main diet of insects and fruit. Furthermore, squirrel monkeys are adapted to seize rapidly-moving, flying insects. Fish, like flying insects, are small, rapidly-moving visual targets. Therefore, we reasoned that although squirrel monkeys undoubtedly do not capture live fish in their natural arboreal habitat, they might nevertheless be attracted to brightly-colored fish swimming in a small pool of water and might attempt to capture them. Fishing by nonhuman primates is not totally unprecedented. Hamilton and Tilson (1985) observed chacma baboons taking fish from desert water holes in Namibia. Fishing by squirrel monkeys in the free-environment rooms was studied by Virginia Landau (1987) as part of her dissertation research.

Before the start of the initial experiments, a group of nine squirrel monkeys was placed in each free-environment room. The other monkeys were placed in individual cages. (These were monkeys serving in experiments requiring individual testing or who did not adapt well to the free-environment rooms.) For three weeks before testing started, monkeys in the fishing experiment were fed a mixture of monkey chow and sardines daily, to accustom them to the taste of fish. After this adaptation period, ten goldfish (*Crassius* sp.) were placed daily in a small (22.4 cm diam) ceramic crock in each room. We had expected that it would take at least a few days before the monkeys tried to capture the fish; but in fact, appearance of the first crock of fish elicited a frenzy of excited attempts to grab them. Ultimately, all but one of the 18 monkeys in the two rooms tried to catch fish at least once and all but two were successful at least once. The two monkeys who did not catch any fish did, nevertheless, eat fish remnants left by others. About 26% of all attempts to retrieve fish from the small crocks were successful. Figure 4 shows squirrel monkeys fishing from one of the small crocks.

To increase the difficulty of fishing, the daily ration of fish was then placed in small wading pools. The first pool was square, 87 cm on a side. As before, ten fish were placed in the pool each day. During the first 22 days, 12 of the 18 monkeys caught at least one fish from the pool. Of all fishing attempts, 57.5% were with the left hand, 17.5% with the right hand, and 25.0% with both hands. About 22% of fishing attempts were successful. Figure 5 shows a left-handed attempt to catch a fish from the wading pool.

Figure 4. (Top) Monkeys fishing from water-filled crock. (Bottom) Monkey in center beginning to eat fish it has just caught.

The monkeys tracked the fish as they swam through the water and frequently ran around the periphery of the pool in pursuit of a particular fish. Individual fish that had become separated from the main school were the usual targets of these pursuits. Caught fish were almost always retained by the catcher. We were impressed by the contrast between the sustained holding of caught fish and the frequent discarding of monkey chow and produce taken from food bowls. Monkeys in possession of a freshly-caught fish were often closely observed and followed by the other monkeys. Theft and attempted theft were fairly common. Figure 6 shows the aftermath of a successful catch.

Squirrel monkeys' fastidiousness toward food item was abundantly displayed in their treatment of fish they were observed to eat. Fish scales were apparently unpalatable; before eating a fish, a monkey typically rubbed it against a tree branch or the side of the pool, which removed some of the scales. Preferred cleaning sites on tree branches soon became covered with a bright layer of scales.

The monkeys were often surprisingly effective at cleaning the fish. They first bit the head off, then removed and discarded the entrails. These rejected portions were usually picked up and eaten by another monkey.

We believe that introducing fishing as part of the monkeys' daily feeding activities offers many interesting possibilities for future research as well as for increasing environmental stimulation for captive squirrel monkeys. (Also see Gilbert & Wrenshall: this volume.) In comparison to taking food from a dish, which makes absolutely minimal demands upon monkeys, fishing offers a challenge; it markedly increases the time, effort, concentration and skill required to procure food.

The most interesting consequence of the daily fishing routine may, however, be its effects on the group's social structure. Dominance, cooperation, imitative learning, food sharing, social roles of skilled fisher-monkeys, and patterns of attentiveness among monkeys are all possible sources of social complexity that may change in interesting ways following introduction of live fish as a food source.

Goldfish can be purchased for less than $10 for 100 medium-sized fish. Tilapia, which are easily bred, can be produced even more inexpensively.

Figure 5. Monkey pursuing fish from the edge of wading pool.

Figure 6. Monkeys on cage stand after a successful catch by the monkey in the center. Notice the gaze of the monkey on the left directed at the fish.

CONCLUSIONS

Our experience with free-environment rooms shows the feasibility of this economical and psychological enriching approach to housing squirrel monkeys. Any room with tiled or waterproof, washable walls, a floor drain and a water source can be converted to a free-environment room with little additional expense. The small size and limited destructive propensities of squirrel monkeys make them particularly suitable for these rooms. A combination of free-environment and individual caging is probably needed because some monkeys will be unacceptably aggressive in groups while others will be perpetual victims of aggression. When adult females are present, the number of adult males in a free-environment room probably cannot exceed one or two. Free-environment rooms offer researchers greater opportunities for increasing behavioral stimulation of captive primates than individual caging. Giving the monkeys regular opportunities for fishing is a particularly effective example.

— 9 —
A Closed Colony of Squirrel Monkeys for Laboratory Studies

Eric A. Salzen

HISTORY OF THE COLONY

This is a retrospective account of an attempt to use a small primate for neurobehavioural study while minimizing the adverse effects that this kind of use can have both on natural populations and on the life and wellbeing of the individuals in captivity. This attempt involved compromises between the demands of controlled behavioural and neurophysiological study in the laboratory and the welfare of the monkeys. The compromises adopted at the time of the establishment of the colony were not unreasonable, but experience with this colony and subsequent developments in primate husbandry have shown that much more can and should be done to improve the lot of such monkeys without imposing unreasonable conditions on the physiological laboratory. Some of these possible improvements will be indicated as "post hoc" findings and "fruits of experience" offerings.

A brief preliminary account of the colony was given by MacRae and Salzen (1977). The requirements to be met at the time the colony was established were to provide a limited supply of a small primate suitable for neurobehavioural work in a small laboratory of a university psychology department. In addition the colony was to be used for teaching in comparative psychology, since only a dual-purpose research and teaching facility could justify the expense of a primate colony. Squirrel monkeys were chosen because they were (at the time) less expensive, easier to handle, less of a health hazard in terms of the Herpes simiae (B) virus, and required smaller scale and less robust housing and experimental equipment than macaques. There was a growing literature of both behaviour and brain studies for this species, triggered principally by the pioneering studies of

MacLean and Ploog (e.g., MacLean, 1959; Ploog & MacLean, 1963; Ploog, Hopf, & Winter, 1963). Because of the cold and inclement north European climate, the housing for such small slender primates as squirrel monkeys had to be entirely indoors and in limited space. Furthermore it was decided to buy one batch of monkeys and breed from them as a closed colony. This would put a once-only demand on the natural population and would involve only one period of the lengthy (six months), strict, and restrictive quarantine required in Great Britain to prevent importing rabies. Furthermore, once screened, tested, and cleared of parasites and disease, the colony would be free from further infections and could be handled and cared for safely by students, research workers and technicians in a more relaxed manner.

Accordingly 40 female and 20 male squirrel monkeys (*Saimiri sciureus*) of the "Gothic" or "Leticia" type (Cooper, 1968) were obtained from a dealer and housed in single-sex pairs in single primate cages (90 cm high x 67.5 cm deep x 60 cm wide) while being treated for parasites and ill-health and becoming accustomed to the laboratory. Dealer-supplied stock is high in aged and immature monkeys and is in poor condition so that many fail to survive. On arrival the males weighed 700-800 gm and the females 500-600 gm. The corresponding weights once established in the laboratory were 1000-1100 gm and 800-900 gm respectively. Monkeys suitable for breeding stock were selected and put into four breeding units, each with four females and two males. The remainder were kept in single-sex groups of from three to six for experimental work or, in the case of immature monkeys, for replacement of breeding stock. The breeding units were maintained successfully with some stock replacement, including laboratory-born females, for nine years, when the colony had to be sold because of financial and staff cuts in the University that left only one interested user of the colony.

HOUSING AND HANDLING

Housing. The colony was housed in five rooms with a total area of approximately 50 m^2. One room held the breeding units, two rooms held stock and experimental animals, one room served as a nursery/sick room, and the fifth held two exercise pens. Artificial lighting was used with a constant 12-hr day because the local day-length varied from 6 hr in winter to 20 hr in high summer. With a ventilation of 12 changes per hour required by licensing authority irrespective of stocking density and humidity needs, it was not possible to maintain humidity above 50%. Room temperature did not fall below 24° C.

In order to reduce disturbances from sudden external noise and to reduce interactions between animals in different rooms, each room had a

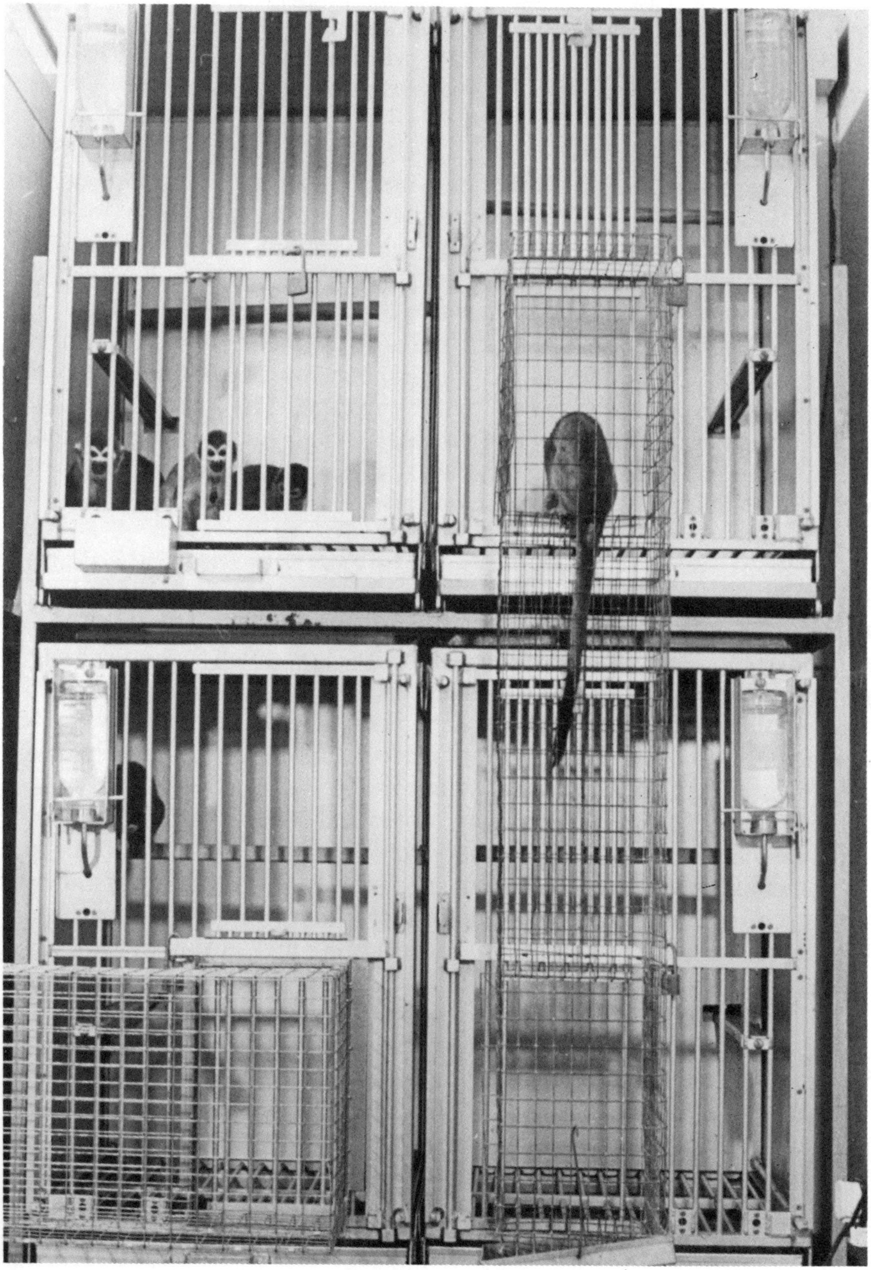

Figure 1. A typical four-cage assembly with vertical connecting passageway as used for breeding units. Two assemblies can be connected by a horizontal passageway (lower left) for larger groups or for studies of group interaction.

loudspeaker so that a national radio program that mixed music with the spoken word could be played throughout the daytime. This also served as familiar background noise in place of masking white-noise during experiments, to the comfort of both subjects and experimenters.

Cage furnishings. The cages were of a standard macaque type, but modified to make them more appropriate for squirrel monkeys while still usable for macaques, if required. The modifications included closer bars (3.75 cm), greater height (100 cm high X 60 cm wide x 70 cm deep), a fixed perch, and a removable side to allow two cages to be united.

In addition each cage was fitted with a trapeze hanging from the squeeze-panel guide bars so that the trapeze could slide forward as the squeeze-panel was moved to the front of the cage. The trapeze proved the most popular part of the cage and provided a vital source of environmental stimulation of the balancing and kinaesthetic systems that would be virtually lacking in the otherwise static environment.

Structuring the living space. The cage floor was of bars, but the dropping-tray beneath was fitted with 1.25 cm wire-mesh to prevent access to the urine and faeces. Cages were in stacked pairs, four to a mobile rack, with adjacent cages joined by their removable sides. Even such paired cages proved to be minimal space for as few as 2 or 3 monkeys and provided a totally limited and uninteresting environment. Therefore, passageways (150 cm high X 23 cm wide X 28 cm deep) made of 2.5 X 3.75 cm wire-mesh were fitted over the guillotine doors of an upper and lower cage so as to connect them (Figure 1). This gave a more structured vertical environment with separate visual areas so that monkeys could get a real change of place and could get out of sight of particular cage-mates. With connection of adjacent cages the vertical passageways allowed for either two- or four-cage units that could hold groups of not more than 6 or 12 squirrel monkeys, respectively.

Normally the upper cages were the preferred ones and in the breeding units it quickly became clear that the females occupied the upper space and the males the lower space, a pattern that reflects behaviour in the natural environment (DuMond, 1968). The external passageways also gave good views of the other cages and their monkey occupants and so proved a great facility for group interactions and displays. This cage arrangement also proved useful for training undergraduates in observing primate behaviour. A full range of individual, social, sexual, and intergroup behaviour could be recorded in a limited and easily definable yet differentiated physical environment of space and objects, with a limited number of monkeys of both sexes and different ages to be identified.

There were, however, unsatisfactory aspects to using standard monkey cages, even with the improvements described. Cages make a barren environment and enrichment with furnishings to make it more complex is difficult because of cleaning requirements. Squirrel monkeys have a habit of rolling and smearing their food on the cage bars, walls, and furnishings (cf. Hopf, Hartmann-Wiesner, Khlmorgen, & Meyer, 1974) which makes frequent cleaning necessary. The extra space provided by the quadruple cage system was still modest for such highly mobile and messy animals.

Exercise pens. Both these problems were alleviated, though not solved, by the provision of larger exercise pens in which each cage group could spend a 24 hr period once a week while their cages received a thorough cleaning. Two pens were built in a single room. Each pen was 180 cm wide X 180 cm deep X 200 cm high, with a roof made of 2.5 X 5 cm wire-mesh, solid walls and a glass-panelled observation front. Each pen was sub-divided by three semi-partitions, two extending from the rear wall and the third placed between them but extending from the front wall. This meant that the monkeys could not see all parts of the pen from any vantage point but the human observers could do so through the observation panel. Rowell (1972, p74-75) has described similar arrangements for caged monkeys. Recently McKenzie, Chamove, and Feistner (1986) reported the use of hanging screensmade of strips of material in enclosures for marmosets and these are an improvement over fixed partitions. The three rear sections of our pens were furnished (left to right) with a sloping ladder and swing; a nylon net suspended horizontally below the mesh cage top at the rear of the centre section; and fixed slatted shelving. Food trays and water bottles were placed low to force foraging on the ground, and a guillotine door was placed higher and opposite the ladder for easy exit of the monkeys into transfer boxes.

Clearly the pens were a compromise between desirable environmental features and ease of cleaning and handling. Perhaps the biggest single improvement that could have been made would have been the use of deep floor-litter. Chamove and his students (Chamove & Anderson, 1979; Chamove, Anderson, Morgan-Jones, & Jones, 1982) have shown how valuable this can be for reducing aggression and stereotyped behaviour, for keeping monkeys occupied foraging for food items scattered in the litter, and for ensuring that subordinate individuals get some of the more favoured food items. A 15-cm layer of litter, such as the conifer bark sold by saw-mills in Great Britain for gardening mulch, would probably have made our exercise pens much more interesting for the monkeys and removed the need for the uniform food-cake that we adopted. Chamove, Anderson, Morgan-Jones, and Jones (1982) found that wood-chip litter need be changed only at intervals of several weeks and that it reduces the odour in the room possibly through anti-bacterial action of ammonia from the urine and the raised temperature in the litter. It may well be possible to use similar litter in standard monkey cages with the addition of deep floor trays and regulations for housing primates in laboratories should be amended to permit this. Even where squeeze-panels are necessary it should be possible to use litter to the level of the floor bars or mesh.

Handling. The squirrel monkeys were handled in the usual way. The squeeze-panels could be used to bring them to the front of the cage to facilitate entry to transfer boxes or catching with gloved hands or handling tongs. The latter, specially made in our laboratory, had a scissor like action with ends that looped comfortably around the monkey's neck (Figure 2). The monkeys quickly adapted to these tongs and seemed to be less stressed in their grip than when held in a gloved hand because they could cling to the body of the tongs, sustaining their weight in a natural way and appearing to be relatively calm and secure. These tongs were particularly useful when monkeys were to be transferred to standard monkey chairs (Lehigh-Valley) where a neck yoke had to be fitted. They also proved quite appropriate for monkeys with implanted electrode assemblies such as that used by MacLean (1972). In our studies the monkeys were introduced to the chair for gradually increasing periods, with fruit juice provided at first as an encouragement and then as reward in appetitive discrete-trials discrimination-learning tasks. Our studies used evoked potentials as markers of neural sites involved in learning, recall, and choice in the manner of John (1967, 1980) with a particular interest in the limbic structures identified by MacLean (1972).

Group composition. Experience in other laboratories had suggested that keeping groups of males could give problems because of fighting. We were able to keep males in two's or three's and even pairs of males in breeding groups. However, some individuals proved too aggressive and caused deep slashing wounds to other males and females with their sharp canine teeth. This damage was greatly reduced by clipping and smoothing the ends of the canines of these aggressive individuals. This clipping was confined to the points of the teeth because of the risk of dental abscess and was done under ketamine anaesthesia (cf. Johnson & Chadwick,1977).

Castration as a means of controlling aggression: Negative results. As an experiment three adult males were castrated under anaesthesia to see if they would be more tractable for caging and experimentation. In the first years of breeding a great excess of males was born and so three year-old males were also castrated for the same experiment. The juvenile castrates were caged with three intact males of similar age and all three began to show the same signs of "fatting" as the intact males when they reached maturity. Coe and Rosenblum (1978) also

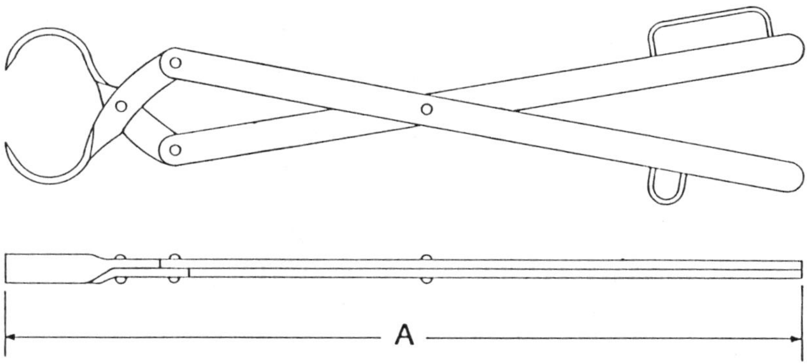

Fig. 2. Handling tongs. Dimension A:58 cm; closed jaw diameter: 4.5 cm. Made from 19 X 3 mm Duralumin. (Drawn and made by P.Bates)

---

noted seasonal weight changes in castrate squirrel monkeys that paralleled the weight changes due to "fatting" in normal males, but their monkeys were castrated post-pubertally. One of our three pre-pubertally castrated males developed large lipomatous lumps scattered over the body and limbs. A similar syndrome appeared in one of the three males castrated as adults. Mendoza, Lowe, Davidson, and Levine (1978) have reported that seasonal "fatting" does not coincide with changes in testosterone levels in male squirrel monkeys. It is possible that the distribution but not the production of the seasonal fattening is under the control of the testes. The possibility of a disturbed physiology in the seasonal "fatting" of castrates warrants further study especially in the light of the report by Portman, Alexander, Tanaka, and Osuga (1979) that male squirrel monkey castrates are high in lipoprotein concentration and are particularly susceptible to gallstones when kept on a high-cholesterol diet.

The behaviour of our pre-pubertally castrated males showed some differences from their normal peers. When 5 yrs old a sample observation of their behaviour showed lower mean frequencies (per 200 min) of genital display (1.7 versus 4.7 in normals), wrestling (0.3 vs 3.7), sexual approach (0.3 vs 1.7) and dominance behaviour (0.7 vs 1.3), and more frequent subordinate behaviour (1.3 vs 0.3). However, one of the castrates was sufficiently aggressive that it was necessary to clip his canine teeth, so it is doubtful if there were any significant differences in aggressive behaviour while caged with other males. There were no obvious differences in the .h2

tractability of either the juvenile or adult castrates. Coe and Rosenblum (1978) could find no differences in the behaviour of their adult-castrated male squirrel monkeys in mixed-sex caging. Their castrates showed normal displays, sexual approach and mounting, but not intromission, and there were no differences in social grouping or in dominance behaviour. Alvarez (1973) has also looked at the effects of gonadectomy in squirrel monkeys and found that castrated males tended to associate socially with the females. Loy, Keifer, Conaway, and Kingston (1984) have since reported a study of gonadectomized rhesus monkeys which showed that castrated males might be more aggressive than normals and that their social and mounting behaviour was directed more to other males. On the whole, therefore, these results do not recommend infant castration as a husbandry practice to control aggression in primate species. Other aspects of our colony management were much as described by Dukelow (1983).

FEEDING AND GROWTH

When the colony was started there was no commercial diet available specifically for New World monkeys, and the later available commercial pellet for marmosets seemed too hard for the normally frugivorous and insectivorous squirrel monkey (cf. Thorington, 1968). Knapka and Morin (1979) noted that high-density pellets are not readily consumed by most nonhuman primate species and that meal and liquid forms have limited application. A special diet was prepared to meet the following specific requirements. It had to be inexpensive, easily prepared and in a convenient form for both staff and monkeys. We wished to avoid the complications of a wholly synthetic diet (cf. Herbert, 1971) and the monotony and hardness of the available commercial macaque pellets, which also lacked protein and vitamin D_3 levels needed for squirrel monkeys (cf. Max Lang, 1968; Portman, 1970). In addition we wished to ensure that even the most subordinate and retiring monkeys in a group would receive their full requirements of the more nutritious and palatable elements of the diet. We fulfilled all these requirements by making an uncooked soft cake, based (75%) on a commercially available dog-meal. Wilson's Dog Meal (Dundee Scotland) is a dry flake preparation of cereals, with 15% fish and meat protein, mineral and vitamin additives, and linseed oil. It keeps dogs healthy and glossy-coated. To this meal we added protein in the form of dried milk or soya meal (18%), and vitamins (6%) in the form of a veterinary vitamin & mineral supplement (Vionate, Squibb) with extra vitamin B_{12} (Feed-grain Type 1000, Roche) and vitamin C (0.3%). The final mix had a protein content in excess of 20%.

The need for B_{12} vitamin is now well established (Nicolosi & Hunt,1979; Rassmussen,Ausman, & Hayes,1980). But there is some question whether squirrel monkeys really need protein levels as high as 25%, a level that has been associated with diarrhoea (Hupp,1972). Adult animals do perfectly well on 5% protein according to Kupper and Britz (1972) and breeding has been as successful with 16% as with 28% protein levels in the diet (Lehner, Bullock, Feldner, & Clarkson, 1967). Later, we changed from dried milk to soya meal as a protein supplement in case the high lactose of milk should cause diarrhoea problems. We found that a cellulose packing (Solka-floc BW 40, as used by Herbert, 1971) was not needed because of adequate fibre in the dog meal.

A different flavour was provided for each day of the week (*"A flavour a day keeps monotony at bay"*) by adding banana, raisins, dates, carrots, cabbage, spinach, apples or other fruits and vegetables according to season. The fruit or vegetable additive was macerated in a kitchen blender, added to the dry food mix, and the whole mixed with water containing the vitamin C. The wet mix was spread into shallow trays and cut into standard size cakes without cooking. The trays could be kept in a refrigerator for several days. Rations of cake were placed on the lower cage floors so that the monkeys would "forage" in the lower cages from which they would take pieces to their favoured places on the perches and trapezes. The cake was sticky and was eaten before it dried enough to crumble into pieces that could fall through the 1.25 cm floor mesh. Exterior food dishes were useless as the monkeys would quickly take the daily food ration into the cages. No other foods were given apart from peanuts, raisins, or pieces of apple which were used as treats at the start and end of each day so that the attendant could entice each monkey to the cage front to check its health and condition. The same treats were used in experimental work so that food deprivation for learning tests was not required provided test sessions were short.

The satisfactory nature of the diet was apparent from the colony weight and breeding records. Figures 3 and 4 show the mean growth curves for males and females born and reared in the colony. The data compare very favourably with data given by Long and Cooper (1968) and Clewe (1969) for captive squirrel monkeys.

We noticed that males become in poor condition at the end of the breeding season and that their coats may become grey with new hair growth. We also saw this greying of coat colour in infants being weaned. Hayes, Fay, Roach, and Stare (1972) have seen this in females following pregnancy. Knapka and Morin (1979) described it in juvenile and mature captive animals at Iquitos and, believing it was due to low mineral content,

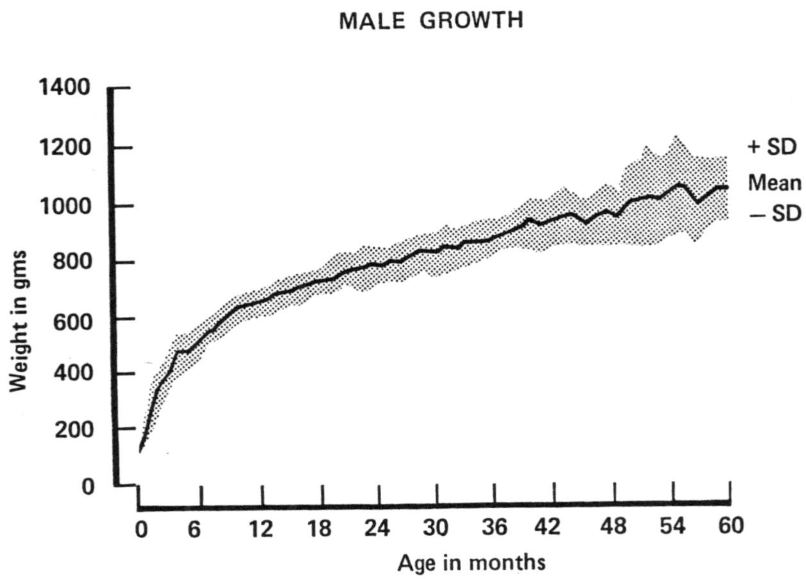

Figure 3. Growth of laboratory-born males weighed every month. Numbers of animals in each age-group were 13, 10, 9, 6, and 6 for 1- to 5-yrs old, respectively.

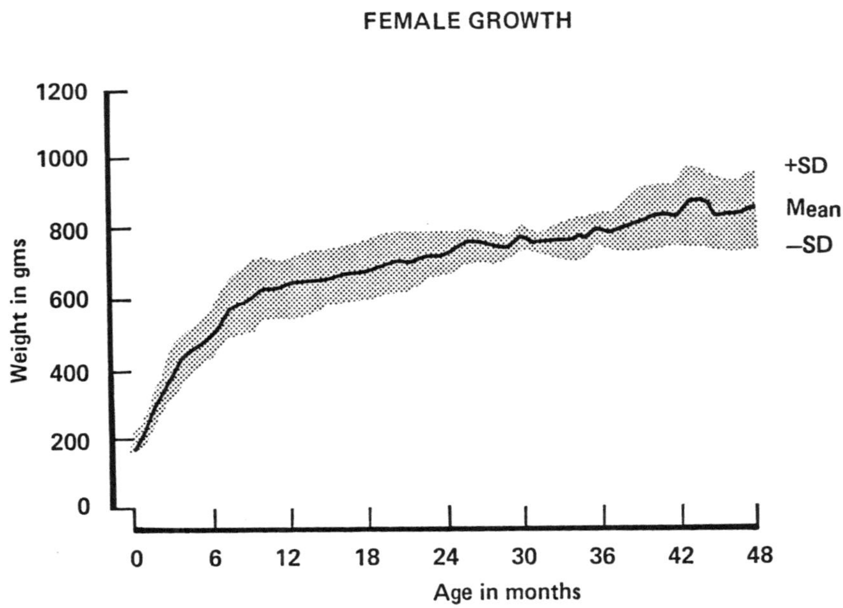

Figure 4. Growth of laboratory-born females weighed every month. Numbers of animals in each age-group were 11, 11, 4, and 4 for 1- to 4-yrs old, respectively.

obtained some improvement with mineral supplements. We think it is a sign of debility and stress but we are unclear as to its cause since our growth and breeding data suggest that the diet we used was adequate in minerals and vitamins as well as protein.

BREEDING

The four breeding units were begun with two males and four females in each unit. After three years, one breeding male in each of the four colonies was replaced with an immature male, which reduced conflict and stress at breeding time while providing male/male social interactions and future succession for breeding in each colony. Three of the remaining original breeding males were replaced by the end of the eight-year period over which the breeding colony was maintained. Of the original 16 breeding females, eight were replaced because of age over the same period. Four of the replacements were laboratory-born females of whom three had pregnancies by their fifth year and before the colony was dispersed. These pregnancies were either abortive or produced stillbirths. Taub (1980) reported that 70% of first pregnancies in his squirrel monkey colony were unsuccessful, due equally to foetal and neonatal death. Both Kaplan (1977) and Taub (1980) have now obtained successful breeding from laboratory-born females and there was every prospect, therefore, that our colony could be maintained with self-replacing breeding stock at least for some considerable time. Young born in the units were left there for one to two years, usually until a second successive birth cohort made the unit too crowded. Taub, Adams, and Auerbach (1978) have given data on the reproductive performance of Brazilian squirrel monkeys that suggest that opportunity to observe sexual and parental behaviour and to interact with infants seems to be necessary for later breeding success. It would seem good husbandry, therefore, to allow at least the females to mature in the breeding units.

Over the eight-year period there was a steady pregnancy rate for the 18 females of a mean 10.3 per annum with a mean of 7.3 live births and a survival to maturity of a mean 3.8 per annum. These figures are comparable with other breeding colonies (e.g., Clewe,1969; Lehner,1967; Kaplan,1977; Rasmussen, Ausman, & Hayes, 1980; and Dukelow,1983). Experience with the first few births suggested that the neonates were vulnerable to heat loss and that the normal colony room temperature (25° C) was insufficient. During the birth period, therefore, the room temperature was raised to 28° C and subsequent neonatal losses in our breeding units were due either to accidents in the cages or inadequate maternal care. There were also some losses of older infants due to too-early removal from the breeding unit until we learned not to do this.

The mean sex ratio calculated from both live and stillbirths for the eight-year period was 0.67 in favour of males (i.e., 41 males : 20 females). But for the first four years it was 0.85 (28 males : 5 females) and in the last four years it was 0.46 (13 males : 15 females). Until the last four years, therefore, it looked as though it would not be possible for the colony to be self-sustaining. This preponderance of male births in colonies newly established with wild-caught monkeys seems to occur commonly in both laboratories and zoos. Van Schaik & van Noordwijk (1983) have suggested that stressful conditions in both captive and wild primate populations may be associated with male-biased sex ratios and that female-biased ratios may be typical of provisioned troops. There is evidence that some ungulates in conditions of poor quality habitat, high population density, or poor nutrition may show high sex-ratios (Clutton-Brock & Iason,1986). Schuster and Schuster (1972) gave evidence that differential parental stress can affect the sex-ratio of offspring in rats and humans. James (1986) has discussed the possibility of, and evidence for and against, hormonal control of sex ratio, with high levels of maternal gonadotrophins being causally related to the production of female offspring. He has also noted that stress reduces testosterone levels in men and gonadotrophin levels in women so that stressed males may father more females and stressed females may well produce more male offspring. Thus the early preponderance of male births in our colony of squirrel monkeys could reflect a greater stress on females by removal from their natural troop. Males do not play an integrated part in the social organization of troops (Baldwin,1968) and may be less stressed by removal into captivity. The subsequent reversion to a more even sex-ratio could reflect the establishment of stable, familiar social relationships among the females in the caged groups over time.

The breeding in our colony showed the seasonal pattern typical of captive squirrel monkeys, including the shift to earlier calendar months that occurs in the Northern Hemisphere (e.g., Du Mond & Hutchinson, 1967; Coe & Rosenblum, 1978). The seasonal pattern is apparent in an increase in weights of the males, who become "fatted" (DuMond & Hutchinson, 1967) in the breeding season, and in a later increase in weight of the pregnant females (see Figures 5 & 6). In the first breeding year after arrival in our laboratory the males became "fatted" in May/June and births were in October/November. The periods for these events moved forward each season and after five years in our laboratory had reached December/January and June/July, respectively, where they stabilized. It is interesting that the seasonal shift occurred even though our colony was kept in artificial light with a fixed day length of 12 hr. However, the relative humidity, although generally low because of the rate of ventilation, did show a seasonal change. In winter the cold outside air was heated to 25-26° C, giving a lower relative humidity (circa 30%) than in mid-summer (circa 50%) when the outside air required less heating (or even none in fine

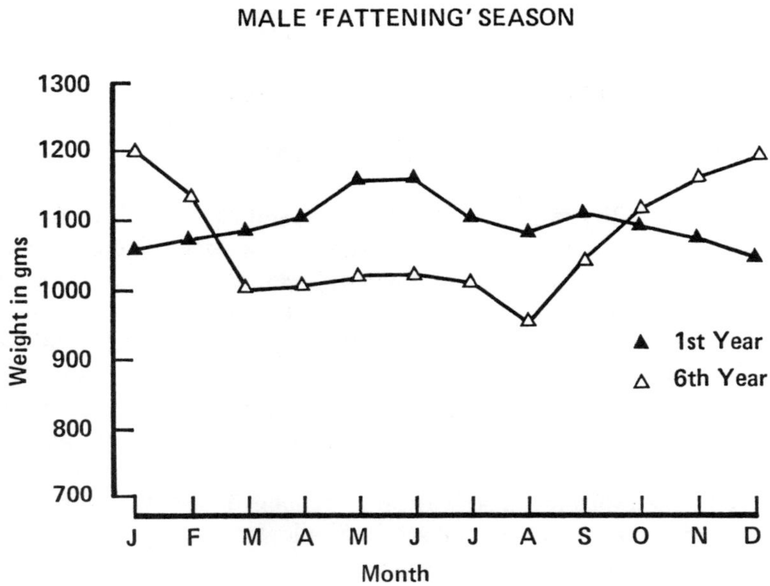

Figure 5. Breeding season shown by a peak in the mean monthly weights of males in the four breeding units due to "fatting". There were 8 males in the 1st breeding year and 4 in the 6th year, when the season stabilized.

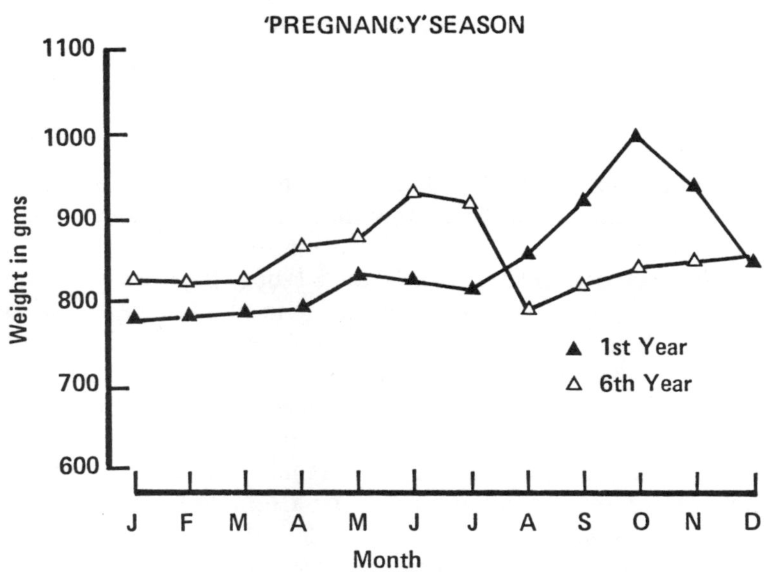

Figure 6. Pregnancy season shown by a peak in the mean monthly weights of pregnant females in the four breeding units. There were 7 females in the 1st breeding year and 10 in the 6th year, when the season stabilized.

weather!). This seasonal change in humidity in the colony rooms corresponds with published data on the relation between humidity and the seasonal breeding of squirrel monkeys (Harrison & Dukelow,1973).

Compatibility of group members. In setting up the breeding units consideration had to be given to personal animosities of particular males and females. When changes in the members of breeding groups were required a new male or female had to be introduced gradually by separating them in one of the four cages in a caging unit using a mesh side, and allowing periods of union while under observation by caretakers. Females would mob an introduced female, vocalizing, seizing and biting, and getting on top of the newcomer and clinging to her back. Often a male would join in, genital display, head grasp, and push the new female and indulge in "back-rolling" display ("Ruckenwalzen", Castell, Krohn, & Ploog, 1969). After this gradual introduction a new female would stay in the lower cages and it could take many days or weeks for full acceptance into the female group in the upper cages. Some introductions just failed to succeed and other individuals had to be tried. Fairbanks (1974) has found that introduced females are rejected by female sub-groups but accepted by adult and juvenile males. DuMond (1968) earlier described the lengthy process of accommodating strange females into a group in a semi-natural environment.

BEHAVIOUR

Social grouping and spacing. The behaviour of the monkeys in this colony was monitored throughout the nine years of its existence. We saw all the behaviour described for captive squirrel monkeys in other laboratories (see especially Ploog, Blitz, & Ploog, 1963; Hopf, Hartmann-Wiesner, Khlmorgen, & Mayer,1974). Some features of the behaviour we saw are of interest in relation to the behaviour in the wild. From the beginning, in our caging units the upper cages and the trapezes were the preferred sites. In the breeding units the females occupied the upper cages and the males stayed separate from them in the lower cages. Later, when there was one mature and one immature male in each unit, the mature males were sometimes accompanied by a female while the younger immature males were tolerated by the females in the upper cages . Subordinate or "unpopular" females were liable to stay in the lower cages. The males were in the upper cages more often in the breeding period. This pattern of social grouping and spacing of males and females is seen in the wild (e.g., Baldwin & Baldwin, 1972).

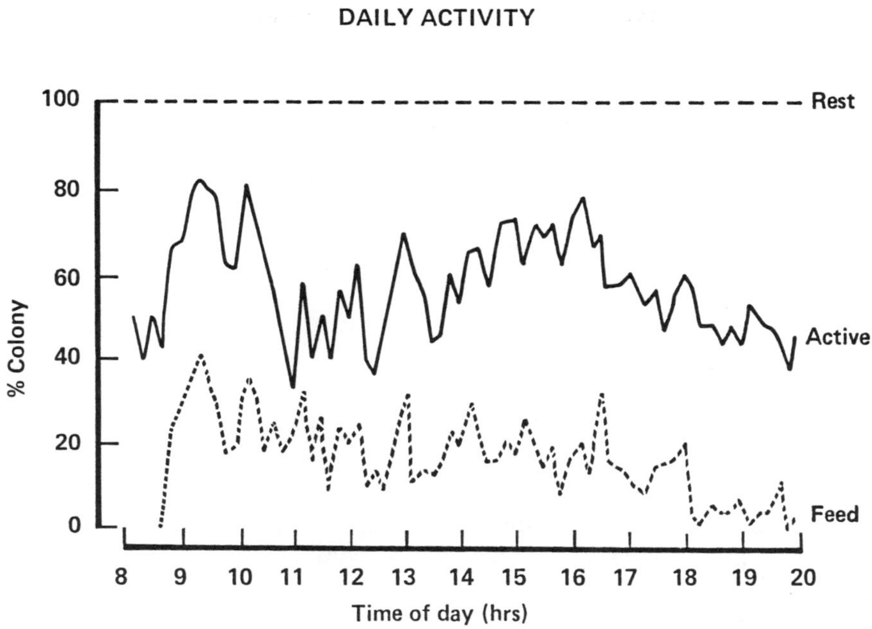

Figure 7. Daily activity in the four breeding units. Food was replenished at 08.30 daily. Each unit was time-sampled every 5-min throughout the 12-hr day in randomly ordered 2-hr periods (two complete days of data on 29 monkeys). Plotted are mean percentage of monkeys feeding, active or resting at 10-min intervals. Graphs are "stacked" to show total activity pattern.

Diurnal activity pattern. The daily pattern of activity in the colony also followed that reported in the wild by Thorington (1968), with a quiet period at mid-day and resumed feeding activity in the afternoon. Figure 7 shows data for the four breeding units, each observed (K. Mearns) for 2-hr periods at all times of the day, to give two complete 12-hr days of observation for each unit. The incidences of monkeys feeding, active, or resting were recorded every five minutes and are plotted in Figure 7, "stacked" to show the typical pattern of these activities throughout the day.

Aggression and displays. Once a breeding unit was established there was little aggressive or display behaviour within the group but it did occur between the breeding units in a room. The displays by the males

Table 1

Incidence of Stealing and Searching For/Retrieving Food From Customary Hiding Places in Cages

Subjects	N	Sex	Hide	Search
Adult (> 9 yr, wild-born)	4	Male	0.0	0.0
	12	Female	0.0	0.3(2)
Adult (5-7 yr, lab-born	0	Male	—	—
	3	Female	1.0(2)	27.7(3)
Juveniles (lab born)	4	Male	0.8(2)	4.5(3)
	6	Female	10.2(5)	39.7(6)

Data are means per monkey for a total of 24 hr. of observations of each of 4 breeding groups in 2-hr continuous-recording blocks. N = number of monkeys observed; (n) = number contributing to scores. Values for Juveniles are significantly greater than the combined Adult values ($p < 0.01$ Mann-Whitney test, 2-tailed).

Table 2

Incidence of Stealing and Hiding Food in Colony Cages and Exercise Pens

		Cage		Pen	
Subjects	N	Steal	Hide	Steal	Hide
Adult (> 5yr)	19	0.83	0.83	1.43	0.22
Juvenile	7	2.40	2.49	2.77	0.77+
All	26	1.25	1.28	1.79	0.37*

Means per monkey per hr from focal-animal observations of 4 breeding groups, each for 3 hrs in cage and 5 hrs in pen. N= number of monkeys observed. The incidence of hiding was significantly less in the pens (+p=0.06, *p=0.049, Wilcoxon test, 2-tailed), while stealing was unchanged.

included "genital display" (cf. Ploog & MacLean, 1963), "standing" vertically to expose the light-coloured ventral surface to other groups, and "rage-shaking" ("Rutteln": Hopf, Hartmann-Wiesner, Khlmorgan & Mayer, 1974) the floor, the cage bars, or the mesh of the connecting passageways. In addition the males quickly learned to shake the trapezes lengthways so that the ends struck the sides of the cages making a great din. This seemed a natural development from branch-shaking and was done in response to males in other groups, as frustration displays to caretakers over delay in presentation of food "treats", and to other interferences. Impressive mobbing displays occurred when infants were removed from cages for attention and were seen in the hands of caretakers. One caretaker was mobbed by these diminutive primates and bitten while trying to free an infant from an entanglement in an exercise pen.

Hiding and stealing of food. An activity that seemed peculiar to our colony was food "storing" or "hiding". It was noticed that some monkeys placed food outside their cage, on top of water-bottle holders, or on the solid roof of the cage, out of sight of the other monkeys in the cage (see Fig. 1). This food would be retrieved and eaten within a short time and other monkeys would be prevented from retrieving it, either by aggression or by pre-emptive retrieval. Marriott and Salzen (1979) described this behaviour and some simple experiments that suggested that the monkeys were deliberately hiding food and desired items and that this might be functioning as temporary storing of food. Kirchshofer (1963) has described "food-play" ("Futterspiele") in squirrel monkeys in captivity and we have seen similar behaviour in which food was dropped over the top of the squeeze-panel and then "looked for" below the panel. The food-hiding described by Marriott and Salzen was done by wild-caught monkeys, all of whom were immature, and it was noted that the habit also developed in immature laboratory-born individuals. It is possible therefore that the food-hiding we observed was another form of food-play.

We have more observations made by K. Mearns during the study of the daily pattern of activity in the colony. In addition to time-sampling the general activity in the colonies, we did continuous recording of specific activities including food-hiding and the related actions of searching customary hiding places, and retrieving hidden food. Table 1 shows the results for the incidence of food-hiding and retrieval (or searching in case no food was actually present in the hiding place). Both hiding and searching/retrieval were characteristic of immature monkeys ($p<0.01$ for all adults vs juveniles, Mann-Whitney tests, 2-tailed). The behaviour seems more common in females than in males, but the differences were not quite significant ($p<0.1$), perhaps because of the small number of males involved.

Three adult females were laboratory-born and their data are shown separately in Table 1. They had hidden food as juveniles but, as Table 1 shows, they now rarely hid food although they still searched their old hiding places. These females were from 5 to 7 yrs old compared with the wild-born adults who were at least 9 years old. Table 1 shows that searching the food-hiding places was very rare in the older wild-born females, being seen only in two of them. The absence of food-hiding in fully adult males and females seems to confirm that it is juvenile behaviour.

Our breeding groups' behavioar was monitored regularly and these records also show a decline in food-hiding with age in three laboratory-born females. The mean frequencies of food-hiding per monkey per hour of observation were 2.3 and 0.48 for mean ages of 3 and 5 years, respectively. Interestingly the corresponding frequencies for food-stealing ("Futter- oder Objekt-stehlen", Ploog, Hopf, and Winter,1967) were 2.2 and 0.4, suggesting that this behaviour is also characteristic of immature monkeys. This suggests that food-hiding could be a response to food-stealing that has arisen among immature monkeys in the confined space of the cages where it is difficult for them to defend against attempts to steal food held in their hand.

We also have data for 5 hr of focal animal observation in the exercise pens and 3 hr of similar observations in the cages for each of the four breeding units. Behaviour, including both food-hiding and food-stealing was recorded continuously and Table 2 presents the data as frequencies per monkey per hour. Clearly food-hiding was greatly reduced in the exercise pens ($p<0.05$ for all monkeys and $p=0.06$ for juveniles, Wilcoxon test, 2-tailed). Food-stealing behaviour, however, was as frequent or even increased in the exercise pens where it was significantly more common than food-hiding ($p<0.001$ for adults and <0.02 for juveniles). The relative absence of food-hiding in the exercise pens might simply reflect a scarcity of hiding places compared with the cages, but it is also possible that in the exercise pens there was adequate room to keep away from other monkeys that might try to take food held in the hand and so there was no need to place food temporarily out of sight of other monkeys. The evidence of differences as a function of sex, age, and cage context, therefore, suggests that this behaviour is unlikely to be true food-storing, as Marriott & Salzen had suggested, and perhaps is best called "food-hiding."

Locomotor stereotypy. Finally it should be noted that some abnormal stereotyped behaviour (cf. Erwin & Deni, 1979) appeared in the caged monkeys soon after arrival in the laboratory. In general this consisted of pacing patterns along trapezes or on the cage floor or in circular routes from floor to trapeze and back. The circuit type of locomotory stereotypy may involve a complete body somersault or a head-toss (cf. Hopf, Hartmann-Wiesner, Khlmorgan, & Mayer, 1974) representing an intention movement for a somersault. In each breeding unit a subordinate female and one or two immature females showed some stereotypies, apparently at times of disturbance, inter-colony displays, and at feeding times. The incidence was higher following the introduction of new subordinate females and juveniles into the breeding units. The virtual absence of stereotyped movements when these same groups were in the exercise pens suggests that enclosures of greater size and complexity would prevent this behaviour.

CONCLUSIONS

Our experience with this closed colony of squirrel monkeys during the nine years of its existence suggests that such a colony could be self-sufficient and could provide monkeys on a small scale for neurobehavioural studies requiring small numbers of subjects in chronic physiological techniques or extensive learning procedures. The economics are acceptable for laboratories or research units of quite modest size, provided use can be made of the breeding units for observational studies, and of the stock animals for learning studies, by undergraduate classes. However, with production of about four surviving offspring per annum from 16 breeding females, more breeding units or more females per unit would be required. We did, in fact, have two more breeding units but these were of the Bolivian strain. Our breeding-cage units could have accommodated six rather than four females so that, with six breeding units, we could have had 36 breeding females in the housing available to us with a potential recruitment to the colony of nine monkeys per annum.

However, for a behaviourally healthy and hence happy existence the squirrel monkey really requires more space than even our quadruple cage system provided. Pens of the size of our exercise pens would avoid stereotypic locomotory behaviour, especially if provided with a deep floor-litter. Such pens would also avoid some of the infant fatalities that occurred in the cages due to falls and entrapment in cage fittings. In retrospect we could have used pens for breeding, rearing, and stock units, though not for experimental animals. We would have saved the time and labour involved in operating the exercise pens on a rotational basis. (Although the monkeys readily learned to enter transfer boxes for transport to their home cages, the exercise pens had to be cleaned thoroughly between groups.) The handling of monkeys in large colony pens does presuppose either well-accustomed wild-stock or laboratory-bred stock. It also requires a little more room space. Given these conditions, it should be possible for laboratories using squirrel monkeys not to have to rely on a constant supply of wild-caught monkeys and thus threaten natural populations, but instead to have their own breeding populations, providing a lively and fascinating teaching and research resource.

There would also be an advantage in having a number of small breeding colonies; exchange of breeding stock between such colonies might avoid the deleterious effects on infant mortality of inbreeding that have been discussed by Ralls & Ballou (1982, 1986). This dispersed small-unit approach to breeding primates in captivity was recommended in 1979 by the Primate Society of Great Britain in an unpublished report prepared by a Working Party on Primate Breeding and entitled *Proposals for a national strategy to guarantee future availability of primates for basic research and biomedical testing procedures,* but the proposals failed to get the support of government agencies in Great Britain.

Finally, there is a further, not insignificant advantage of a breeding unit in a research laboratory. Sustained intimate interactions between monkeys and caregivers, researchers, and students foster caring relationships that ensure that humane requirements will be considered along with the needs of scientific enquiry when experiments (whether for research or for education) are being designed and carried out. Experimental subjects are then no longer code numbers but are named individuals, known and cared for as infants, with a history and personality and often with specific interaction patterns with the individual human caregiver and experimenter. There can be no greater moderating influence on the use of nonhuman primates for scientific research.

ACKNOWLEDGEMENTS

I would like to express my thanks and appreciation for the contributions of the students and technicians who worked with our colony in the course of its ten-year existence. They include Susan Buick, Gail Garnsworthy, Linda MacRae, Kathryn Mearns, Mary Simpson, Loraine Tomon, Clare Trotter. Also to my fellow researchers Alison Williamson and especially Bernadette Marriott who played such a significant part in both colony management and research. The monkeys were given special daily care over the years by Sheila Chalmers, Brian Dick, Norma Morrison, Ron Robertson, and George Taylor, and veterinary care by David Wilson. Caging and housing modifications and equipment were the work of Peter Bates, Dave Cline, and Alex Jaffrey.

Family: Cercopithecidae (Old World Monkeys)

— 10 — A Room with a View for Captive Primates: Issues, Goals, Related Research and Strategies

Peggy O'Neill

PUBLIC CONCERNS

The American public, whose annual attendance at zoological parks and aquaria is greater than at all spectator sports combined ("Major zoos: Attendance figures during 1984," 1987) is starting to sense the need for significant changes to meet our ethical responsibilities toward captive animals. The news media have documented conditions of filth, neglect, discomfort, distress, and avoidable pain as a reality for some captive animals. The public is becoming aware of how much they rely on animal research and testing for their medicines and daily comforts, and of what some animals may be suffering for the sake of these human benefits. The fact that so much animal research and testing is publicly funded makes humane taxpayers feel ethically responsible for the welfare of research animals, and they are pressing for changes in captive animal husbandry.

CONCERNS OF MEDICAL RESEARCHERS

Recent controversies have made medical researchers, too, aware of both ethical issues in the use of animals for research, and scientific questions about the appropriateness of using stressed animals as research subjects. Recent research suggests that psychological and environmental disturbances adversely influence physiological systems, including the immune system (Coe, Cassayre, Levine, & Rosenberg, 1987; Coe, Rosenberg, Fischer, & Levine, 1987; Boccia, 1987; Laudenslager, Ryan, Drugan, Hyson, & Maier, 1983; Laudenslager, & Reite, 1984; Reite, & Capitanio, 1985). In macaques and squirrel monkeys (common species for biomedical research), stressful housing conditions, loud noises, restraint, sensory deprivation, and separation from companions function similarly to electric shock to hamper the antibody response to bacterial and viral infections and possibly to worsen parasitic infestations. Such stresses activate the sympathetic nervous system, leading to an increase in the level of circulating hormones and suppression of immune function. A monkey with suppressed immune function is not a suitable model for most biomedical research, and this has led some researchers to join in the call for better living conditions for captive primates, with the aim of assuring the reliability and accuracy of research findings.

CONCERNS OF CONSERVATIONISTS

Yet others with a concern for the wellbeing of captive primates are conservationists. Wild populations of primates are undergoing catastrophic loss of habitat to the increasing pressure of human agriculture and habitation. Numerous primate species are regularly added to the endangered and threatened lists. To conservationists it is imperative that primates reproduce in captivity, but primates subjected to sensory or social isolation and reared without adequate mothering become depressed and incompetent to breed or rear young (Chamove, Rosenblum, & Harlow, 1973; Goldfoot, 1977; Harlow, Harlow, Dodsworth & Arling 1966; Harlow, Joslyn, Senko, & Dopp, 1966; Rodriguez, 1987; Rowell, 1964; Roy, Wolf, Martin, Rangan, & Allen, 1978). Furthermore, inexperienced primates are ill-prepared to survive and reproduce when released back into a highly competitive and challenging natural environment, as some conservationists hope to do with some species.

Difficulties encountered in initial attempts by the National Zoo to reintroduce the golden lion tamarin to native habitats made it clear that better preparation of animals destined for reintroduction (appropriate environmental enrichment, stimulus learning, and natural rearing conditions) was required if reintroductions were to succeed (Kleiman, 1977). Difficulties also occurred when an attempt to introduce 18 adolescent rhesus who had been surrogate-reared in the laboratory to a monkey island at the Madison Wisconsin Zoo (Harlow, 1962) resulted in total reproductive failure. After three months outdoors, the 16 monkeys who survived remained virgins. None of the males was ever observed mounting a female. My own work in the mid- 1970's with eight surrogate-reared juvenile rhesus had a better outcome. Success required two years of extensive and protected environmental stimulation in elaborately enriched outdoor housing. The resulting social group succeeded in breeding and in parenting offspring, with the assistance of therapist puppies to stimulate social and nurturing behavior. (More on this later.)

Figure 1. These monkeys have just been introduced to an enriched cage. They show poor muscle tone and unsureness with the exercise apparatus.

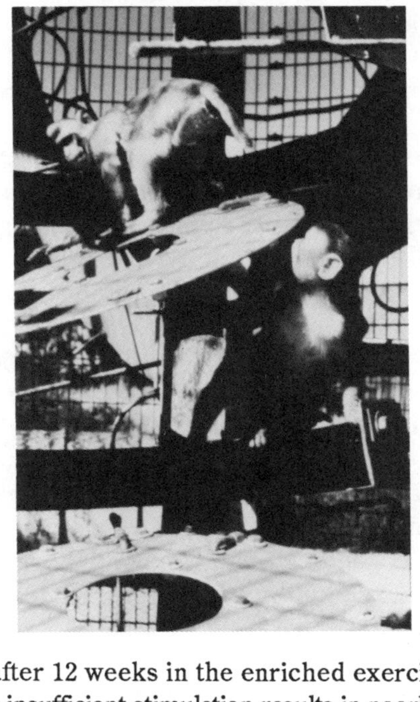

Figure 2. Monkeys after 12 weeks in the enriched exercise cage. Rearing in limited cage space with insufficient stimulation results in poorly developed muscles and poor coordination. It takes time for these young animals to develop the physical skills normal for their species. Photos by author.

A final point from the conservationist's perspective: The rhesus monkey, commonly used in research, is losing out at home in India (Southwick & Siddiqi, 1985). The ranges of all 12 Asian macaque species have been severely reduced (true for the North African Barbary macaque, too). The numbers of rhesus, once between 10 and 20 million in the Indian province of Uttar Pradesh alone, have declined until now there are no more than 80,000 to 100,000 rhesus monkeys in all of India. During the 20-25 year period of this drastic decline, the human population of India doubled from 350 to 700 million. It would not be surprising if the number of rhesus monkeys in captivity exceeds that in the wild within a decade. Hence, it is essential to the preservation of this species that laboratories institute measures to maintain healthy reproduction and rearing in captivity.

CONCERNS OF PSYCHOLOGISTS

Other groups of professionals seeking improvements in the health and stability of captive primates include psychologists and animal behaviorists. "We have become aware that complex, enriched, and more naturalistic settings for captive primates discourage the development of abnormal behavior, induce activity, and facilitate normal social behavior and reproduction" (Clarke, Juno, & Maple, 1982).

Etiology of Psychopathology in Nonhuman Primates

Three decades of research involving denial, restriction, and selective interference with species-typical rearing in macaques have demonstrated that restricting social or sensory experience during early development produces abnormalities in motor activity (Figures 1 & 2), exploration, response to complex stimuli, emotionality (Figure 3), and social behavior (Cross & Harlow, 1965; Goldfoot, 1977; Harlow & Harlow, 1962, 1971; Jensen & Tolman, 1962; Kaufman & Rosenblum, 1966; Mitchell & Clark, 1968; Preston, Baker, & Seay, 1970; Schlottman & Seay, 1972; Suomi, Harlow & Kimball, 1971; Spencer-Booth & Hinde, 1967; Reite & Capitanio 1985; Seay, 1962). We have the cumulative evidence of psychological research to thank for revealing the etiology of pathological behavior in some species of nonhuman primates. Over and above the possible value of this evidence for the understanding of human psychological dysfunction, it has provided us with an excellent foundation for understanding nonhuman primate psychopathology.

Figure 3. Lack of sensory and social stimuli may lead to abnormal emotionality and social behavior.

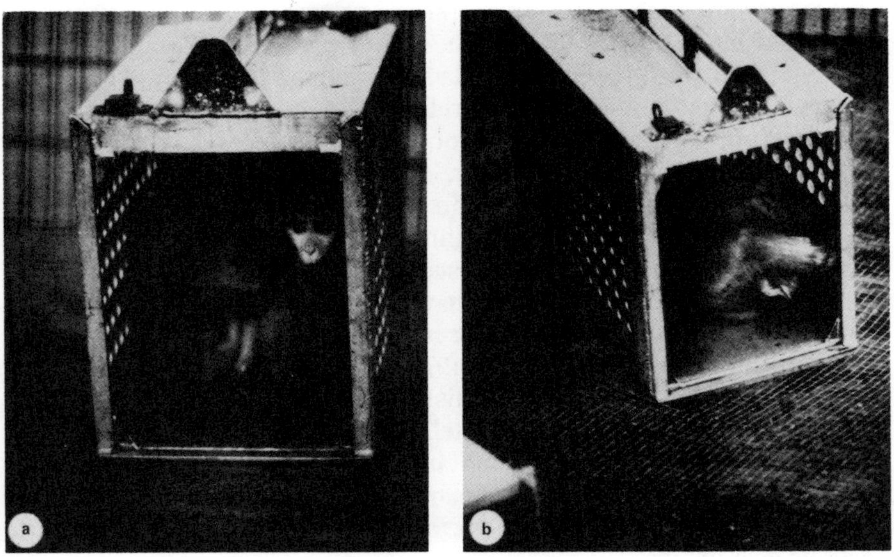

Figure 4a & b. Rhesus macaques raised in restricted environments do not respond to an enriched enclosure at first. Exploration poses a distressing and difficult challenge to inexperienced 1.5-year-olds. Photos by author.

Identifying Psychological Disorders

Abnormal behavior exhibited by monkeys and apes in zoos and laboratories in many ways resembles the behavioral abnormalities seen in institutionalized humans living in similarly restricted and impoverished environments (Cross & Harlow, 1965; Davenport & Menzel, 1963; Erwin, 1973; Erwin & Deni, 1979; Mitchell, 1970; Sackett, 1968). Erwin and Deni (1979) have provided us with an accurate descriptive account of many of the behavioral disturbances that appear in monkeys reared in restricted environments. Among the most persistent behavioral deficits are those classified as maturational arrest, a common disorder resulting from sensory and social deprivation during development. These include infant behavior persisting into juvenile, adolescent, and adult developmental phases such as self-orality (thumb, toe, and penis sucking), body-clasping (clinging to torso or appendages), stereotypic rocking and self-huddling, other distorted and bizarre postures, hyperactivity, hyperaggressivity, excessive masturbation leading to self-mutilation, and appetitive disorders such as hyperphagia and polydipsia.

But abnormalities also arise in monkeys who have been reared by responsive mothers in enriched environments and who exhibit evidence of normal emotional development and emotional security in the form of independent exploration and peer play. When subsequently removed from their familiar and stimulating environment and placed in sterile, solitary cages they are likely to exhibit depressive symptoms not unlike those of a protesting infant rejected by its mother at weaning. This separation-induced depression may be especially devastating in animals too young to have been weaned by their mothers. Their profound depression is characterized by regression, passivity, aimless locomotion, loss of interest in their surroundings, ingestion of body hair, loss of sexual interest, and lack of play. Other manifestations may include hyperactivity, repetitious pacing, stereotypic motor acts such as continuous masturbation and eye-poking, self-biting, and appetitive disorders such as coprophagia and hypophagia.

Size of cage may contribute to the occurrence of such disturbances. In rhesus, smaller cages engender more stereotypic pacing than larger cages (Paulk, Dienske, & Ribbens, 1977). Pigtail macaque mothers raising their infants in smaller cages punish them more than mothers raising infants in larger, group cages (Castell & Wilson, 1977). Cage size is only one factor contributing to aberrant behavior; lack of daily stimulation is another. Many abnormal displays may represent a caged animal's attempts to relieve boredom and sensory deprivation and to exert a degree of control over its environment.

Enrichment or Rehabilitation?

Primates whose abnormal behavior developed only after a normal infancy, and only in response to individual housing in a barren cage, are likely to benefit the most from environmental enrichment. We can restore to them a sense of control and security. On the other hand, primates displaying long-term developmental psychopathology and fundamental behavioral disabilities brought on by inadequate infant rearing require much more elaborate therapy, and even then the outcome is uncertain. It is important to know the precise history and behavioral profile of individual animals one is attempting to rehabilitate. Was the primate weaned by its own mother? Did it have experience with peers and other adults during rearing? Does it respond species-specifically when stimulated? Is the male capable of a double foot-clasp sexual mount with a female conspecific? What form of body language, gestures, and vocal signals does the monkey display toward conspecifics? Do conspecifics respond appropriately?

With animals exhibiting symptoms of permanent emotional crippling originating in infancy, therapy attempts may at first be very distressing. The eight surrogate-reared monkeys that I took on in graduate school were terrified by my initial attempt to help them. I introduced all of them simultaneously to an outdoor pen laden with sources of sensory and social stimulation. They cowered in their transport cages and refused to enter the playpen (Figures 4a & 4b). Eventually the animals accepted, responded playfully, and learned from exploring their new home, but it was not initially a happy experience for any of us.

STRATEGIES: ENVIRONMENTAL CONSIDERATIONS

Scientific evidence as well as my own experience suggest that many of the problems arising in research with confined primates, for the animals themselves, for caretakers, and for scientists can be reduced by selecting as subjects primates who are mother-reared in groups and who enjoy healthy social development (Chamove, Anderson, Morgan-Jones, & Jones, 1982). They display species-typical behavior without resort to therapeutic intervention and they respond with investigation and manipulation to enrichment devices and enriched caging.

Figure 5. Garden hoses are easy to install, remove and replace. They are challenging, and flexible like young tree branches.

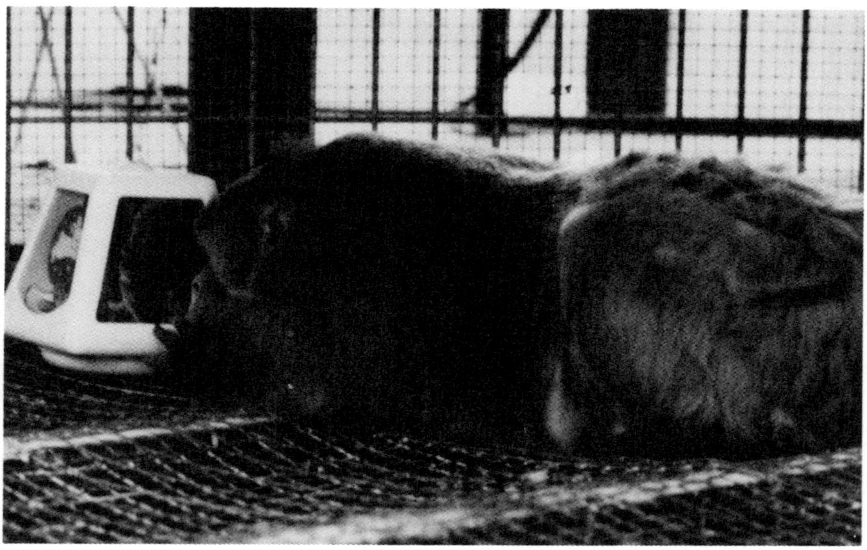

Figure 6. Adult rhesus find more sedentary activities, like investigating child-development toys, to be attractive. Photos by author.

Figure 7. The simple, durable, inexpensive plastic milk crate has appeal with primates.

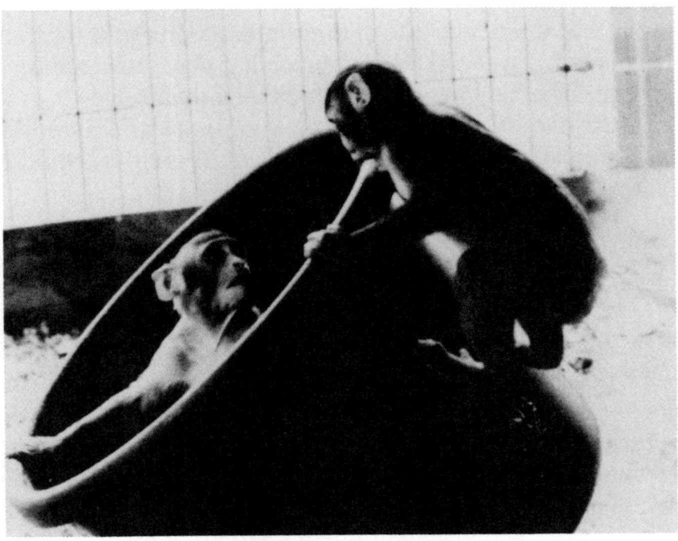

Figure 8. Heavy-duty plastic pails make easy-to-clean play equipment for group pens. Photos by author.

Devices and Cage Enrichment

Different types of floor substrates, greater variety of foodstuffs, as well an array of different stimulus objects, are currently being introduced into group pens and solitary cages with some positive results (Bloomstrand, Riddle, Alford, & Maple, 1986; Chamove et al., 1982; Eterovic & Ferchman, 1985; McKenzie, Chamove, & Feistner, 1986; Novak & Drewsen, 1988; O'Neill, 1988; Reinhardt, Houser, Eisele, & Champoux, 1987; Roy, Hennig, Hill, & Walker, 1981; Westergaard & Fragaszy, 1985; Westergaard & Lindquist, 1987). Some primate species and some individuals respond more positively than others to various methods of stimulation. Enrichment strategies employed in group pens have been a subject for study at the National Institutes of Health (O'Neill, 1988). Infant, juvenile, adolescent, and adult rhesus monkeys show distinctly different patterns of use and social integration of various enrichment devices and apparatus. Activity, both social and nonsocial, involving hanging hoses (Figure 5) outranked eight other activity categories for infant and juvenile rhesus. Adults, on the other hand, when not engaged in social activities, were more inclined to more-sedentary enrichment activities like foraging through sawdust for seeds and investigating child development toys having multiple gadgets, metal mirrors, and sound-producing control buttons (Figure 6). In the same study, infants made considerably more use of hanging milk crates and large plastic pails on the floor of the pens than did the three other age groups (Figures 7 & 8).

Social activities outranked all other available alternatives for adolescents in this study, and 70% of their interactions with toys and other apparatus occurred in a social context involving play or proximity to other group members. The hanging hoses, which lend themselves to group play, got the greatest use by adolescents, as by infants and juveniles. The sawdust and floor toys, selected more often by adults, ranked in the midrange for adolescents.

Because no one solution serves the needs of all primates I suggest rotation of a variety of alternatives from cage to cage. Renewed novelty is likely to maintain interest better than any single object can. At a recent meeting of the American Association of Laboratory Animal Science, Kenneth Pyle, a veterinarian and professor at the State University of New York at Delhi, described the practice of rotating the singly-caged animals in his colony rooms to various positions along the cage racks so that each monkey in turn had an opportunity to reside in the "cage with a view" next to the only window. His staff noted that monkeys displayed less disturbed behavior when in that cage, perhaps not surprising in light of evidence for the positive effects of windows on human health and mood (Moore, 1981; Ulrich, 1979; Ulrich, 1984; Ulrich & Simons, 1986).

Natural Habitats

Primates maintained in well-integrated social groups in environments approximating their natural habitat have been described as happier and healthier than their counterparts in sterile lab cages (Banerjee & Woodward, 1970; Fox, 1966, 1970). There is less disease, mortalities are usually fewer, stress-related disorders are minimal, and reproductive potential is maximized. In addition, research in naturally-housed primate colonies reveals patterns of behavior not otherwise observed in captive populations (Bernstein & Gordon, 1977). The costs of housing captive primates in groups (caging, personnel, and maintenance) are considerably less than the costs of keeping animals in individual cages in colony rooms (Banerjee & Woodward, 1970; Hoffman and Stowell, 1973). (Where animals are already housed individually, there are, of course, one-time costs involved in shifting to group housing.)

Human-Animal Communications and Interactions

The characteristics of the people who interact with captive primates importantly affect the animals' comfort and security. Primates are responsive to the humans who work with and near them (Fox, 1987 & Figure 9). Patience and kindness are essential. Although some captive primates are relatively passive, others may harass the staff, urinating or throwing feces at them. They succeed in getting attention and perhaps a degree of control in their lives, but rarely are the consequences of such actions pleasant. When despair sets in, the animals may attack or injure themselves. Offering the animals opportunities for positive feedback, such as conspecific companionship or playthings to manipulate and investigate, can make both staff and animals more contented (Yerkes, 1925).

Staff stability is very important. Getting acquainted with new staff and their habits can be stressful to confined animals. Caregivers need to show respect for the primates and to be responsive to their needs. Harsh words, sudden body movements, and surprise visits can create distress (Cooper & Markowitz, 1979). Animals may instinctively perceive humans and strangers as threatening intruders. It is important to do nothing to confirm such fears. Responding to a caged animal's aggression with taunting or teasing only aggravates the problem.

Figure 9. Primates are receptive to humans who work with them. Photo courtesy of the Wisconsin Regional Primate Center Library.

Educating staff about the behavior and communication signals of the primates they will be working with is an essential prerequisite for preventing conflicts between humans and animals, and for establishing a sense of wellbeing for everyone. There have been documented reports of infant abuse by primate mothers who ordinarily showed competent maternal behavior but became abusive when their environment was changed (e.g., Troisi, D'Amato, Aureli & Piovesan, 1987). One change that can induce this "maternal anxiety" is bringing in a staff person the monkey mother dislikes. An example I am familiar with involved a seven-year-old female rhesus who exhibited every characteristic of responsive mothering, except when a certain staff member entered the room. At such times she was known to pull her newborn from her abdomen and attack it viciously. This type of problem can be diagnosed and corrected only if it is recognized that individual animals do have personal likes and dislikes. Understanding the relationship between staff members and animals can be vital for assessing psychological wellbeing of captive primates. Although teasing and harassing of caged primates by humans employed to care for them is not a popular topic for discussion among embarrassed professionals, it is a problem that can arise and that must be acknowledged and dealt with promptly.

Another essential is to recognize animals' need to exert some control over environmental events. Giving animals some control prevents the development of the syndrome called "learned helplessness" (Joffe, 1973; Overmier, 1981; Seligman, 1975; Stroebel, 1969). Hal Markowitz (1982) and others (Clarke, Line, Ellman, & Markowitz, 1987) have designed operant devices that allow confined animals some degree of control over their environments. The stresses of confinement can also be reduced by providing signals enabling captive animals to predict potentially disturbing events in advance (Markowitz & Stevens, 1978). Simple routines such as turning on music or lights prior to entering a colony room, and habituating animals to handling through periodic practice sessions, can serve to prevent distress and immunosuppression. I always call to my monkeys prior to entering their room. This has been effective in preventing startle and alarm calls in the colony. I am more comfortable when my animals greet me in a relaxed way.

Social Life for Captive Primates

Volumes have been published on the complex social hierarchies that have evolved in monkeys and apes (Altmann, 1962; deWaal, 1982; Dolhinow, 1972; Goodall, 1986; Hinde, 1983; Smuts, 1987). Healthy social primates seek social interaction and social identity. They are more at ease if given adequate time to develop their social niche, whether they are housed in a group pen or in individual cages within sight of conspecifics. When individuals are interacting in a healthy way, it is wise not to interfere unless absolutely necessary. On the other hand, when they are injuring themselves or others, it is necessary to eliminate serious conflicts by altering cage mates or neighbors. Careful selection of group members whose histories and personalities are known is important (Bernstein, Gordon, & Rose, 1974; Reinhardt, Reinhardt, Eisele, Houser & Wolf, 1987; and Southwick, 1974).

Innovative resocialization research is ongoing at the Wisconsin Regional Primate Research Center. Adult rhesus monkeys displaying maladaptive and bizarre behavior have gradually abandoned their peculiar rituals for expression of species-typical comforting and affectionate behavior, such as grooming and mutual huddling, in the company of a conspecific companion. Under the direction of veterinarian Viktor Reinhardt, weaned infant or adult rhesus have been successfully pair-housed with other adults who have been living alone in barren cages for years (Reinhardt, Houser, Eisele, & Champoux, 1987; Reinhardt, Cowley, Eisele, Vertein, & Houser, 1987; Reinhardt, Cowley, Eisele, Vertein, & Houser, 1988). Other recent research that measured physiological activity while animals were mutually grooming documented the calming effect of social grooming (Boccia, 1987).

Macaques reared in socially-impoverished conditions may be hyperaggressive when placed with cage mates (Mitchell, 1967; Sackett, 1968). They may not have learned to respond to others' signals of submission (Bernstein & Ehart, 1986). Cage space can be designed to enable animals to escape aggressive confrontations. Perches are one simple way to enlarge space and add complexity so animals can avoid physical and visual threats. Visual breaks and barriers are another. Primates in solitary cages may experience less distress if they can shift position to view another, less disturbing, or more interesting part of the room (O'Neill, 1987).

Primate social strategies do not ensure equitable distribution of food. Lower-ranking and young primates frequently must make do with the leftovers of higher-ranking animals, while the higher-ranking and older animals may overeat. Because the food offered to each individual and the amount actually consumed may be very different, it is important to identify inappropriate, inadequate, or poorly-distributed food as a source of stress to captive primates. In the wild, when higher-ranking primates monopolize a feeding site, those who are chased away may search elsewhere for food. They may also opt for high-risk competitive strategies to get to the main food source when escape routes (in case trouble brews) are readily available. In captivity, when there is but a single source of food and no way to escape conflict, lower-ranking animals are distressed.

Problems also arise for singly-caged primates in colony rooms. An individually-housed primate who is threatened or bullied continually by one or more animals living within sight of it gets the same message as it would from higher-ranking animals in a group pen. A timid, lower-ranking animal cannot ignore the others' aggressive gestures and messages and may go without eating the food that has been placed in its own cage, regardless of whether the threatening outsiders are physically able to attack. A perch or visual barrier installed in the cage may provide a means of eliminating the fear, relaxing the animal, and increasing food consumption.

EFFECTS OF ENVIRONMENTAL MANIPULATIONS ON BEHAVIORAL DISTURBANCES OF SURROGATE-REARED JUVENILE RHESUS

To determine the influence of environmental factors such as space, play apparatus, lighting, indoor vs. outdoor setting, and music on the occurrence of four common forms of behavioral disturbance, eight environmental manipulations were studied over a period of 39 weeks in two groups of juvenile rhesus monkeys, each consisting of two males and two females. All had been laboratory-born and reared without mothers in individual sections of quadrant caging (one cage divided into four 76 x 61 x

76 cm. units) until the age of 1.5 years, when this rehabilitation study began. Up to then, their social experience had been limited to five hours per week in a playroom (236 x 249 x 213 cm.) with peers raised the same way, beginning at two months of age (Mears & Harlow, 1975).

The two groups differed only in whether they did (Group II) or did not (Group I) have access to play apparatus in the playroom between the ages of 6 and 9 months. Normally-reared young monkeys first direct their play toward the inanimate environment: swinging, leaping, jumping, and engaging in acrobatics. Beginning at about four months of age, there is a shift toward social play with peers. Mears and Harlow (1977) hypothesized that infants were not ready to engage in more complex forms of play with partners (whose behavior is less predictable) until they had achieved a sense of security playing with the more-predictable physical environment. Harlow and Mears believed that if infants were given the opportunity to cultivate a degree of self-confidence by playful interaction with such responsive features of their physical environment as swings or tree branches, maladaptive behavior induced by nonstimulating caging and lack of mothering would abate.

That is in fact what happened. Group II infants, given the opportunity to engage in positive physical experiences leading to mastery of the leaping, jumping, and swinging motions in a playroom enriched with play apparatus, exhibited consistently and significantly lower levels of rock-huddle and ventral-cling (psychopathological behavior) than did Group I. Group I infants were socialized in the same playroom but, instead of responsive play apparatus, only bare floors and surrogates were available during their five one-hour playroom sessions each week. These studies demonstrated the value of play-stimulating apparatus for development of self-confidence, as well as its therapeutic potential.

I then proceeded to investigate the value of these and other environmental conditions for psychological wellbeing, following Harlow and Mears's subjects over their juvenile, adolescent, and adult stages. Each monkey was observed for 10 min., 3-5 days per week, using a focal-animal sampling procedure. Data were collected at 15-sec intervals using a modified frequency scoring system. Twenty-one behavioral categories were tallied. Four of these were of behavioral disturbance: 1) Stereotypy, 2) rock-huddle, 3) self-mouth, 4) self-clasp and/or ventral-cling. *Stereotypy* was defined as any ritualized, rhythmic, or repetitive pattern of behavior such as pacing or rubbing or brushing some body area at least three times in repetition. The *rock-huddle* category included repetitive forward and backward rocking motions with the subject in fetal position; it might incorporate *self-clasp*, *self-embrace*, and/or lowered head. *Self-mouth* was defined as mouthing any part of the body, not to include grooming. *Self-clasp* and *ventral- cling* were combined as the fourth disturbance category and included clutching any part of the animal's own body or placement of its ventral surface against another animal's body and clinging to it.

Figure 10. Condition 2 enriched playroom. Photo courtesy Bob Dodsworth, Wisconsin Regional Primate Center.

Positive therapeutic intervention was considered to have occurred if any of these categories was significantly reduced during the environmental manipulation. As it turned out, there were no significant differences between the two experimental groups in their response to any of the environmental manipulations, but a number of highly significant behavioral changes were observed in both groups as a function of some manipulations.

Condition 1 (Home Cage) vs Condition 2 (Playroom)

Six weeks of baseline data were collected in the monkeys' laboratory home cage (Condition 1). Animals were housed together in a chain-link rectangular cage (152 x 122 x 152 cm.), raised off the floor, with dropping pans beneath. No perches, play apparatus or manipulanda were available. The overhead room lighting was fluorescent and controlled by a timer. During these six weeks the group was introduced to a separate, larger playroom (236 x 249 x 213 cm.), equipped with bars, perches, ladders, a revolving wheel, and rings (Condition 2; Figure 10). The monkeys were in

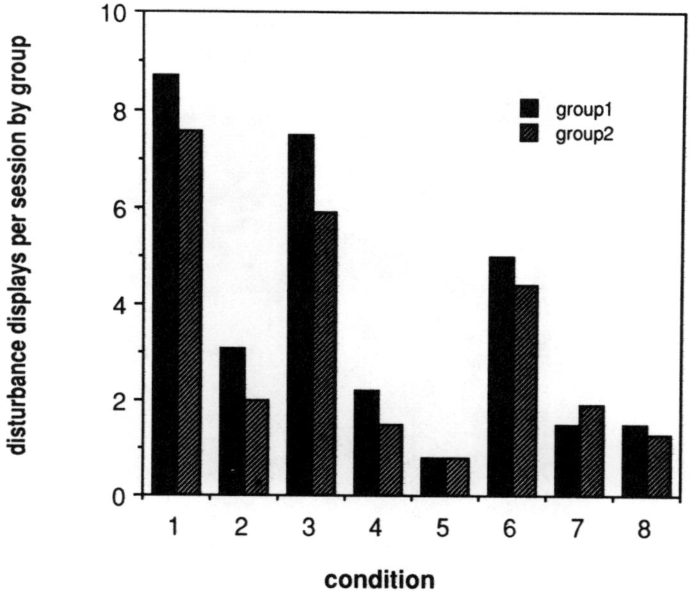

Figure 11. Responses of Groups I and II separately to environmental conditions.

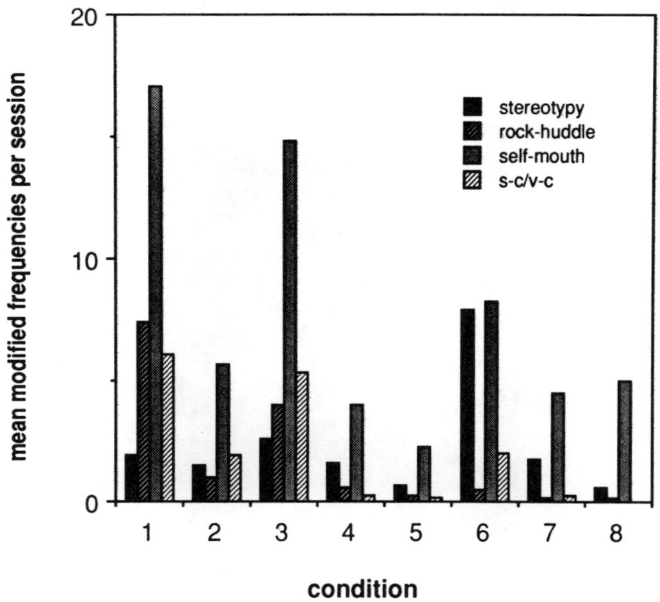

Figure 12. Mean modified frequency of four categories of behavioral disturbance in each experimental condition. Combined data of both groups.

a

b

Figure 13. Two views of enriched corncrib of Condition 5. (a) View of some of the play apparatus. (b) The corncrib with some of the surrounding scenery. Photos courtesy J. Wolf, Wisconsin Regional Primate Center.

this space an hour a day, five days a week. Figures 11 and 12 show that three out of four disturbance categories were significantly lower ($p < .01$ ANOVA and Duncan tests) in condition 2 (playroom) than in condition 1 (home cage). Stereotypy, which was relatively low even in the home cage, also declined in the playroom but not significantly.

Condition 3: Natural Lighting in the Home Cage and Condition 4: Concurrent Behavior in the Playroom

Because plans were being made to house the group outdoors, natural lighting was the next variable studied (Condition 3). For the next six weeks the monkeys continued to spend one hour a day in the playroom, but back in the home cage the fluorescent lights were switched off and the wooden covers were removed from the thick glass blocks that made up the north wall of the windowless room, allowing in diffuse daylight. Figures 11 and 12 (Condition 3) show the monkeys' reactions to natural daylight. There was a significant decrease ($p < .05$ ANOVA and Duncan tests) in rock-huddling in the home cage relative to the baseline home-cage value (Condition 1). No other behavioral disturbances were significantly affected. During this six-week period the frequency of disturbed behavior in the playroom was in general lower than in the previous six weeks but not significantly (Figure 11 and 12: Condition 4).

Thus far the monkeys had only experienced the enriched playroom for limited times. Although in the playroom both groups displayed significantly less of three of the four classes of disturbed behavior, these therapeutic results were not transferring back to the home cage. Natural lighting alone did not substantially lessen disturbed behavior in the home cage.

Condition 5: Outdoor Apparatus-Enriched Housing

Next the monkeys were moved outdoors to a spacious circular corncrib (12-ft diam, 14 ft from floor to metal roof), made of wire mesh and enriched with apparatus, that was located at my farm in rural Columbia County Wisconsin. Not only did the monkeys have a hanging 10-ft mobile with moving wings and parts to swing, climb, and jump on (Figure 13), but they had other stimuli to respond to such as the sounds of insects, amphibians, birds, and mammals, weather fluctuations, and many new sights. All eight monkeys spent nine weeks (September-November) in the corncrib. Although frequency of the four types of behavioral disturbance continued to decline relative to playroom levels, they had already reached such a low level in the indoor playroom that further declines relative to that setting were not statistically significant. The monkeys exhibited very little disturbed behavior during the nine weeks they were in the corncrib, and in

11 cases a specific type of disturbance dropped out of an animal's repertoire entirely.

Condition 6: Return to Indoor Housing

After nine weeks in the outdoor corncrib the monkeys had to be brought indoors for the winter. Their old home cage, enriched with two swinging milk crates and a hanging chain, was moved to a quarantine area where they were held for nine weeks. There was no significant increase in the frequency of rock-huddle or self-clasp/ventral-cling in this enriched home-cage setting, but stereotypy and self-mouth increasd significantly ($p<.01$ ANOVA and Duncan tests). In fact, stereotypy, previously infrequent, reached its highest level (Figures 11 & 12, Condition 6).

Condition 7: Music Stimulation

Following the fifth week of quarantine, the animals were given classical music in their room between 0700 and 1600 hrs. Music was piped into all animal rooms in this building, where research was conducted primarily on domestic livestock. Music has been used for environmental enhancement in the dairy industry for some time, and it has been reported to have therapeutic effects on human anxiety (Smith & Morris, 1976) and mental illness (Charlesworth, 1982). It was thought that the music might relax the monkeys and reduce their distress at being confined after nine weeks in the outdoor corncrib. There was no significant effect on the already-low categories of rock-huddle and self-clasp/ventral-cling but the two categories that had been elevated significantly in Condition 6 now showed a decline [ANOVA and Duncan tests: stereotypy ($p<.01$); self-mouth ($p <.05$). Figures 11 & 12: Condition 7].

Condition 8: Return to Original Laboratory Room

The final manipulation, lasting nine weeks, was to move the old home cage, still modestly enriched, out of quarantine and back to the original laboratory room. This room had no music. Disturbed behavior stayed about as low as in Conditions 5 and 7. Self-clasping/ventral clinging had entirely disappeared (Figure 12).

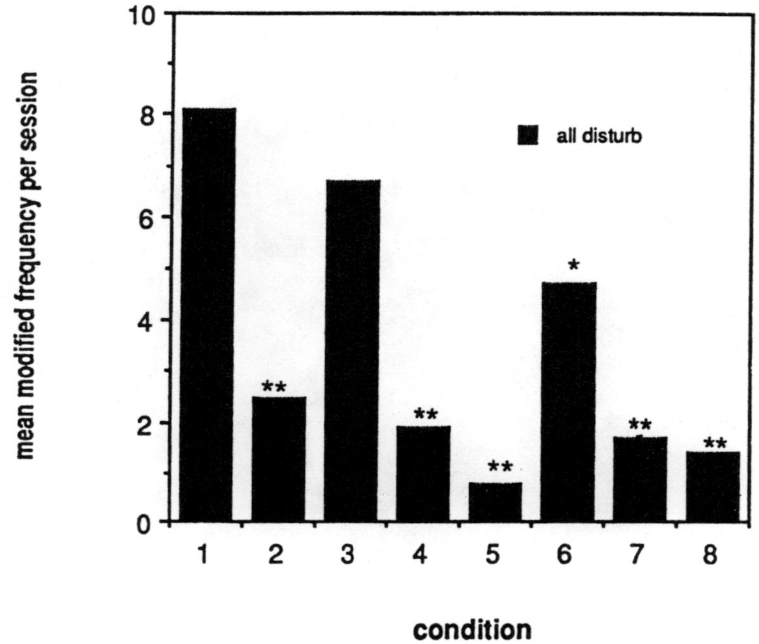

Figure 14. Mean modified frequency of disturbed behavior in each condition, combined over behavior categories and subjects. * Different from Condition 1 at $p<.05$; ** different from Condition 1 at $p<.01$.

CONCLUSIONS

This study lacked a separate control group to gauge the influence of time and maturation alone on the frequency of disturbance displays. There were, however, internal controls: the significant difference, over the same span of time, between disturbed behavior in the group home cage devoid of playthings, and in the larger enriched playroom (Condition 1 vs 2, Condition 3 vs 4, Figure 14); the increase in disturbed behavior when the monkeys were moved back indoors to the much-smaller (albeit now modestly enriched) indoor home cage after the glorious time outdoors in the relatively-vast enriched corncrib (Condition 6 vs 5, Figure 14). Even after the novelty of the corncrib wore off, disturbed behavior remained at low levels, very likely because of the variety of stimulation that the corncrib provided. The significant increase in a previously-infrequent type of disturbed behavior, stereotypy, when the monkeys were returned indoors perhaps reflects increasing maturation. Stereotypy might be a stress response more typical of older primates and perhaps supplanted stress responses more typical of monkeys when they are younger. Perhaps it was directly related to their perception of reduced living/play space. The quarantine cage, although familiar to the animals, had only 1/12 the

Figure 15. Group II animals continue to thrive in an outdoor research setting. Photo by author.

Figure 16. Rehabilitated surrogate-reared rhesus mother still shows self-mouthing characteristic of motherless infants, yet this does not interfere with her own mothering. Photo by author.

Figure 17. Puppies seem to stimulate nurturant behavior in rhesus monkeys that did not receive parenting during their own rearing. Photo by author.

volume of the corncrib. One behavioral abnormality (rock-huddle) declined significantly when natural light was provided in the home cage (Condition 3), perhaps indicating that natural light is preferable to artificial lighting. This, too, could have contributed to the therapeutic effects of the corncrib.

Thus, while not definitive, these data illustrate the possibilities for simple, effective environmental enrichment to reduce abnormal behavior in rhesus macaques, as well as illustrating some of the possibilities for research on enrichment and rehabilitation.

Figure 18. Surrogate-reared infant Sleepy clings rigidly to his unresponsive cloth-covered surrogate mother for security. Photo by author.

Figure 19. Sleepy, aged 12, reverts to his rigid, self-mouth posture during an aggressive interaction between other group members. Photo courtesy June Mathwick, Earthwatch Foundation.

Comment: The Long-Term Outcome

Something should be said in closing about the lack of difference between Groups I and II in this study. At the outset, Group I monkeys lacked the experience with play apparatus that Group II monkeys had. Although their behavior was not significantly different from that of the more sophisticated Group II monkeys across conditions, there is reason to believe that there were deficiencies. Three of the four Group I animals died within 6 months after this study. After returning to the farm the following year, one Group I male fell to his death and the other Group I male and a Group I female died following a fight between the remaining monkeys of both groups. The only Group I monkey that survived for the long term was injured in the same fight that killed her peers and she eventually assumed the status of lowest-ranking female and continued in that rank until she died ten years later.

All Group II animals are alive and thriving today, 13 years after environmental enrichment therapy began (Figure 15). They have reproduced successfully and reared two generations of offspring. These four monkeys have exhibited very low levels of disturbed behavior throughout adolescence and adulthood while residing in a five-acre outdoor naturalistic habitat in the company of their first and second generations of offspring. All surviving females were successful first-time mothers (Figure 16), perhaps owing to the practice in social and nurturing behavior they enjoyed in pre-puberty when they had the company of therapist puppies (Figure 17). All surviving males exhibited species-specific mounting, and paternity tests gave evidence that each male fathered offspring.

Although none of the group was exposed to any rhesus monkeys but themselves during development or adulthood, the group has organized itself along species-typical matriarchal lines with separate dominance hierarchies between and within families. It appears that in rhesus monkeys basic behavioral repertoires and social organization owe much to the species' genetic code; whether this is equally true of other primate species remains to be seen (Suomi, 1986).

Other observations on these monkeys are of interest from a conservationist perspective. Only the progeny of the original rhesus group consistently exhibit species-normal behavior in response to stress. The surrogate-reared parent-generation monkeys sometimes revert to their previous rituals, stereotypies, and disturbance patterns when placed under stress (Figures 18 and 19). Demanding challenges of group life such as antagonistic interactions with conspecifics, rivalries between families, and conflicts arising in the dominance structure trigger the return of such displays and replace the species-typical and dominance-rank-specific responses observed in their offspring. It is safe to predict that years of extensive therapy and efforts to enrich the lives of macaques who have histories of deficient rearing may allow them to mate and to parent normal, healthy offspring. But such parents are unlikely to meet the intense challenges of the wild successfully, so reintroduction to the wild would appear to be ruled out for them.

THE PRESENT UNFOLDS INTO THE FUTURE: RESEARCH PROVIDES HOPE

A number of professional and lay groups have dedicated themselves to improving psychological conditions for the creatures most closely resembling ourselves, the nonhuman primates that have contributed so much to improve the quality of our health and welfare. We now have the experience and wisdom to prevent psychological disturbances associated with captivity, rather than having to resort to rehabilitation treatments after the fact. We can appreciate the work of psychologists like Harry Harlow that enables us to understand why enriched rearing conditions for healthy infant and adult primates are a crucial beginning to their lives.

Let us take more seriously the need to know individual animals in their specific situations. Let us be more aware of staff habits and practices that may influence the wellbeing and safety of the primates in their care. By offering all staff the opportunity to participate in research and helping them to understand the communication signals and displays of their captive charges, we open the door to more positive attitudes and better research. And let us understand that the psychological wellbeing of captive primates is the product of a complex interactive system whose proper functioning depends on the contribution of each and every part. If we will invest the time and energy, we can achieve more successful husbandry with monkeys and apes and reap greater rewards from our research as well as greater respect for ourselves.

ACKNOWLEDGEMENTS

I would like to express sincere gratitude to the University of Wisconsin Primate Laboratory; Past Director, Dr. Stephen J. Suomi; and the people of Leeds Township in Columbia County Wisconsin for making my primate rehabilitation efforts possible.

— 11 —
Enriching the Lives of Captive Primates: Issues and Problems

Melinda A. Novak and Karla H. Drewsen

Scientists and members of the public have been increasingly drawn into a serious debate over the use of animals in research and teaching. At the core of this interchange are issues pertaining to the moral justification of animal use and concerns about the quality of life for animals in laboratory environments. This latter point, quality of life, is important to all scientists for practical as well as humane reasons. Animals in distress can confound research outcomes. Primatologists are particularly concerned with this issue in light of new federal regulations requiring researchers to promote the "psychological wellbeing" of captively-housed primates. While the intent of the new regulations is laudable - we must perceive our subjects as living organisms and not as mere research tools, implementation of these regulations represents a formidable challenge.

Questions such as "What is quality of life?" and "How do we define and measure psychological wellbeing in nonhuman primates?" are not easily answered. The lack of direct communication between nonhuman primates and their human caretakers is one stumbling block, but even if we could communicate verbally with our nonhuman primate subjects, this would not guarantee that we would understand their psychological needs. Animals, like humans, do not always make choices that promote their psychological health (Curtis, 1985). Nor can we simply assume, because of the genetic heritage we share with other primates, that what produces wellbeing in ourselves will promote wellbeing in monkeys and apes. Instead, we have to seek out methods and measures that demonstrably promote wellbeing for our primate subjects. The task before us, then, is to determine what constitutes an optimal range of laboratory environments for each species of captive primate. This requires: 1) defining and operationalizing such phrases as "quality of life" and "psychological wellbeing," 2) identifying the kinds of environmental changes that may improve the quality of life for captive primates, and 3) devising assessment strategies to monitor the effects of environmental change. We will discuss these three requirements and then examine their implementation in several of our studies in which we *attempted* to enrich the environment of captively-housed primates.

DEFINING PSYCHOLOGICAL WELLBEING

There is currently no consensus among scientists as to what constitutes psychological wellbeing in nonhuman primates, and it is unlikely that a simple, widely applicable, definition will emerge. Different primate species have different needs, different modes of locomotion, come from different habitats, and have different species-typical responses to conspecifics and to the environment. Situations that could promote wellbeing in one species might have no benefit whatsoever in another. Thus, prescriptions for wellbeing will have to be uniquely formulated for each species, taking into account our current knowledge of ecological and behavioral requirements.

Although detailed programs for each primate species will require a high degree of specificity, it may still be possible to devise a general framework for defining and assessing wellbeing that will be applicable to a broad range of subjects. One such formulation has been suggested by Moberg (1985). He indicates that it is impossible to define wellbeing directly because it is subjective and each person brings unique personal experiences to the task. Instead, he argues for an approach based on illbeing or stress because most people would agree that stress compromises an animal's wellbeing. In the absence of chronic stress, then, the animal's psychological health is not at risk.

Although this stress approach has merit, we must consider several problems. First, stress itself is not easily defined. Despite extensive research on this topic, scientists still have no precise meaning for stress, nor can they necessarily measure it accurately in individual organisms (Mason Maher Hartley Mougey Perlow & Jones 1976; Moberg, 1982; Ursin, 1982).

Second, stress is not a unitary concept; some stressors may actually benefit organisms. Selye (1974) proposed the term "eustress" to denote such beneficial sources of stimulation and "distress" to denote potentially injurious stimuli. More recently, Frankenhaeuser (1982) has examined the positive and negative aspects of psychological arousal using the terms "effort" and "distress". Given this dual function of stress, it may be appropriate to characterize wellbeing more specifically as the absence of distress. One advantage of this characterization is that distress is more conveniently assessed. Whereas stress is primarily measured by physiological changes, distress is also behavioral in nature and can be directly observed by caretakers and researchers. There are reliable guidelines (clinical signs) for the recognition of distress in primates (Morton & Griffiths, 1985), including hunched posture, high-pitched vocalizations, facial grimace, abnormal breathing patterns, and tremors. CARE, the American Psychological Association's Committee on Animal Research and Ethics, has also developed a list of examples of behavioral disturbance in primates. These include:

1) apparent inability of non-aged primate to move quickly without strain or difficulty when appropriately stimulated to do so
2) chronic self-mutilation or abnormal pattern of repetitive motor acts of the limbs or whole body which fit within the generally accepted definition of stereotypic behavior
3) persistent pattern of avoiding or ignoring novel stimuli within the immediate living space
4) failure to respond in species-typical fashion to social partners
5) repeated species-specific fear or avoidance behavior in the absence of noxious stimulation

A third problem with Moberg's approach involves the presumed relationship between distress and wellbeing. Even if we could eliminate distress in our captively-housed primates, would this suffice to assure their psychological wellbeing? Or must the elimination of distress also be accompanied by active efforts by researchers to enrich the environment of captive primates? While these two processes are linked in some definitions of wellbeing, one can conceive of conditions in which a primate might neither experience distress nor wellbeing (a moderately bored monkey might fit this category). Thus, we prefer to define psychological wellbeing in substantive terms rather than simply as the absence of something else.

Another approach to defining psychological wellbeing is in terms of physical health. To assess wellbeing, one might monitor: 1) physical characteristics such as the condition of the skin (flaky, scaly, rashes, smooth), quality of the coat (sparse, patchy, full), color and consistency of urine and feces, appearance of eyes (swelling, crusts, clear, opaque etc.), gait pattern (shuffling, limping, stiff walk, normal), and rate of growth or aging, 2) various physiological variables such as hormones typically measured in blood, urine, and feces and 3) immune function. However, we question how tightly linked psychological functioning and physical condition actually are. Although poor health would impair psychological wellbeing, it is not clear that excellent physical health guarantees enhanced psychological functioning. Physical health, like the absence of distress, may be only part of what constitutes psychological wellbeing in animals.

Another part of the picture may come from studying the animal's behavior to try to identify behavioral markers for wellbeing. Given the differences across primate species, it is unlikely that any universal list of behavior patterns can be developed, but a general framework for behavioral assessment might be established. In one scheme, psychological wellbeing is viewed in terms of the range of species-typical behavior that can be expressed by captive primates (Markowitz & Spinelli, 1986). Animals experiencing enhanced psychological functioning would be those that displayed the full behavioral repertoire shown by the same species under free-ranging conditions. Although this approach has the advantage of easy assessment (i.e., simply measure the number of activities expressed by captive primates), several problems are apparent.

First, the phrase "range of species-typical behavior" is vague and unspecified. Is it sufficient to see one occurrence of grooming per day in captive monkeys, even though free-ranging monkeys groom much more frequently through the day? Should the frequency and duration of species patterns be considered?

Second, must all species-typical activities occur before we can say an animal exhibits psychological wellbeing? Or do we consider other factors? For example, do single-sex groups of monkeys (which can be found under free-ranging as well as laboratory conditions) lack wellbeing just because breeding is prevented?

Finally, the species-typical repertoire includes patterns such as antipredator behavior and aggression, which are often associated with distress. Highly socialized rhesus monkeys frequently express mild, and sometimes more severe forms of aggression toward group members. This occurs in both field and laboratory settings so that it is unlikely that changes in the laboratory environment will reduce this response to negligible levels. Does this mean that distress is sometimes acceptable and desirable in captive primates as long as it occurs in a species-typical context and is not an artifact of the lab? Or should we exclude part of the species repertoire from our definition of psychological wellbeing?

Using another scheme, we can examine how competently an animal reacts to social and nonsocial stimuli. This would include assessing: 1) how responsive is the animal to environmental events? 2) how appropriate is the response, given the species and the stimulus? 3) how successful is the animal in obtaining goal objects? (e.g., does it feed efficiently? Do mating attempts succeed?) 4) does it have effective strategies for deterring antagonists? and 5) how quickly does it recover from noxious events? Although these questions can be difficult to answer, it should be possible to adapt this approach to different primate species.

As we have seen, psychological wellbeing can be examined from a number of different perspectives. Perhaps enhanced psychological functioning is the product of all the factors proposed in these differing viewpoints. Thus, an animal may experience wellbeing if it is free from distress most of the time (some distress may be unavoidable), is in good physical health, exhibits a substantial range of the species-typical repertoire, and is able to deal effectively with environmental stimuli. While this is not a simple definition, it can be operationalized for each species of nonhuman primates that resides in research labs.

TYPES OF ENVIRONMENTAL ENRICHMENT

There are many features of laboratory environments that might be altered to try to promote captive primates' wellbeing. These features can be classified as physical, social and psychological. We can expect these classes to be interdependent; changes in one sphere may differentially affect how the animal responds to changes in the other spheres. Before examining such potential interactions, we shall look at the features in each class.

Physical Environment

Physical features can be divided into those pertaining to the building or room and those that are a part of the animal's living cage. Regulations and guidelines govern room features such as light intensity, temperature/humidity range, and air exchange (see Animal Welfare Act, 1985; NIH Guide for the Care and Use of Laboratory Animals, 1985), although there is some question as to whether these guidelines can be applied uniformly to all primate species. Little is known about other potentially important room variables such as light on/off transitions, the spectral density distribution of light, and the color and texture of floors and walls.

Regulations also cover the animal's living cage, but only with respect to size and construction materials. In recent years, researchers have increased the complexity of the cage environment by changing the floor substrate, introducing manipulable objects or puzzle boxes, dividing the space with screens and panels, and providing branches and ledges. These changes frequently produced beneficial effects. For instance, using deep woodchip litter as a floor covering reduced inactivity and fighting in both arboreal and terrestrial primates (Chamove, Anderson, Morgan-Jones & Jones, 1982; McKenzie, Chamove & Feistner, 1986; Chamove & Anderson: this volume). The introduction of manipulable objects was associated with increased activity in captive orangutans (Tripp, 1985), increased manipulation in capuchin monkeys (Westergaard & Fragaszy 1985), lion-tailed macaques (Westergaard & Lindquist, 1987), and squirrel monkeys (Eterovic & Ferchmin, 1985; Fragaszy & Mason, 1978; and Roy et al., 1981) and a higher incidence of species-typical behaviors in chimpanzees (Bloomstrand, et al., 1986). Placing branches in cages of individually-housed rhesus monkeys stimulated manipulation and activity (Reinhardt et al., 1987). Not all changes are beneficial, however. The introduction of Nylaballs (small nylon balls) into the cages of individually-housed rhesus monkeys at one primate center produced little reaction; only 10% of the animals explored them in 11 months of exposure (Line, 1987). (See Markowitz & Scott, this volume, for discussion of what it may take to produce beneficial changes in the behavior of individually-housed macaques).

Social Environment

Social interaction is vital in some primate species. Infant rhesus monkeys reared in total or partial isolation and thus deprived of social experience display profound deficits in social behavior (Cross & Harlow, 1965; Harlow & Harlow, 1965; Mitchell & Clark, 1968; Suomi, Harlow & Kimball, 1971). Some of these deficits can be reversed when isolates are exposed to much younger monkey "therapists" (Novak, 1979; Novak & Harlow, 1975; Suomi & Harlow, 1972), but others appear to be permanent. For troop-living primates such as rhesus monkeys, the best way to promote their wellbeing in the laboratory may be to rear them with partners or in social groups.

Socially-reared rhesus monkey infants sometimes end up living in individual cages as adults. Here, too, wellbeing may be most readily promoted in these animals by removing them from the individual cages and placing them in compatible social groups. Although there are significant health risks, as a certain amount of tension, aggression and wounding occurs when new rhesus groups are formed (Bernstein Gordon & Rose 1974; Southwick Siddiqi Farooqui & Pal 1974), these usually dissipate with time. Careful selection of group members may reduce this initial trauma. A group formed of individuals all of whom are strangers to one another may be less aggressive to each other than one formed by merging existing groups (Erwin, 1986). Animals that have been reared in total or partial isolation and housed individually throughout their lives may not benefit from social interaction unless some rehabilitation is undertaken. Although social interaction is important to rhesus and pigtail macaques, we should not make the mistake of generalizing these observations to all species of primates without further study.

There is more to consider about the social environment than rearing and housing animals in groups. Composition of the social group may also affect psychological wellbeing. For instance, does it take a minimum number of animals to comprise a psychologically healthy group that allows for the expression of a wide range of species-typical behavior? Is there an optimal sex ratio or range of ages? Group continuity may also affect psychological functioning. Constantly changing group membership produces stress, aggression, and an increased incidence of atheroschlerosis in cynomolgus monkeys (Kaplan Manuck Clarkson Lusso Taub & Miller 1983). On the other hand, constant group pairings over years have been associated with a decline of reproductive function in rhesus macaques (Michael & Zumpe, 1978).

Another important feature of the social environment is the relationship between laboratory primates and the personnel who care for them. Although too little emphasis has been placed on the caretaker in the past, these individuals may represent a major source of stimulation for isolated and individually-housed primates. Caretaker characteristics such as mannerisms, voice patterns, and time of visits and departures (Cooper & Markowitz, 1979) may calm or stress laboratory animals. Thus, the nature of this interaction may be critical for the animal's wellbeing and should be studied further.

Psychological Environment

We are only beginning to understand what is meant by the psychological features of the laboratory environment, but several elements may be important. These include not only the amount but also the predictability of environmental stimulation and the extent to which animals can control it.

There is a commonly-held view that laboratory animals are best served by moderate levels of stimulation. Low levels are thought to produce boredom, while high levels lead to tension and distress. If so, researchers should strive to create an environment that stimulates the animal moderately. However, it is not a simple matter to quantify the amount of stimulation in the animal's environment. Stimuli are not uniform and there is no standard way to group them. Furthermore, they have both a physical value (intensity, frequency etc.) and a psychological value (the significance to the animal), the latter being difficult to assess except in a very general way.

Predictability, like amount of stimulation, is thought to be beneficial at moderate levels. Total predictability is assumed to be boring while a high level of unpredictability would be stressful. Some laboratory environments may be too predictable; there is a standard husbandry routine that never varies, a standard chow that is never supplemented, and a standard experimental procedure that goes on for months or years. Other laboratory environments may be too unpredictable. Animals may be used in a variety of studies with no common thread; they may put into and taken out of social groups frequently, husbandry practices may be variable; and there may be considerable turnover in caregivers.

Laboratory animals have only a modest degree of control over their environment and are subject to the experimental protocol of the researcher and the husbandry practices of the caregiver. Control, or the perception of control, over environmental events appears to have powerful effects on social, emotional, and cognitive functioning in human children (Gunnar, 1980; Lewis & Goldberg, 1969; Watson, 1979). More recently, giving infant monkeys control over appetitive stimuli was found to decrease fear and increase exploratory and coping responses (Mineka, Gunnar, & Champoux, 1986). Thus, control over environmental events may enhance psychological wellbeing in laboratory primates.

Interactions

Although we can classify features of the laboratory into physical, social, and psychological divisions, these are not independent of each other. Changes in the social environment, for example, may alter an animal's reactions to physical features. At present, we cannot be sure which variables or combination of variables both within and across divisions are important to primates nor do we know how their responses might vary across species. The effect of a combination of variables may not be readily predictable from their effects taken one at a time.

ASSESSMENT STRATEGIES

Evaluating the impact of experimental changes in the laboratory environment is one of the most difficult aspects of the task before us. In an area where we often presume to know in advance what is beneficial for captive primates, it is important to maintain objectivity and establish meaningful assessment procedures. For example, we need to ascertain that monkeys will respond to potential enrichment devices such as Nylaballs rather than assuming they will. Otherwise we waste effort and money and the monkeys are no better off than they were before.

Several different assessment strategies have been employed, including: 1) monitoring behavioral changes (ranging from a focus on one activity to surveying the full behavioral repertoire), 2) measuring physiological or anatomical variables (e.g., hormone concentrations, immune function, brain and bodily growth, or joint mobility), and 3) recording primate preferences. Each of these strategies has its strengths and limitations.

Behavioral change can be observed and recorded from a distance, making this the least disruptive strategy to the animals being studied. On the other hand, different classes of behavior are interdependent; increases in one category will most likely be associated with a pattern of increases and decreases in others. Because beneficial changes, such as an increase in activity, can sometimes covary with negative changes, such as a decrease in sociality and an increase in stereotypic behavior, it is advisable to monitor a number of different activities and evaluate the overall pattern of response rather than to concentrate on one or two target acts. This complicates data collection and interpretation.

Physiological and anatomical measures are problematic from two standpoints. First, they are typically invasive and therefore potentially disturbing to the animal. If the animals are not thoroughly habituated to the invasive procedure, the assessment process itself may mask or exacerbate the effects of the environmental manipulation. Fortunately, less invasive procedures for monitoring hormone concentrations are now being developed (see Rissler Wasser & Sackett 1987 for one example). A second problem involves variability. Physiological measures (especially those used to monitor stress) can be quite variable both across and within subjects. On the other hand, physiological measures are sometimes more sensitive than behavioral measures to environmental change.

Studying preferences has a certain intuitive appeal. It may seem as though the animals are actually informing us of their wants and needs. Such preferences can be misleading, however, because subjects may not choose optimally with respect to long-term consequences. Previous history can create wants that are independent of or contrary to the animal's needs. Even the nature of the choice test itself can influence the outcome. Researchers must distinguish between the benefits derived from the act of choosing and the significance of the choice itself.

Effective assessment requires selecting measures that are suitable and appropriate for the proposed environmental manipulation. Consideration should also be given to the number of measures to be employed, keeping in mind that too few may bias the results and too many may make the results uninterpretable given our present understanding of how measures interact.

Effective assessment also requires comparison between and/or across subjects. When we introduce a change in the environment and measure an animal's response to the change, how is the response to be evaluated? Given a between-subjects design, should the comparison be merely with a group of animals maintained in their typical lab environment or with several groups of animals housed in a variety of ways, some of which may even be stressful? In the first case, the significance of the change may be underestimated, especially if the standard laboratory environment is already stimulating to the animals. The more groups to which the manipulated group is compared, the better our understanding of the effectiveness of the manipulation.

Finally, the length of the assessment period, whether it be two weeks, two months or two years, needs to be determined. There are no general guidelines available at present, but it is important to note that animals have both short- and long-term reactions to environmental change. Primates may initially withdraw or even show an increase in aggression or other adverse reactions after an environmental change but these responses may dissipate or reverse within a few short weeks.

Figure 1. A female from the S group and her male infant.

ENRICHMENT STUDIES IN SOCIAL GROUPS OF CAPTIVE MONKEYS

Over the past several years, we have examined the effects of environmental manipulations on captive groups of rhesus monkeys. Here we will describe three separate manipulations involving cage size, lighting, and exposure to music.

The subjects were 12 adult rhesus monkeys maintained in 3 permanent groups along with their infants (Figure 1). The Q group consisted of two adult males, four adult females, and their infants. The S group contained two adult males, two adult females, and their infants. The B group included one adult male, one adult female, and their infant. All monkeys had early social rearing experiences and had been maintained in the same groups all their lives. They were housed in floor-to-ceiling chain-link and stainless-steel pens.

During the course of these studies, each monkey was observed for 5 or 10 min per day, 4 to 5 days per week, using a focal-animal sampling procedure. Data were collected using a modified frequency scoring system with 15-sec epochs. Our Behavioral Profile System is designed to record the daily activities of rhesus monkeys in social groups. The following composite categories were measured: 1) locomotion, 2) tactile-oral exploration, 3) passive alone, 4) passive social contact, 5) social grooming, 6) self-grooming, 7) social play, 8) stereotypy, 9) dominance-submission, and 10) sexual behavior. In addition, coat quality, reproductive condition, fecundity, wounding, illness, and gait pattern were noted daily. Urinary cortisol values were obtained during the lighting and music studies, and preference for music was measured by means of lever pressing during the music study.

Cage Space Manipulations

Although there are numerous federal regulations pertaining to the amount of space required for the housing of captive monkeys, none of these rules appears to have been based on empirical data. Instead, experimenter convenience, economic factors, and some vaguely defined standard of humaneness seem to have been the governing variables (Fox, 1986). The problem of adequate cage space becomes even more complex for group-housed animals. More or less space per animal may be appropriate depending upon the species and the social composition of the group.

Although amount of living space is thought to play an important role in aggressiveness, recent studies have not clarified this relationship. Increases in aggression have been associated with a reduction in cage space in rhesus monkeys (Southwick, 1967), Japanese monkeys (Alexander & Roth, 1971), baboons (Elton & Anderson, 1979) and galagos (Nash & Chilton, 1986). In the first two instances, however, other variables were changed along with cage size, thereby weakening the conclusion. In other cases, there was little, if any, change in aggression when space was reduced. Chimpanzees were not more aggressive when moved from their summer quarters to smaller winter quarters at the Arnhem zoo (Nieuwenhuijsen & de Waal, 1982). In still other cases, aggression may actually decrease along with cage space. In a detailed study with several replicates, Erwin (1979) demonstrated that aggression in pigtail macaques surprisingly *declined* when the living space of harem groups was reduced from two rooms to one. Higher levels of aggression returned when the harem groups were once again maintained in two rooms.

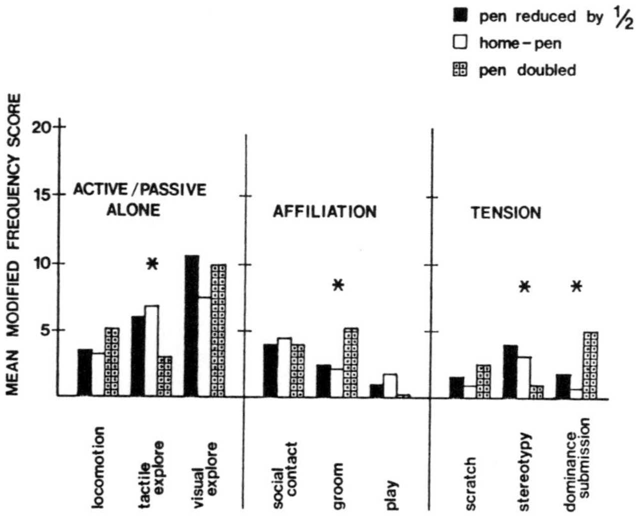

Figure 2. Mean modified frequency scores for the behavioral responses of the S Group to cage reduction and expansion. (* denotes significant differences with ANOVA at the $p < .01$ level.)

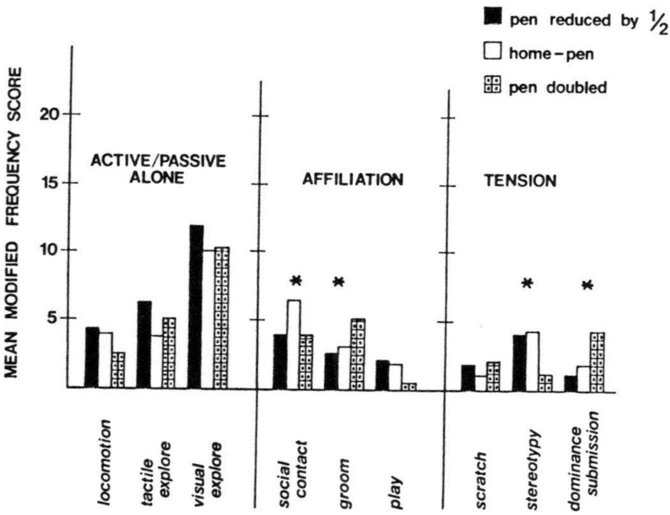

Figure 3. Mean modified frequency scores for the behavioral responses of the Q Group to cage reduction and expansion. (* denotes significant differences at the $p < .01$ level.)

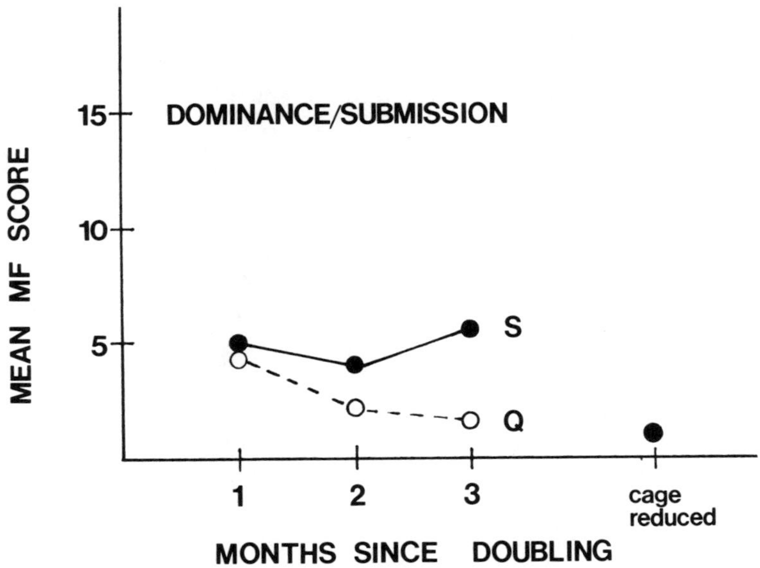

Figure 4. Changes in the levels of dominance/submission for the S and Q Groups over a three-month period following cage expansion.

The following study on cage space contraction and expansion came about fortuitously. Our goal was not to sort out the relationship between space and aggression but rather to do something that we thought would enhance the monkeys' wellbeing. About 11 years ago, we had the opportunity to enlarge the pens of two of our social groups. To accommodate construction, the original pens had to be reduced by half for a short time. Thus, each group's living space was halved for a month, returned to its original value for a month, and finally doubled. At the time, we assumed that "bigger must be better" and that the quality of life of our animals would be enhanced through cage expansion.

Figures 2 and 3 show the reactions of groups S and Q to these space manipulations. Both groups were more affected by cage doubling than reduction, perhaps partly because expansion was a novelty. A novelty effect has also been noted for galagos (Nash & Chilton, 1985). Although cage doubling produced a significant decline in stereotyped behavior in both groups, agonistic behavior increased. After one month, agonistic behavior declined to preexpansion levels in the Q group, but remained elevated in the S group for a full three months. Aggression returned to baseline in the S group only when their cage was reduced to its former size (Figure 4). Both groups showed an increase in grooming but not in passive contact or other forms of affiliation when the cage space was doubled. Thus, the change in grooming may represent reconciliation in response to increased tension (Terry, 1970).

In this study, monkeys showed an increase in aggression when their cage was enlarged. These results were certainly surprising and are contrary to most people's intuitions about living space. But this should reinforce the view that there is no simple relationship between aggression and living space in captive primates. Other factors, some yet to be recognized, must play a significant role. Our findings are similar to those of Erwin (1979) who suggested that certain aspects of group dynamics (e.g., presence of a male) might explain his atypical results. Group dynamics, or the personalities of individual group members may partially explain our outcome as well, especially since one of our groups eventually adjusted to the larger cage while the other did not. This group difference may be attributed to the number of animals in each group (six adults in the Q group that adjusted vs. four adults in the S group that did not) or to social relationships. The dominant male of the S group had always been more wary, active, and more likely to displace companions than the dominant male of the Q group. The S group was also much more neophobic than the Q group in response to novel foods and objects.

This study is important for three reasons. First, it demonstrates the folly of *a priori* assumptions in enrichment work. We cannot presume to know in advance what is beneficial for captive primates; bigger is not always better. Second, it shows that reactions to enrichment may be quite variable. Two groups of rhesus monkeys, similarly reared and housed, nevertheless responded differently to cage expansion. Finally, monkeys make both short- and long-term adjustments to environmental change. One of our groups showed a decline in aggression within a month following cage expansion and thus adjusted to and benefited from the manipulation.

Lighting Manipulations

Laboratory primates are commonly housed indoors under artificial light, usually fluorescent. Standard fluorescent lighting differs from natural light in several ways including spectral density distribution. Sunlight has more ultraviolet radiation than fluorescent lights and there is a peak in the blue-green rather than yellow-green region of the visible spectrum (Thorington, 1985). The ramifications of these differences for captive primates are not yet understood, but one might assume that natural light would be better for monkeys and apes than artificial light.

Researchers may not be able to house their primates outdoors or in indoor settings that make use of natural light, but lights that more accurately simulate sunlight are now available commercially (e.g., Vita-Lite, Lustra Ultra-D60, VL Ultra-D). Vita-Lites, manufactured by Durotest Corporation, have been tested in rodent colonies and appear to facilitate sexual maturation in rats (Wurtman & Weisel, 1969) and to reduce the incidence of dental caries in hamsters (Sharon, Fellner, & Burney, 1971). These changes have been attributed to the increase in UV radiation provided by the Vita-Lites. UV radiation has also been reported to reduce the incidence of metabolic bone disease in some marmoset colonies (Hampton, Hampton & Landwehr, 1966) and to improve immune function (Parrish, 1983).

We examined the effects of Vita-Lites on the behavior of two groups of rhesus monkeys housed in separate rooms. We used the same Behavioral Profile System as in the previous study but certain categories were combined to form the new categories of affiliation and nonsocial activity. In collaboration with Dr. Jerrold Meyer, we also obtained measures of urinary cortisol excretion in response to the moderate stress of removal from the group cage and individual housing for six hours. Adrenocortical activity can be measured by assaying either plasma or urinary cortisol concentrations. We chose the urinary assay because it is difficult to obtain representative blood values when testing multiple subjects living in social groups. As subjects become familiar with the blood-drawing routine, animals still waiting their turn show artificially elevated cortisol concentrations (see Meyer, Novak, Bowman, & Harlow, 1975). In addition, there is a potential advantage in examining urinary rather than plasma or serum concentrations because the urinary values represent integrated as opposed to point estimates of pituitary-adrenal activity.

Total cortisol in the urinary samples was determined by the method of Clark, Rubin, Kales, and Poland, (1970; 1971). To ensure that our urinary cortisol values accurately index stress, other factors contributing to cortisol excretion, such as an animal's body weight, rate of urine production, etc., must also be taken into account (Schneeburg, 1970). This was accomplished by measuring the amount of another substance that is known to vary with body weight (viz., creatinine) and relating cortisol to creatinine output. Urinary cortisol was measured twice for each animal at the end of every phase of the experiment.

A pretest-test-posttest design was employed. In pre- and posttest phases standard fluorescent lights were used, whereas in the test phase, the animals were exposed to Vita-Lites. To distinguish the effects of increased UV exposure from possible changes due to altered color rendition, Vita-Lites were installed in both colony rooms but special diffusers selectively blocked the transmission of UV light in one of the rooms. Behavioral and urinary cortisol responses were measured in every phase.

Figure 5 depicts the behavioral responses of the UV and NONUV group during the 10-week pretest, test, and posttest phases. Both groups became more affiliative when housed under Vita-Lites; however, the NONUV group returned to baseline while the UV group continued to show an increase in affiliation during the posttest phase. Although sexual behavior significantly increased in the UV group during the test phase, the level of aggression (dominance/submission) was likewise elevated. These changes were not manifested by the NONUV group. Furthermore, each animal in the UV group showed a heightened urinary cortisol stress response during the test phase compared to pre- or posttest, a pattern that was not present in the NONUV animals (Figure 6).

UV radiation had both positive and negative consequences for our monkeys. Affiliative and sexual behavior increased, but so did aggression and stress responsivity. This pattern is probably best understood from a breeding perspective. Both groups were sexually active throughout most of the study and breeding is known to increase aggression and tension in rhesus monkey groups (Teas, Richie, Taylor, & Southwick, 1980). However, the length and intensity of this breeding activity were greater in the UV-exposed group and thus it is not surprising that this group showed an increase in both aggression and stress reactivity. Desirable outcomes (such as increased breeding especially in groups where reproduction is low) may also be associated with undesirable effects. Thus, researchers may often have to weigh the costs and benefits of their environmental enrichment before deciding on its use.

Musical Stimulation

Captive primates are frequently housed in static environments where novel events are rare. The purpose of the present experiment was to determine whether exposure to, and control of, a novel stimulus such as music would promote psychological wellbeing in rhesus monkeys.

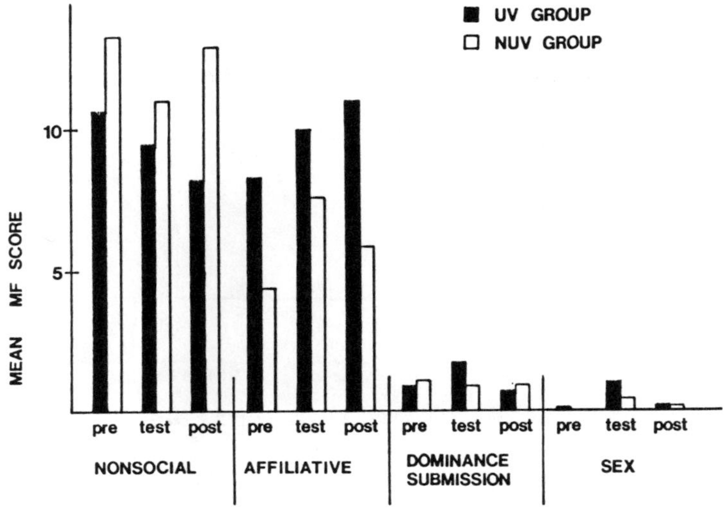

Figure 5. Mean modified frequency scores for the behavioral responses of rhesus monkeys to ultraviolet radiation present in Vita-Lites. UV = group exposed to ultraviolet radiation present in Vita-Lites. NUV = control group exposed to Vita-Lites through special diffusers that blocked the transmission of UV radiation.

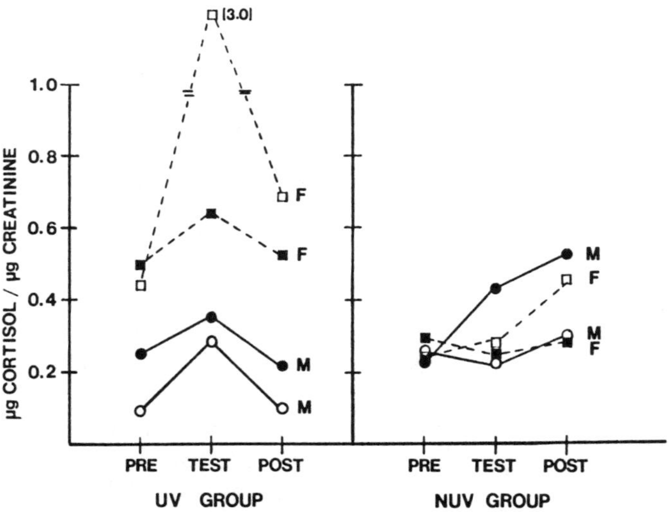

Figure 6. Urinary cortisol responses (measured as ug cortisol/ug creatinine) to an increase in ultraviolet radiation contained in Vita-Lites. UV = group exposed to ultraviolet radiation present in Vita-Lites. NUV = control group exposed to Vita-Lites through special diffusers that blocked the transmission of UV radiation. (M = males, F = females)

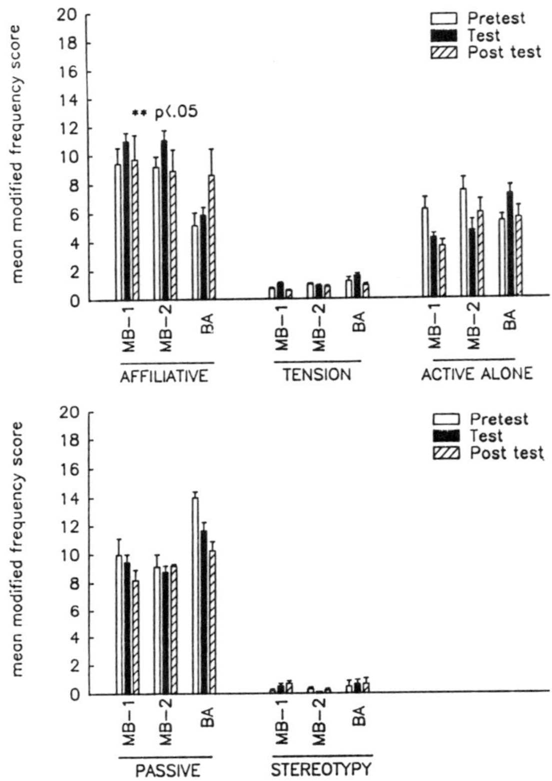

Figure 7. Mean modified frequency scores for the behavioral responses of rhesus monkeys across the three phases of the musical stimulation study. (** denotes significant differences with ANOVA at the $p < .05$ level.)

Little is known about the potentially beneficial effects of music on animals except for anecdotal reports, but there is an abundance of human data indicating that music has both psychological and physiological benefits. It has a therapeutic effect in the treatment of mental disorders (Alvin, 1965; 1978; Charlesworth, 1982; Ishiyama, 1963); it reduces anxiety (McFarland, 1978; Smith & Morris, 1976); and it has other physiological effects (Dainow, 1977; Landreth & Landreth, 1974; Schullian & Schoen, 1948).

We expected that music, and control over its presentation, would enhance the wellbeing of our monkeys, as manifested by behavioral changes, an increased ability to cope with the mild stressor of cage removal, urinary cortisol values, and by operating manipulanda to control music onset and offset.

The manipulanda were mounted in an aluminum box (34 x 13 x 26cm) that could be temporarily attached to the front of the pen. There were two levers on the box, one for activating a tape recording of light jazz music (George Winston, piano) and one for turning the music off. During adaptation trials the monkeys were observed to press the levers. Throughout the test phase of the experiment, the monkeys had access to the box for 2 hours per day, 4 days per week.

Using a pretest-test-posttest design, the effects of two conditions were evaluated. In the pre- and posttest phases both groups were maintained under the standard laboratory regimen (baseline). During the test phase, the groups received one of two treatments:

1) Music Box (MB) - These monkeys were given the opportunity to turn the music on and off by pressing the levers.
2) Box Alone (BA) - These monkeys had access to a deactivated lever box, and therefore never received musical stimulation.

To assess the impact of these treatments, behavioral and physiological responses were monitored across the four-week pre- and posttest phases and the nine-week test phase. The Behavioral Profile System was also used in this study, and certain categories were combined to form the grouped categories of affiliation, tension, active alone, passive, and stereotyped behavior. Urinary cortisol values were measured in the same way as previously described, and each subject's cortisol levels were evaluated twice at the end of every phase. Operant behavior measures (number of presses of either lever and total duration of the musical stimulation) were recorded in the test phase.

Figure 7 presents the behavioral reactions of the groups across all three phases of the experiment. There was a significant group-by-phase interaction (ANOVA) for affiliation, with the MB monkeys showing the highest levels of affiliative behavior during the phase when the music box was present. This is in contrast to the BA group where the differences were difficult to interpret in light of the manipulation.

To satisfy ourselves that this was not merely a result of group personalities, the BA group was retested under the Music Box condition (MB-2) with music now contingent upon their lever-presses, and a similar increase in affiliative behavior was seen. There were no other significant interactions for the other behavior categories. Cortisol values did not vary significantly across groups or phases (data not shown). Thus, musical stimulation did affect the levels of one target class of behavior and appeared not to subject the animals to appreciable amounts of stress.

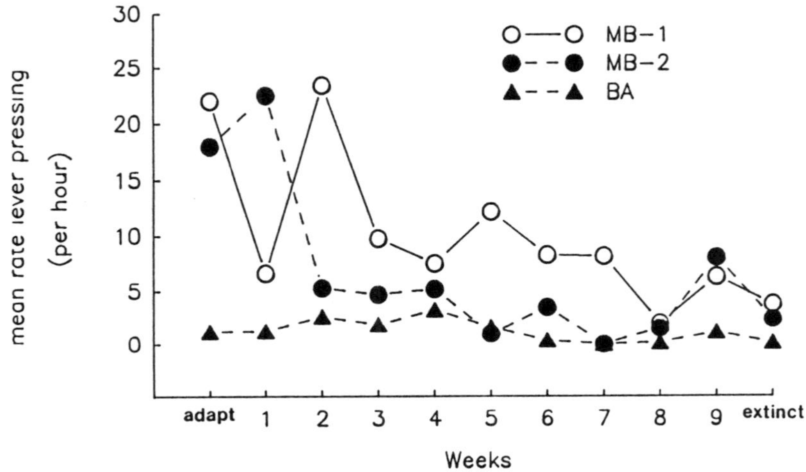

Figure 8. Mean hourly lever pressing response rates for the MB-1, BA, and MB-2 groups during the test phase. Each test session was 2 hours in length. Adaptation (adapt) and extinction (extinct) phases are also shown here. (MB-2 denotes the retest of the BA group when music became contingent upon their lever presses.)

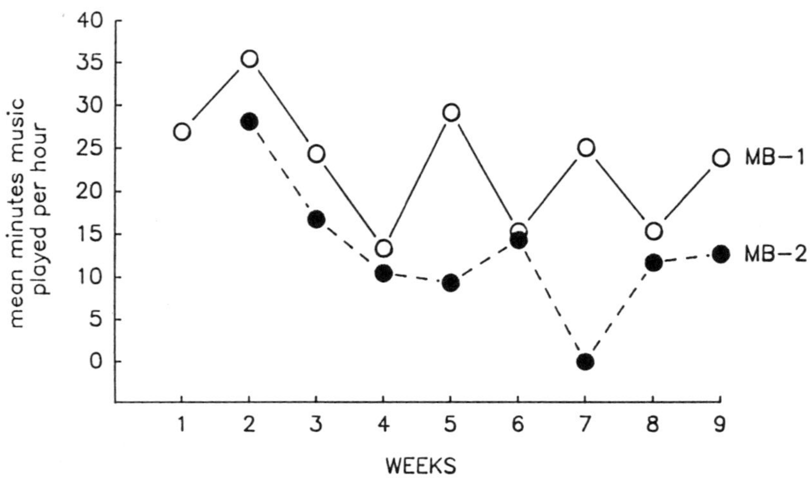

Figure 9. Average duration per hour (for a 2-hour test session) of musical exposure for the MB-1 and MB-2 groups during the 9-week test phase.

The preference data reveal that Music Box monkeys spent a considerable amount of time playing the music. Figure 8 shows that these monkeys repeatedly pressed the levers to activate and deactivate the tape, whereas lever-pressing rates were extremely low for the nonmusic group (BA). When the Box Alone group was subsequently retested under the Music Box condition (MB-2), this group also showed a high rate of lever-pressing when lever-presses controlled the occurrence of music. Both groups played the music for about 45 to 60 min of the two-hour observation period. As the experiment progressed, the rate of lever-pressing declined. But only one group (MB-2) showed a corresponding reduction in the total duration of music exposure (Figure 9).

This study demonstrates some of the problems of interpretation when different assessment procedures yield different outcomes. Music did have an effect on choice of stimulation and levels of affiliative behavior of these monkeys. On the other hand, other portions of their behavioral profile were not affected, nor were urinary cortisol levels. Given these results, how do we rate the effectiveness of music in promoting psychological wellbeing in rhesus monkeys? Does the fact that monkeys played the music necessarily make this a form of enrichment? In this case, we think it does. The monkeys could turn the music on and off and yet left it on for substantial periods. Other types of music or other types of stimulation (e.g., slides or videotapes which could also be controlled by manipulanda) might be even more enriching to these monkeys. Before we conclude that music is a form of enrichment for our colony groups, it would be prudent to compare it with other types of stimulation that monkeys can turn on or off.

Comments

Obviously, one of the limitations of the research described above is that only a few social groups of captive primates were studied. Testing a treatment on a greater variety of groups would give us a clearer picture of its effectiveness. These data were not presented with the aim of recommending types of enrichment for rhesus monkeys. Rather, they are detailed here to illustrate some of the pitfalls and problems that may ensue from this kind of research. These include: 1) defining psychological wellbeing adequately, 2) constructing assessment schemes that will be sensitive to the treatment to be assessed, 3) determining the appropriate length of the assessment period , 4) recognizing that treatments can have both positive and negative effects, thus requiring judgments about a treatment's effectiveness and 5) recognizing that diversity exists both across and within species with regard to the psychological needs of individuals.

The last point, about diversity, is an important one. Is our goal to identify, in a general way, what is enriching to captive primates or is it to understand the particular animals we work with and attempt to enrich their lives based on what we know about them as individuals? Hopefully, we will develop some broadly applicable procedures for improving the psychological wellbeing of primates, but even the most broadly applicable enrichment may not be beneficial for all animals of a species. Primates are housed in such a variety of ways and used in such diverse studies that it may be unreasonable to assume that the same treatment will serve to enhance the quality of life for all of them. Furthermore, primates have unique personalities and can be expected to respond to enrichment manipulations in an individual manner. As with humans, perhaps one monkey's caviar is another's rotten fish. Thus, the burden may still fall to each researcher to evaluate the effects of recommended treatments on his/her own subjects.

CONCLUSIONS

Although each of these studies was quite different, a few general conclusions about the reactions of our animals to environmental change are warranted. First, different groups of similarly-reared rhesus monkeys can and do respond differently to the same environmental treatment. This was evident in the space study where one group eventually adjusted to an increase in cage size while the other group did not. Second, manipulations were observed to have both positive and negative consequences for the animals. This was particularly surprising in the cases of doubling cage size and introducing Vita-Lites - both manipulations we had intuitively assumed would enhance primate wellbeing. Such results point up the necessity of using objective assessment procedures to evaluate the potential benefit or cost of any environmental change. Finally, some manipulations were found to have both immediate and more long-term consequences for the monkey subjects (e.g., increases in space and presentation of music). Thus, the assessment period has to be long enough to yield an accurate picture of primate responses to environmental change.

— 12 —
Examining Environmental Enrichment

Arnold S. Chamove & James R. Anderson

INTRODUCTION

When the need to confine animals leads to an environment that is less complex and more predictable, the degradation often creates a downward spiral:

unstimulating environment -> suboptimal behavior-> reduced use of environment -> worsening behavior, etc.

Many researchers are increasingly interested in interrupting this trend. Most of the published research toward this goal has been with animals living in small groups in zoological gardens, but some has been done with individually-housed animals (Markowitz 1982), and some in laboratories (Erwin & Deni 1979). Concern has centered on improving the welfare of captive animals, yet until very recently there has been little interest in operationally defining environmental improvement or enrichment, or evaluating the cost-effectiveness of various enrichment strategies, or developing theoretical models to guide research and integrate divergent studies (but see Chamove 1986; Dawkins 1980; Hughes 1980).

The study of enrichment can be viewed as a conceptual extension of past research on the developmental effects of early experience. In that earlier work, questions about the role or importance of various factors (e.g., vision – Anderson & Chamove 1984b; Bovard 1959; Chamove 1984b; Sackett 1966) in social development were studied by withholding or limiting sensory or social stimulation. In enrichment studies, on the other hand, such questions are addressed by adding a variable to a relatively-impoverished standard environment, and assessing its consequence for behavior. These studies often have two goals: (1) to assess the importance of a given factor for the subject's behavior and (2) to improve the captive environment. Both are achieved by adding to an impoverished environment that is presently the status quo.

PROBLEMS OF CONTAINMENT

Confinement can amount to environmental impoverishment in a number of ways. Issues to address include (i) cage size in relation to group size and structure, (ii) cage design, and (iii) cage furnishing. One can think of the last two as ways of increasing the psychological size of the cage (discussed at length in Chamove in press).

Cage size

One common problem is a cage that is too small. There is evidence that small cages are stressful enough to inhibit even that most robust measure of general wellbeing, reproduction (Doolittle Wilson & Geisking 1976; Snowdon Savage & McConnell 1984). The mechanism for this might be the adrenocortical stress response (Kalin Carnes Barksdale Shelton Stewart & Resch 1985). Several other behavior patterns are adversely influenced by cage size and its related variables of crowding and group size (Adams & Babladelis 1977; Archer 1979; Chamove Hosey & Schatzel in press; Dantzer & Mormede 1983; Kaplan 1986; Siegel 1980). Of course group size also has effects on behavior in the wild where space restrictions are less salient or even absent (Boelkins & Wilson 1972; Garber 1986; Kinzey 1986; McFarland 1986; Singh 1969).

Small cages are known to increase the incidence of stereotyped movements as well as other types of abnormal behavior not involving locomotion (Draper & Bernstein 1963; Hediger 1968; Paulk Dienske & Ribbens 1977). It is hard to imagine a case where simply increasing usable cage space would not constitute an enrichment procedure, even though gross activity levels might not change dramatically (see Line 1987).

Even when it is impossible to provide more space, and cages are quite small, enrichment is still possible. Reinhardt and his coworkers have shown that compatible pairs of rhesus monkeys can be formed from unfamiliar, individually-housed adults, allowing "the expression of a hitherto inhibited basic need: the need for a companion and for social interaction" (Reinhardt Cowley Eisele Vertein & Houser 1987a p.8).

Where social grouping is not possible, even visual contact with other animals is better than nothing. The presence of a full-length mirror in the cage reduced the development of abnormal behavior in young stumptailed macaques otherwise reared alone (Anderson & Chamove 1986). Group-living monkeys rapidly habituate to a large mirror just outside their cage (Anderson & Bayart 1985), but small, portable mirrors were repeatedly manipulated and remained a source of fascination over months (Anderson

1983).

For isolated individuals with no possibility of social interaction, one can at least provide varied objects that elicit manipulation. Food-acquisition games and devices such as those designed by Markowitz (1982) improve the quality of the animal's environment as well as the quality of the animal (Markowitz & Spinelli 1986). (Also see Rumbaugh Washburn & Savage-Rumbaugh: this volume.) Simply installing a branch in an otherwise barren cage gives a monkey an object to manipulate, hold, shake, gnaw, perch, and stand on (Reinhardt Houser Cowley & Champoux 1987).

Cage design

Commonly, enclosures are designed by architects whose primary considerations are engineering, maintenance, cost, and human comfort and convenience. Bare walls allow paintings to be hung, straight walls are best for human furniture, conventions for location and level of lighting are for reading. Bare, flat, smooth walls are easy to build and to clean, but most animals cannot make use of them – they do not hang pictures on them and, more importantly, they cannot climb them. One often sees zoo primates housed in spacious cages, but unable to make use of the large open spaces. One problem of enrichment is how to use these valuable areas most efficiently (Berkson Mason & Saxon 1963).

Cage furnishing

Floor covering. An area of the cage that is often neglected or ill-designed is the floor. The floor and areas just above the floor are often kept clear for cleaning and for humans to work. In eight callitrichid enclosures in our laboratory, the floor makes up about 40% of the total usable surface area and 66% of the horizontal surface area. In the wild these arboreal monkeys rarely come to the ground, only occasionally descending to forage in leaf litter or to cross open areas (Heymann 1987; Neyman 1980; Sussman and Kinzey 1984). In captivity these monkeys almost never visit a bare floor (1% of the time), but ground time increases ten-fold if the ground is covered with some leaf-like substrate

Branches. Another aspect of primate cage design that has been virtually neglected is branch placement One field study suggests that callitrichid monkeys primarily use horizontal branches moving on vertical substrata only 4% percent of the time, and this mainly on large tree trunks (Garber & Sussman 1984). Nevertheless, many of the cages housing these monkeys have few horizontal tree branches but many hanging vertical branches, perhaps because it is easier to attach these vertical clumps. Rough tallies (personal observations) indicate that about 25% of the branches are horizontal, 30% oblique, and about 45% vertical, with vertical branches usually being thinner than horizontal ones. In the wild, thin vertical branches are almost never traversed except to reach food. In large outdoor areas (Chamove & Rohrhuber in press) and in the wild (Garber 1980), the monkeys spend most of their time (89%) in dense networks of thin, flexible, nonwoody tangles. (Also see Wright Haring Izard & Simons: this volume.)

We are currently looking at branch use and branch preferences in caged callitrichid monkeys (Chamove & Goldsbrough: unpublished). Our results make it clear that these monkeys prefer a larger number of unevenly spaced horizontal branches; widely spaced ones are especially used by the more-active juveniles, and narrowly spaced ones by the less-active adults. These results are for preference, and it should be noted that measures of preference and measures of enrichment are not necessarily the same. Animals' preferences are not always or necessarily in their best interest (Chamove 1979; Dawkins 1976; Duncan 1978; Hughes 1976).

Range of behavior

Small cages, poor cage design, and impoverished furnishing combine to limit the range of behavior shown by an animal. The most obvious of these, and perhaps the most important, is the opportunity to move about and forage. Primates spend an average of over 50% of their day searching for and processing food. In cages this is dramatically reduced. Some animals continue to perform relatively normally, commonly exemplified by hamsters making piles of food; some perform species-typical behavior in inappropriate ways, like raccoons washing soluble food in their water dish, or stereotyped pacing; some display food-related behavior not seen in the wild, like regurgitation and coprophagy in primates (Fox 1986; Gould & Bres 1986).

It is often suggested that repetitive stereotyped movements "can be seen as the animal's attempt to increase its sensory input" when it is too low (Bareham 1972; Duncan 1977; Wemelsfelder 1985 p.143). This view presumes that the animal is receiving input that is inadequate in amplitude, variability, or quality. Some stereotyped movements may be an expression of arousal or stress, perhaps due to boredom, frustration, or merely stimulus change (Berkson Mason & Saxon 1963; Chamove & Anderson 1981; Chamove Bayart Nash & Anderson 1985). The ability of stereotyped behavior to reduce adrenocortical hormone levels supports this interpretation (Anderson & Chamove 1981; Blackshaw & McVeight 1984; Dantzer & Mortmede 1983). Stereotyped movements are probably best considered as habitual patterns of stimulation which, by virtue of their familiarity and repetitive nature, calm the individual when input is too high or arouse it when too low.

Cages may be designed so that it is impossible to perform certain actions at all, like tool-making and tool-using in apes and capuchins (McGrew 1981; Nash 1982), wood-gouging in marmosets (McGrew Brennan & Russell 1986), and, indeed, foraging in most primates (Rosenblum & Smiley 1984). Several lines of research have shown that increasing the range of foods eaten is healthier (Kirkwood 1983), increasing the range of muscles used is healthier (Turnquist 1985), and gnerally increasing the range of stimuli is both healthier (e.g., Bollhorn 1980; Ekesbo 1981; Fox 1986; Gross 1972; Kendrick 1972; Renner & Rosenzweig 1987; Riley 1981) and less boring (Wemelsfelder 1985). Consequently, enrichment efforts should try to increase the range of behavior that an animal can express as well as the range of stimulation available to the animal.

GOALS OF ENRICHMENT

Two complementary aims of enrichment studies re to increase "desirable" behavior and to reduce "undesirable" behavior. Activities such as coprophagy, regurgitation, hair-pulling, self-injury, or tereotyped movements re generally judged undesirable, while active exploration, play, affiliation, and foraging are judged desirable. Other behavior is classed as one or the other depending on its frequency: withdrawal or avoidance, reproductive behavior, aggression, inactivity, displacement, and feeding are judged desirable if moderate in frequency but undesirable if very frequent (e.g., Blackshaw & McVeigh 1984; Chamove Anderson & Nash 1984; Goosen Fransen & Gommers 1986; Gould & Bres 1986; Reinhardt Reinhardt & Houser 1986; Tripp 1985). An increase in desirable behavior is taken as an index of improvement in physical or psychological health.

Criteria of Desirability

There are at least four bases for assessing desirability of behavior: (i) normality, (ii) observer/caretaker acceptability, or (iii) theoretical or (iv) practical considerations.

(i) Normality. Probably the most common, implicit criterion is normality of behavior: Behavior that approximates what is found in the wild is held to be desirable, while behavior uncommon in the wild is undesirable. In other words, the form and frequency of behavior should fall within the range of values seen in the wild (McGrew 1981; Snowdon in press; but see Wemelsfelder 1985 p.137). One problem is that typical behavior patterns in the wild vary, sometimes widely, with group size, habitat, season, and even the daily weather (Chapman 1985; Harrison 1983; Oates 1987; Richard 1985).

(ii) Public acceptability. The second approach takes into account what is acceptable to the public -- often meaning the most outspoken elements of the public. It appears that the public does not like to witness stereotyped or sexual behavior, or "unpleasant" behavior such as regurgitation, reingestion, and coprophagy by captive gorillas (Akers & Schildkraut 1985; Gould & Bres 1986); they like animals to be active and clearly within view; and they do not read or necessarily believe posted notices (Cherfas 1984). These assertions need testing by data, but in the meantime they appear to be assumed by zoo curators and to influence the housing of zoo animals (Adams & Babladelis 1977).

(iii) Theoretical considerations. Some enrichment goals are chosen because they have a theoretical rationale. Examples might be setting goals for exercise or calorie expenditure for captive animals based on their activity budgets in the wild; or providing problem-solving opportunities that simulate challenges encountered in the wild, for example burying food in woodchips. Swimming pools and fishing ponds are further examples of innovative ways to increase activity in the laboratory (Gilbert & Wrenshall: this volume; Molzen & French: this volume).

(iv) Practical considerations. Finally, the practical need to improve husbandry often influences enrichment studies. Goals to be tackled include reducing aggression and self-injury, reducing food and other resource wastage, and reducing costs in general. One technique to reduce food wastage while making better use of space is to mix species (O'Neill: this volume; Waser 1986). Ground living rodents, birds, or fish eat fallen fruit in bird, bat, or primate exhibits in Edinburgh, Milwaukee, Sydney, Copenhagen, Stockholm, and San Diego Zoos.

Of course, though we may discuss them independently, the various bases for assessing desirability of behavior are related, and what satisfies one criterion will often satisfy others. As an example, management practices aimed at reducing stereotyped movements, say by increasing food variety (a practical measure), may lead to behavior that is more like that seen in the wild (normality approach), increases the similarity of the activity profile to that observed in the wild (theoretical approach), and probably also increases public acceptability along with improving the animals' physical health (practical). When hiding food in floor coverings led to increased activity in zoo orangutans, the visitors rated the exhibit more favorably (Tripp 1985).

PSYCHOLOGICAL FACTORS IN CAPTIVITY

Control

Cages usually reduce the degree of control that individuals have over environmental stimuli. When rooms are visually divided by opaque barriers, male macaques cannot control female aggression through posture and facial expression, and so female aggression increases (Erwin 1971).

The importance of control is also exemplified by the effects of visitors on the behavior of zoo animals (e.g., Hosey & Druck 1987; Glatston Geilvoet-Soeteman Hora-Pecek & van Hooff 1984). One can postulate that zoo visitors are beneficial stimuli in an otherwise-barren environment. Certainly some primates interact with visitors both positively and negatively (Fa: this volume). An alternative view, supported by an increasing amount of evidence (Chamove Hosey & Schatzel: in press), is that visitors cause the same kind of changes in the behavior of display primates as other stressors do; increased locomotion is the most consistent effect. The data suggest that the more visitors there are, the more behavioral change there is; noisy and active visitors cause more change; and visitors viewing monkeys from above (rather than at a lower level) cause more change. This research adds to the evidence for the importance of providing animals with privacy refuges (see Fouts Abshire Bodamer & Fouts: this volume, and Miller-Schroeder & Paterson: this volume) from which they can control visitor contact (Adams & Babladelis 1977). Other evidence for the importance of control is the surprising preference animals have for working for part of their food in the presence of free food (discussed later). Of course hidden choice titbits can also stimulate persistent searching (Figure 1; Anderson 1986).

Figure 1. A young adult male *Macaca tonkeana* uses a mirror to guide his hand movements as he searches for a hidden raisin stuck onto the outside of the cage at position 4. The mirror is in the lower left corner of the photograph.

The importance of control to the psychological health of animals is currently a topic of great interest. In the wild, animals have more control over certain stimulus variables and most response variables than they do in captivity. Cages restrict animals' control, particularly with regard to their distance from stimuli outside and inside cages, as well as by limiting opportunities to act effectively on the environment. The discrepancy between control of stimulus and response variables in captive vs natural environments, first elaborated by Hediger (e.g., 1968), may be the most fundamental issue in enrichment.

An extensive study of the effects of early control over environmental events on socioemotional development showed that young rhesus monkeys that could operate devices dispensing food, water, and other treats showed less self-directed behavior and more exploration than monkeys receiving the same commodities independently of their own behavior (Mineka Gunnar & Champoux 1986). Thus, in order to produce behaviorally competent individuals, animals should be allowed to act on their environment to produce significant consequences. Extreme loss of control, especially over aversive events, leads to syndromes such as "learned helplessness" (Seligman 1975).

Complexity/Predictability

On a more conceptual level, cage environments can be thought of as reducing the complexity or as increasing the predictability of stimulation. These two are inversely related; if one declines the other usually rises. In an environment where predators are nonexistent or highly predictable (e.g., laboratory or zoo), or where only one type of food is offered, options are fewer, stimuli more repetitive and predictable, and decisions are simpler.

Stress

Stress is another theoretical construct in enrichment studies. It is believed desirable to eliminate stress, so as to increase desirable and decrease undesirable activities. Poor cage design, cramped cages, crowding, etc. are believed to stress animals, and this is supported by studies on physiological and behavioral measures of stress (also Bollhorn 1980; Clark 1980; Eisen 1966). So one challenge for enrichment studies is to reduce stress, or better, to identify and strive to achieve an optimal level of arousal rather than a constant minimum. In the normal animal arousal varies with the ordinary challenges of life (e.g., Beuving 1980). So any optimal level is bound to vary, and to include occasional brief peaks. We now have results to suggest that such brief challenges, such as birds passing over tamarin enclosures, are beneficial (Moodie & Chamove: in preparation).

Some researchers view "stress" as equivalent to "distress", although this is not a necessary interpretation. Garmezy (1982, p.238) defines stress as "any action or situation that places special physical or psychological demands upon a person–anything that serves to unbalance an individual's equilibrium or homeostasis." Wemelsfelder (1985) uses "distress" for the effects of understimulation, "stress" to those of overstimulation. If stress is any event that is destabilizing or that taxes the system, then positive events, too, are of interest.

Beneficial effects of stress? The literature suggests that stress levels that are markedly and persistently below those likely to be found in the wild lead to individuals who do not deal well with subsequent stressors (over-reacting both in intensity and duration to mild stimuli as well as to stressors -- e.g., Stolba & Wood-Gush 1980), and who appear to have difficulty inhibiting habitual responses (Chamove 1984a; Sackett 1970). In animals, some stress early in life enables adults to cope better with stressors (Chamove 1980; Denenberg 1964; Denenberg & Whimbbey 1963; Ewbank 1973; Fox 1986). Captive animals also appear to seek challenge by choosing to work for food despite the availability of free food (Anderson & Chamove 1984; Carder & Berkowitz 1979; Harlow 1950; Havelka 1956; Inglis & Ferguson 1985; Kear 1961; Markowitz 1982; Neuringer 1969; Wood-Gush 1973).

EVALUATION OF EFFECTIVENESS OF ENRICHMENT

Evaluation should be seen as crucial to studies of enrichment; we need answers to questions about the real (not merely presumed) benefits of enrichment efforts, the reliability of the effects, and whether the benefits justify costs.

Cost

Costs are most often discussed in terms of time or money, but can also be measured in terms of alternatives forgone. One form of enrichment may preclude others, or may mistakenly lead caretakers to believe they have done enough. If a ball is chosen as the enrichment device (e.g., Renquist & Judge 1984), this may be at the expense of other possible devices; investing in one expensive device may exhaust the budget for enrichment. Evaluating an enrichment technique can be likened to evaluating a new drug. If there is no other drug available for treating an illness, the drug is compared with a placebo (unless its benefits are so great or so obvious as to make such a trial unnecessary – e.g., vaccination for smallpox). If the drug is to supplant an existing drug, it is compared with that drugs.

Benefit

Just as cost can be seen from several aspects, so can benefit. The short- and long-term effects of intervention both need to be measured, and criteria for the duration of effect need to specified. For example, a small, controlled fire in a large chimpanzee enclosure produces a dramatic response the first time, but the apes lose interest after the second time (personal observation). It is usually long-term changes in behavior that are of interest and measured in competent studies of enrichment (e.g, Westergaard & Fragaszy 1985). Changes in behavior need to be maintained for long periods or to show up frequently during the day to be deemed significant and worth the effort.

Finally, to measure benefit one needs to decide on the importance of the behavior that is altered. Two extremes will illustrate the difficulty of such decisions. Certain behavior may be considered important because it occurs frequently in the normal situation; examples are foraging, locomotion, and huddling. King and Norwood (this volume) consider the opportunity to leap important in squirrel monkeys, as over 40% of their travel is accomplished by leaping. Other behavior may be important because, although its normal frequency is low, it has known value for the individual or species. Dust-bathing in fowl is an example (Dawkins 1976), or food-offering and sharing in cotton-top tamarins (Feistner & Chamove 1986). And other behavior, initially rare, may be important for its potential to supplant common behavior. The use of tools enables monkeys to process certain nuts four times as rapidly (Visalberghi & Antinucci 1986). This tool use reduces feeding time but at the same time it represents an increase in the range of behavior available to meet future environmental challenges (Figure 2; Anderson, in preparation).

Two studies illustrate the benefits of proper evaluation of enrichment procedures. Rosenblum and Smiley (1984) found that providing a foraging task reduced such behavior as self-aggression and abnormal posturing in high- and low-ranking members of a group of isolation-reared bonnet macaques, but increased these actions in intermediate-ranking monkeys. Others (Bloomstrand Riddle Alford & Maple 1986) reported behavioral deterioration in some members of a chimpanzee group that had access to a food puzzle, while other members showed improved behavior. It seems likely that such negative reactions stem from social tensions arising when an enrichment task stimulates (even if inadvertently) competition. Such enrichment devices may be better suited to individually-housed animals; or competition may be averted by providing several devices to the group.

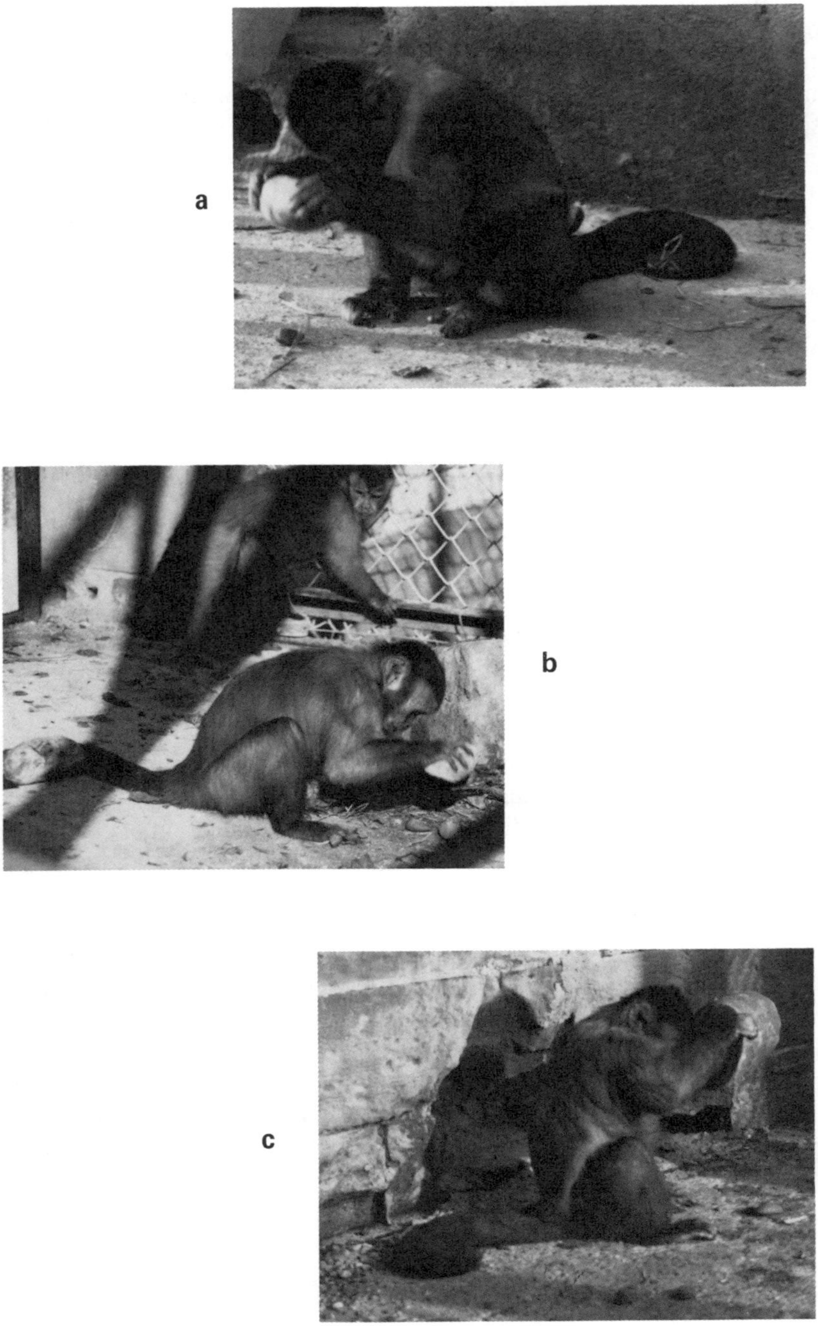

Figure 2. A male *Cebus apella* uses a stone (a,b) or a block of wood (c) to break open an almond.

AN EXAMPLE OF AN ENRICHMENT TECHNIQUE

To illustrate several of the points raised in this chapter, we will describe a series of studies, conducted between 1979 and 1986, that assessed the effects, in eight different primate species (listed in Figure 4), of covering the floor of large group cages with a deep absorbent litter (Anderson & Chamove 1984a; Chamove & Anderson 1979; Chamove Anderson Morgan-Jones & Jones 1982; Chamove Anderson & Nash 1984; McKenzie Chamove & Feistner 1986). The goal was to improve behavior, especially to increase activity and to reduce abnormal and undesirable behavior that some of these species were displaying.

The problem. In the wild, primates may spend 25% to 90% (mode about 55%) of their waking hours searching for and eating food (reviewed in Clutton-Brock & Harvey 1977). The pattern of foraging and feeding varies between species and with time of day; often highly digestible foods that can be rapidly harvested are eaten early in the day (Oates 1987). In captivity, however, search time is decreased almost to zero, and eating time varies from a low of only about 5% in callitrichids (personal observation; also see Molzen & French: this volume) to a high of 10% in the great apes (Kirkwood 1983; see also Rosenblum & Smiley 1984). When, as is common, food is offered in only one or two discrete locations, the frequency of conflicts at feeding times may exceed non-feeding values by a factor of up to ten (Anderson & Chamove 1984a; Belzung & Anderson 1986), and there is a concomitant 50% increase in blood cortisol levels, suggesting considerable stress (Chamove & Bowman 1978).

The technique. Our method of enrichment was to cover the floor to a depth of 4 cm with woodchips. Items of food were sometimes mixed into this litter. The aim was not to emulate–to copy nature, as this is impractical for almost all captive environments, but rather to simulate–to provide certain possibly-critical components of the natural environment. The food that was mixed into the floor litter varied: one to a mixture of eight different types of cereals [diameter 1.6 mm (millet) to 8.3 mm (maize)]; or insect larvae (mealworms); or large pieces of fruit.

The hypothesis. We worked from the hypothesis that enrichment would be more effective when it targeted behavior that is common in the animal's natural setting. We predicted (i) that changing the opportunity for an important and common behavior pattern like foraging would lead to substantial changes; and (ii) that arboreal species would be less affected by floor covering than would more terrestrial species, since the former are less likely to forage at ground level.

Effectiveness. The effectiveness of the woodchips, with or without food, was measured over several months to determine costs and benefits and the stability of the behavioral effects. Technician workload was assessed, cage cleanliness and odor were evaluated, and changes in the animals' behavior were recorded. Bacterial vigor in the woodchips was also measured.

Results

These studies clearly showed the high cost-effectiveness of the floor covering in improving behavior. The desired goal was achieved in the initial studies on stumptailed macaques (*Macaca arctoides*); self-aggression (Anderson & Chamove 1981) was reduced by more than half and aggression was reduced by a factor of two in adults, and ten in juveniles. Similar reductions occurred in all but one of the species subsequently studied in which aggression had been seen prior to introduction of litter (Figure 3). When, prior to introducing woodchips, fruit was scattered over the bare floor (as is commonly done), aggression levels were high, as dominant individuals tried to monopolize the food. When we buried the fruit under woodchips, fighting decreased by half, and the fruit was more evenly distributed among the group members. To our surprise, when all the fruit was placed in a single, easy-to-monitor pile, aggression levels were as low as when the fruit was buried: the dominant animals controlled the pile of food and other animals simply waited their turn; food distribution was highly skewed across subjects (also Belzung & Anderson 1986; Brennan & Anderson 1988). When the piled-up fruit was frozen, food-processing time increased (a piece of frozen fruit takes six times as long to eat as a piece that has not been frozen); aggression also increased, because dominants now controlled the pile for longer. However, when frozen fruit was scattered, distribution over group members was more even, and aggression was as low as when unfrozen fruit was covered with woodchips. Increasing foraging demand can increase or decrease agonistic behavior, depending on how foraging demand fits in with group social structure (also Rosenblum & Smiley 1984).

When the floor of the enclosure was bare, macaques, of course, spent almost no time searching for food items on it. When the floor was covered with woodchips only, the animals spent about 5% of their waking time sifting through this floor covering. When grain was added to the woodchips, the monkeys' searching increased to almost 15% of waking time, even when grain was also freely available elsewhere. When free grain was removed, so that the only grain available had to be found and extracted from the woodchips, the monkeys spent over 30% of their day foraging.

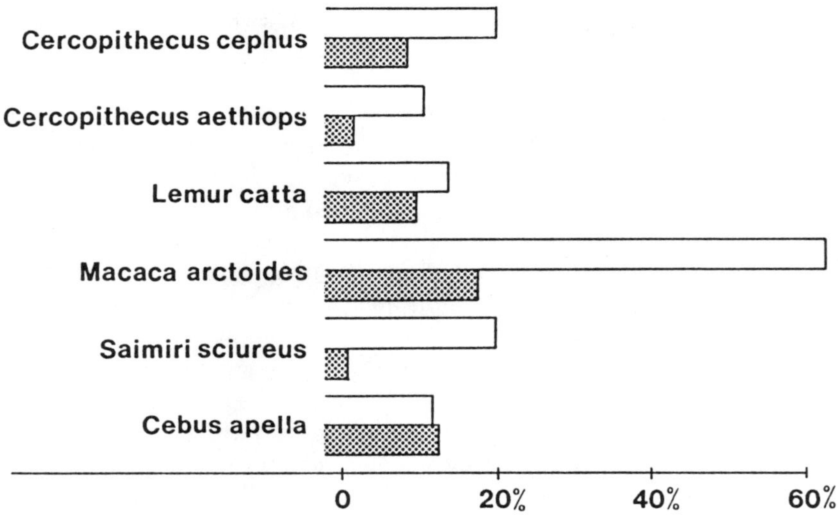

Figure 3. Percentage of time engaged in agonistic behavior when the floor is bare (open bars) or covered with floorcovering (filled bars).

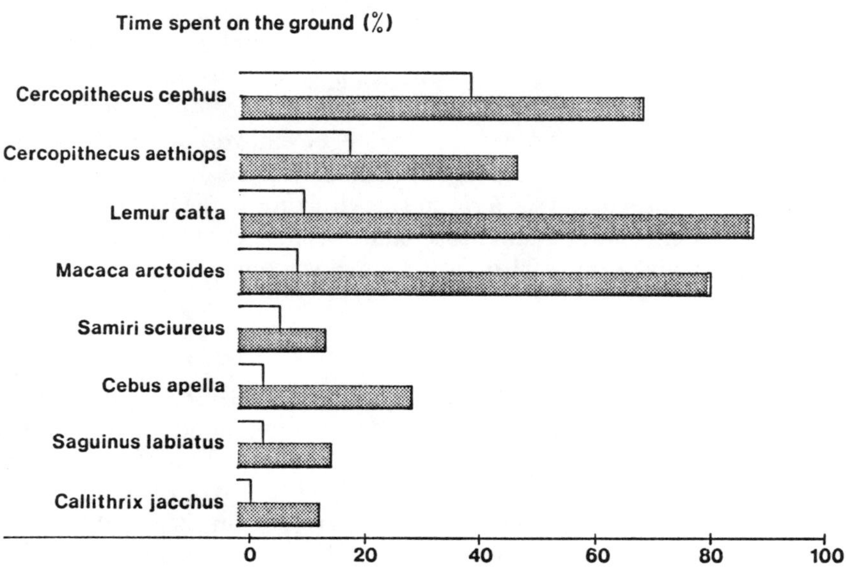

Figure 4. Use of the floor when uncovered (open bars) or covered with woodchips (filled).

All of the primate species assessed showed increased use of the floor when it was covered with woodchips (see Figure 4). We were surprised to find no support for our prediction that arboreal primates would benefit less than terrestrial primates; there was no systematic relation between the amount of increase in time spent on the floor and whether a species was primarily terrestrial or arboreal.

All monkeys spent more time looking for food hidden in the woodchips, taking over four times longer to locate a hidden item than the same item on the bare floor. Monkeys of every species moved around more on the floor when the floorcovering contained food. They took longer to exhaust the smaller grains, while collecting these items almost three times as fast as the larger grains. The arboreal monkeys were less hesitant in going to the ground to retrieve food from the covered floor. The increased time spent foraging was at the expense of other behavior: aggression and abnormal behavior were reduced, but so were play and affiliative behavior, and so too was time inactive. Westergaard and Fragaszy (1985) also reported decreases in quiet social contact and proximity, and increased exploration, when a straw floorcovering was given to a group of capuchin monkeys. [Opportunity to forage also reduces abnormal behavior and increases activity in nonprimate species: chickens (Bareham 1972; Hughes 1976); pigs (Ekesbo 1981; Koomans 1981; Vestergaard 1981); and cows (Albright 1982).]

The cost of this procedure was assessed in several ways. Woodchips reduced the time required for cleaning each enclosure from five hours per week to two hours. The enclosures, especially the walls and windows, were judged to be cleaner, and they were rated as smelling less after four weeks with woodchips than after one day with a bare floor that was cleaned daily with water, detergent, and disinfectant. To assess the potential for spread of disease by litter, samples of woodchips were removed periodically over eight weeks and either (a) tested for the presence of bacteria, or (b) inoculated with *Salmonella* bacteria and survival time of the latter measured. The results clearly showed that the longer the litter was in use, the greater was its bactericidal activity (also Turnbull & Snoyenbos 1973).

Conclusion

From these and other studies (e.g., Bloomstrand Alford & Maple 1987; Westergaard & Fragaszy 1985) one can recommend certain general techniques for enrichment in primates. One popular and successful approach employs food that is desirable, perhaps prized or rare (Feistner & Chamove 1986). To ensure adequate distribution, it should not be presented massed but widely distributed, and in such a way that animals can express at least part of their repertoire of gathering/extraction techniques. To reduce aggression, food is best chosen that naturally requires long processing times, such as artichokes, *Aframomum*, nuts, millet, browse; or that has been modified so as to increase processing requirements, for example by freezing fruit, kibbling or grinding maize, or by hiding food in holes or under woodchips or straw. Any food-containing devices should be inexpensive and easily maintained and baited. Figure 5 illustrates such a device.

CONCLUDING REMARKS

We should continually be looking for ways to improve the treatment of animals in captivity. As far as nonhuman primates are concerned, the reduction of imports and the trend toward self-sustaining captive populations (Benirschke 1986) mean that many captive primates face a lifetime in conditions of spatial restriction. Research aimed at perfecting environmental conditions promises multiple benefits, including improved quality of the animals and even preparation for their eventual reintroduction to protected wild areas (Snowdon: in press). Considerate design of housing, paying attention to space, temperature, cover, etc, can to a large extent take care of an animal's physical comfort. Where compatibility can be achieved, social housing should be a priority, given the highly-evolved social potential of most primates, and the relative fragility of this potential in the face of restriction (Erwin & Deni 1979; McGrew 1981). Environmental enrichment is not a simple, all-or-none concept. The issues are complex and deserve dedicated consideration, as we have tried to convey in this chapter.

ACKNOWLEDGEMENTS

The following made helpful comments on the text or in discussion: R.Cooper, S.Evans, A.T.C.Feistner, W.C.McGrew, E.M.Moodie, E.Price, C.T.Snowdon, and especially E.F.Segal. Thanks also to Mangustan, Angus, and Star.

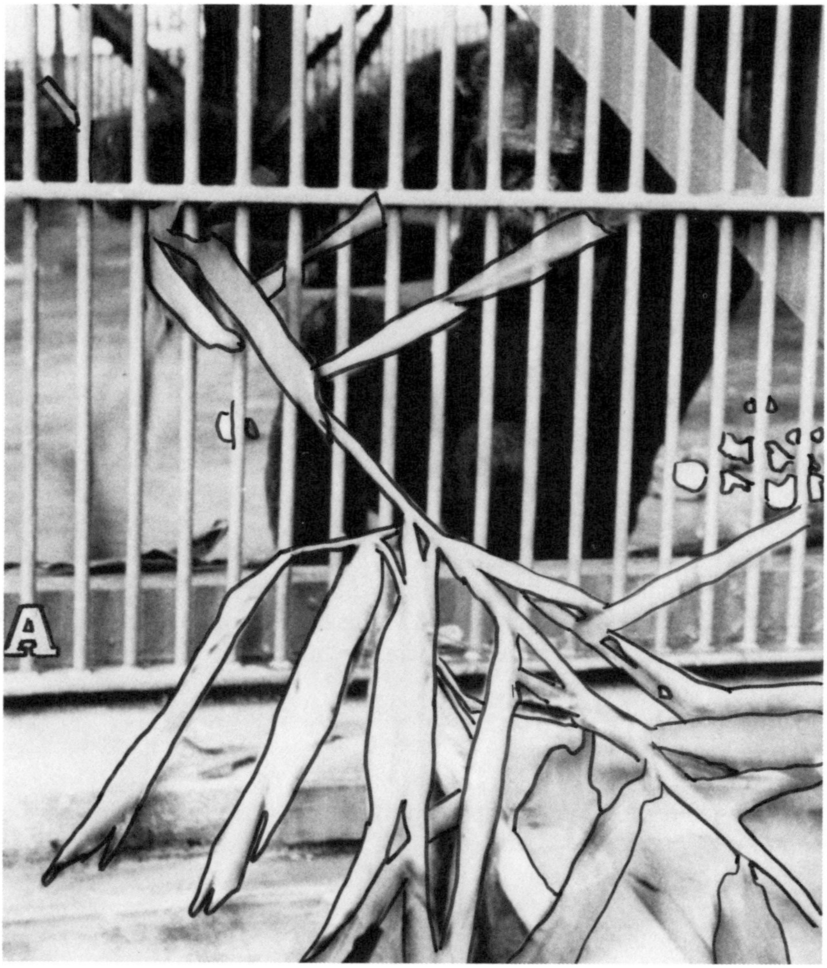

Figure 5a. This chimpanzee uses an *Aframomum* stem with leaves intact (outlined in pen) to retrieve peanut kernels from a long distance (90cm). A pile of food (outlined in pen: bread, tomato, avocado, pineapple, mango, papaya) lies uneaten behind the chimpanzee.

Figure 5b. The chimpanzee uses the *Aframomum* stem with leaves intact to retrieve peanut kernels from an intermediate distance (60cm).

Figure 6. For close work (30cm), Mangustan strips off the leaves and uses the bare stem to guide the kernels through the obstacle of nails. Near retrievals took about 70 sec and far retrievals about 400 sec. Individuals spent about an hour a day working this simple, inexpensive, easily-baited device to obtain prized peanuts.

— 13 —
Primate Research Models and Environmental Enrichment

Hal Markowitz and Scott Line

INTRODUCTION

We offer two major points for consideration in this chapter. *First,* there is a clear need to document with care the behavioral and physiological changes that may follow attempts at environmental enrichment. This is time-consuming and sometimes expensive work, but it is essential that we search for methods that *demonstrably* improve the lives of animals in our care.

Second, careful planning of enrichment activities may greatly enhance research in a number of ways. Animals that are active, exert some control over their daily schedules, are challenged to solve problems, and have daily positive interactions with humans are almost always better research subjects than those that have lived under typical behaviorally-impoverished caging conditions.

Several general findings have emerged during our current efforts to develop effective environmental enrichment methods for primates caged in typical research facilities:

1. For enrichment procedures to be effective without the need for intensive training, they should begin with younger animals, ideally prior to their long-term residence in individual cages.

2. For rhesus monkeys, in spite of all the obvious difficulties involved, finding ways to provide the animals with social opportunities is of critical importance in maintaining their behavioral fitness. Interactions with conspecifics also lead to increased use of enrichment devices which become integrated into such behavior as dominance displays and sharing of food.

3. Rather than searching for panaceas in the form of universal devices that will serve the behavioral needs of all members of a species, we should provide a variety of opportunities and anticipate individual differences in the

animals' choices of enrichment activities.

4. It is clearly possible to find methods by which environmental enrichment can be combined with research protocol to enhance both. Consultation between behavioral enrichment specialists and other researchers may result in new research protocols that make the animals better research models, *and* make the research an enriching experience for the subjects. This may reduce or eliminate the need to coerce the animals' participation in research activities. Brief examples will illustrate these points.

EFFECTS OF LONG-TERM CAGING WITHOUT ACTIVITY

Research at the Animal Care Facility of the University of California at San Francisco (UCSF) began with the use of a number of simple game devices that could be attached to the cages of singly-housed rhesus monkeys. These games were similar to ones that had been used successfully with a variety of primates in zoos (Markowitz 1982). Markowitz and Spinelli (1986a) intentionally selected animals that had been in the facility for a long time, showed clear signs of abnormal behavior, and were part of a project where the investigators were especially anxious to see something done to enrich the static daily environment of their monkey subjects.

The outcome of these first efforts was rather startling. Unlike animals that had lived in more complex milieus, these monkeys almost never used the offered opportunities to entertain themselves. Nevertheless, as the months passed, technicians and investigators reported remarkably healthy changes in their animals' general behavior. The monkeys no longer constantly turned somersaults or banged their cages whenever research personnel entered the room. Excessive vocalizing and grimacing were greatly reduced, as was self-abuse.

After considerable evaluation we were able to come to only two straightforward conclusions. The first was that these adult animals had been so negatively influenced by their long-term stay in an unresponsive environment that it would take strenuous efforts to teach them to use new opportunities. This finding was recently corroborated by a research group in West Germany led by Bruno Preilowski. They also found that inexperienced adult rhesus monkeys required considerable training before they would use enrichment devices placed in their home cages.

The second conclusion was that the presence of researchers who did nothing but greet the animals, give them occasional fruit treats, and watch and evaluate their behavior regularly, by itself significantly reduced apparently fearful responses to humans. While this will surprise few of our readers, it is an important reminder to researchers that however busy their schedules and however overburdened they may be by paperwork and meetings, it is critically important to habituate their animals to the humans who will work with them.

THE IMPORTANCE OF SOCIAL OPPORTUNITIES

Our second major point is the importance of providing social opportunities for rhesus monkeys. Even adults that have been housed singly for long periods can be resocialized. Researchers at the Wisconsin Regional Primate Research Center successfully paired adult female rhesus monkeys with either juveniles or other adult females (Rheinhardt Cowley Eisele Vertein and Houser 1987; Reinhardt Houser Eisele & Champoux 1987). An example from our work at UCSF illustrates one of the benefits of expanded social activity. We placed a simple color-and-sound game in the monkeys' home cages. When the monkeys initially showed little in the way of positive responses to the device, it was embellished with ropes (see Figure 1), making it appear irresistible to the researchers. Nevertheless, it was weeks before most of the monkeys touched the device, and even after months they made only limited use of it. An identical game, ***without ropes or other embellishments,*** was tried with Peggy O'Neill's rhesus monkeys living in social groups at the Poolesville, Maryland research station directed by Steven Suomi of NICHD. Within ten minutes these monkeys were actively using the game, ***and*** showing dominance displays, bouncing and vocalizing, as they caused changes in lights and sounds. (Also see O'Neill: this volume.)

In earlier work with diana monkeys (Markowitz 1982), we had observed that older primates may learn some of their most interesting responses, and thereby reduce stereotypies, by watching younger conspecifics. For example, the adult male that had sired all of the youngsters in the diana monkey group at the Portland Zoo showed a remarkably rigid routine in earning tokens during the first several years of behavioral work at the zoo (see Figure 2). He always took the same turns in the same lengthy path, requiring minutes to move from the first to the second station to obtain a token, and then to move on to deposit the token for food. One day he was seen watching one of his youngsters for a long time, as the young diana took a variety of shorter paths that gained him food in a few seconds. For the first time in all the years we had been noting that adult male's behavior, he suddenly moved quickly to the first station, took a short loop around the top bar, and closely emulated his youngster's performance. Both human and monkeys stared in surprise as he practiced this new routine, vocalizing occasionally (Markowitz 1982).

With animals like rhesus monkeys that have complex social organizations, we believe it is crucially important to give them opportunities to watch conspecifics perform complex solutions efficiently. In our current work with singly-caged animals we have found that one effective method for teaching rhesus monkeys to use new equipment is to rearrange cages so they can observe others that have already mastered the devices.

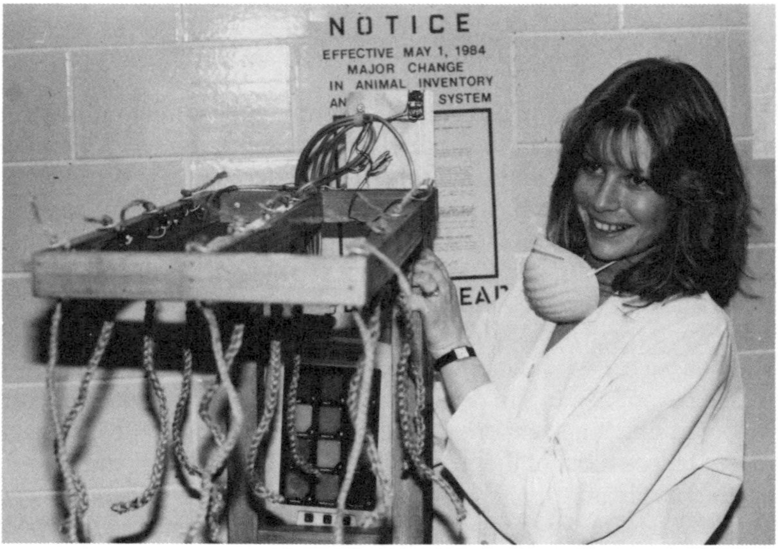

Figure 1. Sue Stern displays the rope-embellished color-and-sound game. When attached to a monkey's home cage, the monkey could produce colors and sounds by pulling the ropes.

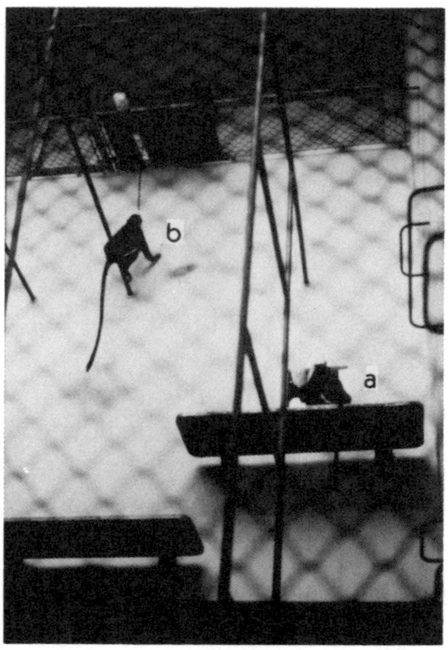

Figure 2. Male diana monkey responds at topmost station (*a*) while juvenile male waits at platform (*b*) where tokens are delivered.

PRIMATES ARE INDIVIDUALS

The next point, that (as with other things in life) the search for panaceas is doomed, is most easily illustrated with a simple anecdote from our work in the Animal Care Facility at San Francisco. In a room with four individually-caged rhesus monkeys of approximately the same age and experience, one animal showed no interest in splashing colors across a television monitor, pulling ropes to play the color-and-sound game, or working with any of the equipment we provided. One day a device was introduced that allowed the monkeys to turn on music. As it turned out the first monkey scheduled for observation with the new apparatus was in the cage adjacent to the reluctant animal. Immediately upon hearing the music its neighbor had turned on, this monkey became quite animated, even apparently moving in rhythm with the music. When the observation session with the first monkey was over, the device was moved to her cage, and in the first 45-minute session she turned the music on and off several times. We all suffer from a tendency to lump all animals of a species together (unless they are humans); yet anyone who has worked closely with a complex species is acutely aware of individual differences and personality quirks. We have also noticed large and consistent individual differences in the use of enrichment devices at the California Primate Research Center. In designing enrichment equipment it is important to provide sufficient variety to benefit potentially all of the animals.

It is also important to design equipment that can sustain interest. Simple, inert toys in great variety have been suggested as enrichment devices, but once their novelty wears off (often within hours), most animals ignore them. Results from our work at Davis demonstrate that prolonged use of enrichment devices is indeed an achievable goal.

Sustaining Interest with a Responsive Device

Figures 3 through 6 depict results with an apparatus that allowed monkeys to turn a radio on and off, and to operate a food dispenser to get banana-flavored pellets. The device was a metal box, approximately 25 x 30 x 10 cm, attached to the front of the cage. Three contact detectors, extending approximately 2 cm into the cage, provided independent controls for turning the radio on and off, and for activating the food dispenser. Automatic counters recorded how often the monkeys operated the food dispenser and turned the radio on. Once a radio had been turned on, it continued playing until the monkey turned it off. Fourteen weeks after the boxes were introduced, we fitted them with digital timers to record how long the radios played.

The subjects were five adult female rhesus macaques (age range 4-10 yrs), all colony-born in large outdoor multi-male, multi-female groups at the

California Primate Research Center. During this study they were housed indoors in nonadjacent single cages (.60 x .66 x .80m) in two colony rooms housing about 50 monkeys each. Three subjects were kept in one colony room and two in the other. The rooms were illuminated 12 hours daily, and cleaned daily between 0900 and 1000 hours. The monkeys were fed commercial monkey chow at 0700-0800 and 1400-1500 hours.

Most rhesus monkeys will readily work to obtain banana-flavored pellets. So that the pellets would not become a major part of the diet, the amount of work required to obtain pellets was increased over time. The food dispenser operated on a fixed-ratio reinforcement schedule, the ratio initially being set to one touch of the control switch per pellet. After each monkey had touched the control a total of 2,000, then 5,000, 8,000, and finally 10,000 times, the ratio of touches required for each pellet was increased to 2, then 5, 8, and finally 10 touches, respectively. At no time were the monkeys deprived of their twice-daily regular feedings.

Figure 3 represents the mean number of times the subjects touched the control that operated the pellet dispenser. All monkeys learned to operate the device without any training by the investigators, and three of five continued to obtain banana pellets throughout the five-month study. Of the remaining two, one made very few responses for the first 13 weeks, then became the highest responder for the last 7 weeks; the other made a few responses in the first week, and only occasional responses thereafter.

Figure 4 depicts the mean number of pellets earned by the subjects. The number earned per week increased rapidly at first, declined temporarily each time the ratio requirement was raised, and then increased again during the last part of the experiment. All monkeys used the food dispenser, but variably over individuals. Total pellets earned over the 20-week experiment ranged from a low of 24 for one monkeys to a high of 66,000 for another. A Friedman analysis of variance of the total pellets earned per month showed no significant differences over time. These data suggest that this type of device might be readily utilized by most adult rhesus monkeys. While there was considerable week-to-week variation, there was a generally increasing trend in responses to the apparatus over the experiment, suggesting that the subjects did not become bored with the feeder.

Although we do not know how much attention the monkeys paid to the radio when it was on, Figure 5 shows that they continued to turn it on *and off* throughout the study. The continued use of the radio and the large amount of on-time leads us to conclude tentatively that the radio was a source of positive stimulation for the monkeys. Use of the radio varied from week to week, but the mean number of times the monkeys turned on the radio each week was as high in the 20th week as it was in the first.

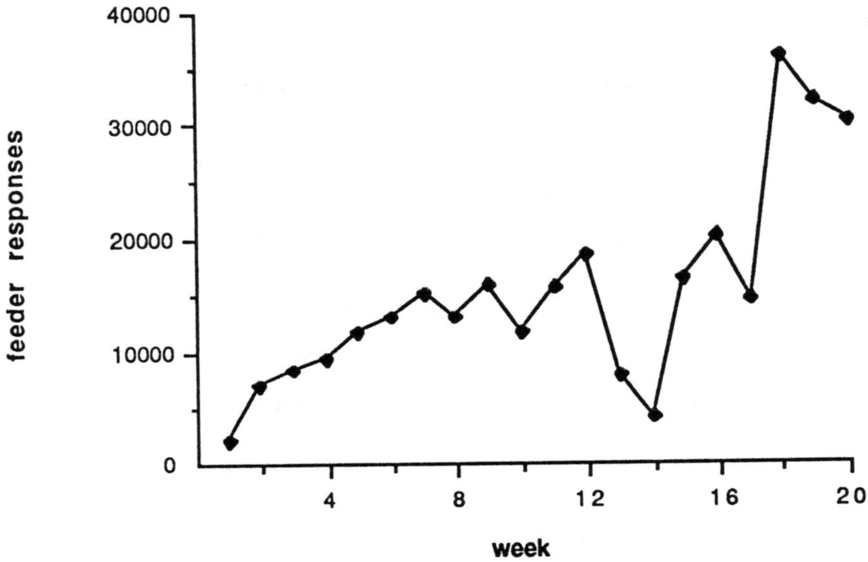

Figure 3. Mean weekly feeder responses. Mean touches per monkey per week of the control that operated the food dispenser (N=5). Individual monkeys touched the control from 0 to 130,000 times per week.

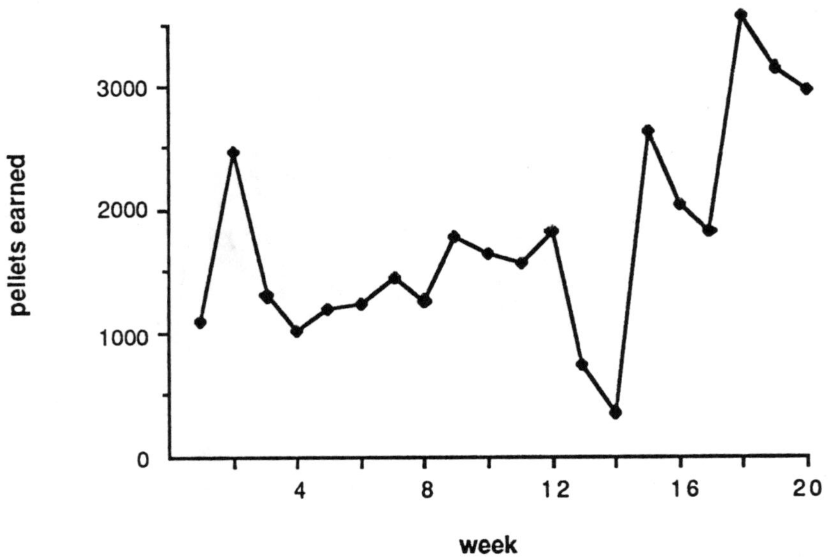

Figure 4. Weekly pellets earned. Mean banana pellets earned per monkey per week. Individual monkeys earned from 0 to 12,000 pellets per week.

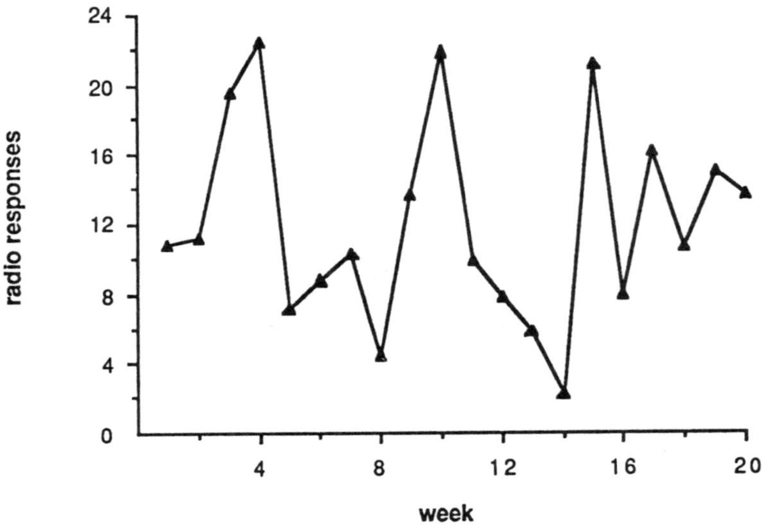

Figure 5. Mean weekly radio responses. Mean touches per monkey per week of the control that turned on the radio. Individual monkeys turned on the radio from 0 to 85 times per week.

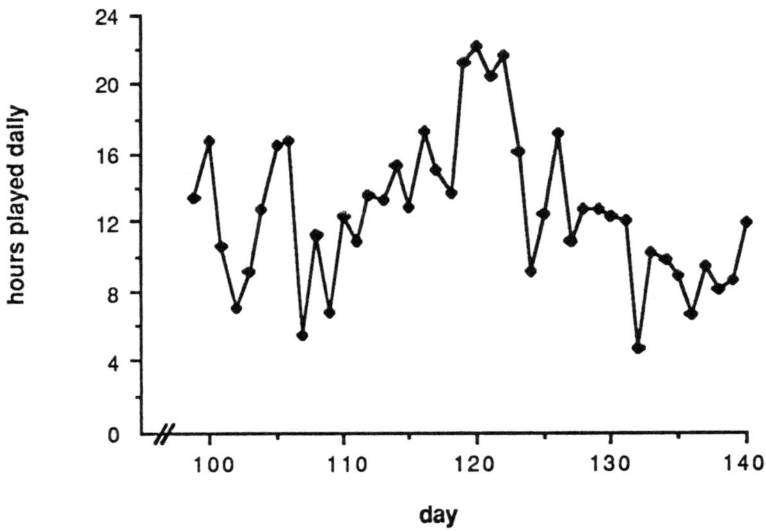

Figure 6. Mean daily radio playing time. Mean hours per day that the radio was played by each monkey during the last seven weeks that the apparatus was attached to the cage. Individual playing time varied from 0 to 24 hours per day.

Another indicator of sustained interest appears in Figure 6, showing the number of hours the radio played each day over the last seven weeks of the study. Again, playing time varied widely, but mean daily hours of playing time at the end of the project were about the same as when we first began recording playing time seven weeks earlier.

Our experiences with this and other enrichment devices suggest that *an apparatus that the monkey can control, and that responds to the monkey in some way,* will be used by a larger proportion of animals, and for a longer period of time, than devices or toys that are not actively responsive.

COMBINING RESEARCH AND ENVIRONMENTAL ENRICHMENT

A final point is of potentially greatest importance to many laboratory scientists. In previous papers we stressed the importance of behavioral enrichment for providing healthy, representative animal models (e.g., Markowitz and Spinelli 1986b). Preilowski's research group in Ravensburg, Germany has recently produced convincing data showing that incorporating research apparatus in the home cage for the animals to use on their own schedules is a very effective means of collecting research data *and* reducing abnormal behavior in individually-caged rhesus monkeys. Preilowski's work involves a complex response: the monkeys must squeeze together two manipulanda, maintaining a constant pressure within very narrow tolerances, for a period of one and a half seconds. Preilowski's group is investigating laterality of brain function and hemispheric transfer with and witht ut the intact corpus callosum. These investigators have shown that our goals of making better research models while at the same time pretraining subjects for later participation in complex research were too modest. It is clearly possible with laboratory primates not just to pretrain and prepare them, but actually to conduct complex research itself, on a continuing basis, without greatly disrupting the lives of the primate subjects (Preilowski Roger & Ngele: in press).

CURRENT RESEARCH

To date we have relied largely on behavioral measures to evaluate the effectiveness of our enrichment procedures. For example, we have calculated the frequency of various stereotypies and other types of abnormal behavior, drawing on work such as that of Erwin and Deni (1979) for the categories of quantitatively and qualitatively abnormal behavior. Beginning in November 1986, a battery of physiological measures was assembled in order to assess the effect of enrichment procedures on several indicators of stress responsiveness. In one group of animals used in regular research at

UCSF the measures involve urine sampling by means of metabolism pans. We are looking at factors such as patterns of free cortisol levels and correlating these with behavioral observations.

At the California Primate Research Center, where we are using rhesus monkeys specifically assigned to the enrichment project, we are monitoring a larger array of physiological variables. Blood samples are taken regularly and are analyzed for several hormones, including cortisol, serotonin, and glucagon. A biotelemetry system allows us to sample EKG, temperature, and movement. We record direct behavioral observations as well, in preparation for assessing correlations between behavioral and physiological variables. It is critically important to ascertain whether behavioral changes evoked by environmental enrichment are paralleled by changes in physiological variables. Strictly speaking, without such data we can only say that we have altered gross behavior in directions that superficially look healthier. We must now proceed to demonstrate that environmental enrichment positively affects the internal states of these laboratory primates.

We are dedicated to the idea that well-conceived basic research is essential and justifiable for the enhancement of the lives of humans and other animals. We also believe that all of us have a responsibility to see that the animals used in such research are treated humanely, that they are protected from unnecessary injuries and, as well, from unnecessarily impoverished lives.

ACKNOWLEDGMENTS

Preparation of this chapter was supported in part by NIH grant RR0169 from the Division of Research Resources to the California Primate Research Center.

— 14 —
Long Term Animal Studies

Cynthia L. Bennett and Roger T. Davis

Long-term studies of animals in laboratories and zoos require special care. One of us (RTD) studied the behavior of a group of rhesus monkeys over a period of 29 years. The other author (CLB) took part in what became a study of aging in that rhesus group, and later, at the Santa Barbara, California Zoo, prepared a group of lesser spot-nosed guenons (*Cercopithecus petaurista*) for display in social groups. The two studies convinced us that the same special care is needed to assess behavior as to insure long-term survival. The study of aging, having been reported elsewhere (Davis 1978 1985a 1985b), will be described only briefly as a background for dealing with factors related to longevity.

ANIMAL MODELS OF AGING

Every being manifests biological and psychological changes with age. While biologists have studied the fundamental processes of aging at length (Comfort 1959), psychologists have done so decreasingly. Proportionally less space is devoted to basic research on aging in animals in each of three successive editions of a standard handbook on the psychology of aging (Birren 1959: 9 of 24 chapters; Birren & Schaie 1977: 5 of 30 chapters; Birren & Schaie 1985: 2 of 32 chapters. Our knowledge of psychological changes, being thus narrowly based on just one species, *Homo sapiens*, has less generality than do biological theories of aging. A further limitation of studying aging in humans is that there are few longitudinal studies because people age more slowly than most other animals. Researchers resort to cross-sectional studies, which confound age and cohort and require more subjects, as the basis for psychological gerontology. Our longitudinal study of aging in 15 rhesus monkeys (*Macaca mulatta*) (Davis & Leathers 1985) is in many respects unique.

A LONGITUDINAL STUDY OF AGING IN MONKEYS

Our 29-year study of a single group of rhesus macaques enabled us to answer several basic questions about aging in this species and to compare our findings with those on old people.

How long do rhesus monkeys live in a laboratory?

In the wild, monkeys of the family *Cercopithecidae* who are past their prime lose their competitive advantage, often leave their troops, become solitary, and die. Remains found in the wild indicate that monkeys typically do not survive much past maturity. Before Bowden and Jones (1979) surveyed life spans of approximately 600 nonhuman captive primates and reported maximum life spans for genus *Macaca* in captivity between 30 and 40 years, it was widely assumed, based on the wild data, that they lived no longer than 20-25 years. Early records of longevity from many zoological parks are suspect because, before the modern-day demand for accurate records, caretakers often substituted new animals for a dead favorite without recording the substitution (Bowden & Jones 1979).

The group of rhesus monkeys RTD and his coworkers studied in the laboratory for 29 years must have been at least 31 years old, because they were estimated to be at least two years old on arrival. However, postmortum studies of racemization reactions in their teeth and optical lenses (Bada & Brown 1985) suggested that, quite possibly, they were several years older than we had estimated. Even with our conservative estimate of their age, they were very old by laboratory standards of the day, certainly the oldest of their species ever studied longitudinally.

How do changes in behavior compare in aging humans and nonhumans?

Analogous psychological changes occur in the aged of both species, suggesting common causes. The monkey data showed the following:

- Difficulty shifting set from one type of problem to another (Davis 1978 1985b).
- A decline in short-term memory (Davis 1978 1985b; Medin 1969 1972), apparently related to brain pathology (Price et al. 1985).
- Greater susceptibility to the effects of visual masking (Anderson 1970; Davis 1974 1978 1985b).
- Slower reaction times (Davis 1978 1985b; Davis & Ruggiero 1973).

- o In the oldest animals: a loss in long-term memory for mistakes as compared to correct responses in paired-associates recognition learning (Davis 1985b; Medin O'Neal Smeltz & Davis 1974).
- o In the oldest animals: partial hearing loss at frequencies between 125 Hz and 16 kHz, complete loss at higher frequencies. There was also reliable hearing loss in animals aged 24 years when compared to animals aged 8 years (Bennett Davis & Miller 1983; Weisenburger 1979).
- o Visual changes: a loss of efficiency at discriminating saturations of color (Senner 1962), a decline in acuity (Davis 1985b; Lovelace & Davis 1963), and senile changes in the optical properties of the eye (Leary Bayne & Bennett 1985).
- o Social homophily was evident in both young and old monkeys (Davis 1978 1985b; Fitts 1982): Old animals associated with old animals and young with young.
- o Old monkeys were more cautious (Davis 1978 1985b).

When federal funding for our behavioral studies stopped, we euthanized the animals in our colony and performed comparative pathology studies, especially valuable in the case of old animals whose ages were known with certainty. These studies offered the opportunity to determine if pathologies seen in older people occur in other primates. Fifteen monkeys were euthanized and their tissues studied by 18 teams of investigators. Three monkeys were estimated to be at least 31 years old, three were their offspring, aged 24, and three others from our laboratory were aged 8. To complete the age sampling, three monkeys aged 13 from the Johns Hopkins Medical School and three aged 4 from the University of California: Davis were euthanized as well.

Pathology of old monkeys

The oldest animals showed little gross pathology in the heart, lungs, muscle, kidneys, viscera, and breast, and no infectious, inflammatory, or neoplastic lesions (Leathers, 1985). The animals appeared to have been very healthy. Other pathological studies on the sense organs, brain, lungs, endocrine glands, heart, and bone suggested that the old monkeys had aged under nearly optimal conditions of diet, exercise, and shielding from infectious disease. The pathology in many organs resembled that seen in old people except that effects of toxic agents, accidents, and disease were lacking. Notably, investigators found that:

a) Unlike young animals, old animals had a wide distribution of neuritic plaques in their brains. One old animal had plaque distribution like that seen in Alzheimer's disease (Price et al. 1985). In earlier tests, this animal had been unable to remember at brief retention intervals (Davis Bennett & Weisenburger 1982). We interpreted this deficit as a failure to initiate rehearsal necessary for information to progress from short- into long-term memory (Davis 1985b).

b) Cells from the aortas of old monkeys showed less replicative potential than analogous cells from younger monkeys (Ogburn & Martin 1985). This finding supports a theory of cellular aging predicting that, with age, cells lose the ability to replace themselves.

c) Age-related degeneration of sensitive hair cells occurred at both the apex and basal turns of the cochlea suggesting an uncomplicated sensorineural presbycusis (Hawkins Miller Davis & Rarey 1985). In people, pathology of the inner ear shows much more extensive damage related to accidents, ototoxic drugs, and disease states.

d) The visual apparatus underwent age-related changes (presbyopia) as determined by opthamological biometrics (Leary et al. 1985).

e) Studies of endrocrine functioning showed a positive correlation between age and cholesterol levels, but glucogon concentrations were related to weight, not age (Howard & Palotay 1985), as well as a multiple-endocrine syndrome, less extensive than is seen in aged people (Blumenthal 1985).

f) A smaller loss of cortical mass was seen in the ribs of old monkeys than is seen in old people (Przybeck 1985).

g) Little blockage occurred in the coronary arteries of old monkeys. This is unlike the findings in older people (Leathers 1985). Further, Weick and Ritter (1985) reported that biochemical changes in heart tissue in these monkeys did not parallel those in human beings.

h) On the other hand, Engle and Farrell (1985), studying blood constituents and lung tissue, found that plasma serum and fatty acids in old monkeys parallelled those of older people. Also, there was lowered phospholipid content with advancing age. Chemical analysis suggested that the animals had been fed a sufficient diet to maintain normal levels of lung protein, lipids, and fatty acids.

i) Collagen cross-linkage in the monkeys' lungs was related to advancing age (Reiser Rucker & Last 1985).

j) There was little age-related change in the skeletal muscle of the older animals, suggesting they had had enough exercise and good care (Hegreberg & Hamilton 1985).

In summary, many age-related pathologies were analogous in our aged rhesus monkeys and old people, but there were notable differences, many of which underline the good health our sample of old monkeys enjoyed. For this reason, we think we can offer suggestions to other investigators who plan to do longitudinal studies of aging animals.

Designing studies of aging using animals

Our study lasted a long time and included many experiments on the same small number of animals. We took many steps to ensure the wellbeing of the animals so that our data would reflect learning, memory, and perception precisely, and not the pathological consequences of trauma. From our experiences, we suggest that gerontologists planning longitudinal studies of animals heed the following:

Estimate life span of a species as accurately as possible. Monkeys and possibly other mammals can live longer than was previously thought.Perhaps some procedures used in laboratories and zoos in the past did not favor long life, but with optimal care we now know some rhesus can live at least 31 years.

Define aging with well-chosen behavioral and biological markers. Those mentioned in the previous sections provide a good start for future studies of aging in animals. Undoubtedly we don't know all such markers, but the present set is a beginning. Gerontological studies in animals will aid not only in the wellbeing of pets and domestic animals but also will assist in the understanding of general problems of aging and long-term care.

Establish baselines of behavior and health. An investigator should know each animal's social experiences and disease and accident histories from birth, and have a record of the animal's blood chemistry, trauma, and exposure to drugs. For example, animals that were ever treated with the antibiotic drug streptomycin, which is ototoxic, should never be used in studies of hearing. (Our animals never received this drug.)

At the beginning of our 29-year study (Davis & Leathers 1985), we spent more than a year establishing behavioral baselines; in retrospect, that was not enough time. Also, we did not have baselines on many common biological measures. An investigator studying aging should obtain *noninvasively* a wide variety of physiological, behavioral, and biochemical baselines.

Optimum conditions of housing

Optimum conditions of care and housing are essential in longitudinal research. When our study began, we adopted some basic principles of care and handling, and adopted some more as we went along, but further investigation is needed to determine whether (or which) of our guiding principles are essential.

Treating the animals humanely was our cardinal rule. We took the time to train each animal to go into a carrying cage voluntarily for daily transport to a room where it was given choice foods for solving problems. We never punished an animal for making a mistake. Although we occasionally netted the animals during the first 25 years of the study to give tuberculin tests, we avoided indiscriminate netting or restraint (one reason was that we did not know the behavioral consequences of various traumas).

Disease control. Some colonies are frequently tested for tuberculosis since it is an ever-present danger in laboratories using Old World monkeys and spreads rapidly, forcing sacrifice of the entire colony if one monkey becomes infected. It would be economical of animals and money to keep spare monkeys, not used as subjects in an aging experiment, in the same room as the experimental animals, to be used for routine tb testing. This would avoid the trauma of netting, restraining, and injecting the experimental animals. Also, very tight security on the health of people and animals entering a laboratory is essential. All medical and maintenance personnel, reporters, and others should be barred from the laboratory if they cannot provide a recent medical certificate that they are free from tb. Veterinarians especially may have had contact with tuberculosis in other animals, so they should be screened carefully. Likewise, new animals should be quarantined before being introduced into a colony. These precautions should apply when any airborne disease is a threat to the survival of any species of animal.

Human contact. In the case of Old World monkeys, another aspect of humane treatment is prohibiting laboratory workers from touching, petting, caressing, staring at, or harassing the monkey unless the monkey initiates contact. Monkeys are not domesticated animals, and they generally respond aggressively to human staring or contact. It is best to respect their wildness and not give in to caretakers' wishes to pet and caress them. Since our animals were housed in pairs, they had ample contact comfort from each other.

Feeding is important in the care of monkeys. Commercial monkey chow provides essential vitamins and minerals but lacks the texture and taste of fruits, nuts, and vegetables. We provided apples, oranges, carrots, and cabbage as dietary supplements and rewarded correct problem solving with peanuts and raisins. A common fault in feeding captive animals is providing too much food. Obesity favors longevity in neither animals or people, yet fat is common in zoos and laboratories. We avoided this problem by feeding one animal in a spare carrying cage and the other in its indoor living cage. Each animal received a controlled amount of food based on its expected body weight, allowing for normal preadult growth.

Exercise. Another principle we developed in caring for our monkeys was that they, like us, need ample living space and exercise. For 12 years of the project, when the animals were particularly active, each pair not only had an inside cage measuring .9 m on three sides but was able to pass through a short tunnel with gravity-operated doors at either end into an outside cage that measured .9 m x 1.8 m x 1.8 m. This gave the animals opportunity for exercise and to experience a variety of climactic conditions. This housing was available during most of the time the colony was located in South Dakota, which has a summer climate as hot as 105° F and a winter climate as cold as -20° F. In bad storms we locked the animals into their warm inside cages, but otherwise they had access to the outside. This freedom to go outside seemed to contribute to excellent coats of hair. Also, early experience with cold may prolong life. Regrettably, when the colony moved to Washington State University, the laboratory had the usual small inside single monkey cages, so for the final 11 years of the project, the animals were not housed optimally.

Socialization, breeding, and feeding. Our laboratory at South Dakota provided for social interaction. Rhesus monkeys, and many other animals, spend a large part of their time in social behavior. Recognizing this need, but also wanting to be able to feed and train animals individually, we compromised between housing the animals singly and housing them in family groups by pairing them. When the animals arrived in 1952, we tried various pairings to determine ones that minimized fighting. Most of the animals were female, so we rotated males among the cages of receptive females so that the male would be in unfamiliar surroundings and less dominant. There were 14 offspring.

Medication. Overmedication should be avoided in chronic studies. Weigh the risks of restraint and trauma while injecting an animal against the benefits of giving a drug. Also, consider the following:

(i) In many primate laboratories, rectal temperatures have been taken daily. This is not necessary if an animal is active and apparently healthy. Monkeys' temperatures can be taken by training to grasp a thermometer.

(ii) Superficial wounds heal rapidly in monkeys, and lesions are less likely to become infected if there is no bandage to become soaked in urine.

(iii) Antibiotics are often unnecessary, especially when all you are doing is sensitizing the animal to the drug and destroying its usefulness in combatting major diseases.

In summary, our experiences in a long-term study of the behavior of rhesus monkeys offer some lessons for doing longitudinal aging studies in other species. If we want to arrive at general laws of aging, we should examine the behavior of several species as they age so as to have a broad perspective on behavioral and pathological changes. It is absurd to think that one can arrive at general laws of aging just by studying *Homo sapiens*.

PREPARING LESSER SPOT-NOSED GUENONS FOR DISPLAY IN SOCIAL GROUPS

Guenons (*Cercopithecus*) include approximately 20 to 25 species and 60 subspecies (Napier & Napier 1985); they are the largest and most varied primate genus. Most of the species are readily distinguishable by their striking external characteristics including boldly patterned and contrasting coats, colorful genitalia, and a wide variety of facial masks including beards, moustaches, spots, dots, and lines. Despite these differences in appearance, the many guenon species are anatomically indistinguishable except for minor differences in body size, wrist joint structure, and certain cranial characteristics (Kingdon 1971).

Guenons, like *Macaca* and *Homo sapiens*, have a diverse range of habitats including grass and woodland savannas, margins of the desert, mangrove swamps, and a variety of forest habitats. Most inhabit forests and are arboreal (Bourliere Hunkeler & Bertrand 1970). Within the forests, they spend most of their time in mid to lower canopy regions but occasionally in the upper canopy or on the forest floor.

The guenons' appearance, great adaptability, and diversity make these species interesting subjects of study, yet the social behavior of only three has been studied in detail: *C. aethiops* (Bernstein 1971; Dunbar 1974; Gartland & Brain 1968; Hall & Gartland 1965; Struhsaker 1967 1971); *C. ascanius* (Haddow 1952; Struhsaker 1978 1980); and *C. talapoin* (Gautier-Hion 1966 1968 1970). Neither the behavioral repertoire nor the performance of guenons on cognitive and sensory tasks has been investigated thoroughly.

This situation may change soon because various aspects of guenon behavior are now recognized as important exceptions to ideas about the evolution of primate morphology and behavior. First, there is evidence that guenons represent exceptional movement for a primate away from the savanna and back to an arboreal habitat. Second, guenons, unlike other primates, can have from 42 to 72 chromosomes and have a tendency to create and pass on new chromosomes at a phenomenal rate (Dutrillaux 1986). Finally, although guenons can interbreed and live in multispecies groups, hybrid forms are not reported at a rate to be expected from their opportunity for interbreeding. Kingdon (1971) believes their flamboyant external characteristics, combined with species-typical head movements, may serve to separate species behaviorally as well as to enhance intraspecific communication in the trees. Guenons may represent an ongoing speciation event. The behavioral and visual differences presumably act to isolate each group genetically. If this is the case, comparative investigations of behavior patterns, communication, and reproductive strategies take on new significance (Maranto 1986).

Opportunities to study guenons are not hard to come by. The International Species Inventory System (ISIS) sponsored by the member institutions of the American Association of Zoological Parks and Aquariums (AAZPA), American Association of Zoo Veterinarians (AAZV), and various foundations lists some 20 guenon species, with over 800 specimens, in North American zoos. This is undoubtedly an underestimation because many zoos neglect to list all their specimens in ISIS.

Some aspects of behavior can be studied very profitably in zoos. Zoos do by charter what neither laboratory nor field workers can do. Individual animals can be identified reliably and their behavior monitored extensively. Histories, including health, medication, reproductive status, social hierarchy, institutional history, age, and genealogy are often a matter of record. Zoos also provide the opportunity for experimentation. Specific environmental, social, and management variables can be manipulated systematically. Routine practices such as additions to existing troops, multiple-species exhibits, births, deaths, and changes in the design of exhibits offer excellent opportunities for study. Many animals that a field worker knows only as a blur retreating into the bush can be examined closely for extended periods of time. Finally, most zoos try to exhibit their primates in social groups and try to simulate natural habitat.

Yet zoos remain underutilized in the study of primate behavior, perhaps because the behavior of captive animals is thought to differ radically from their wild counterparts. Several investigators have compared the behavior of captive and wild primate troops and found that the range of behavior was similar. Some activities, like sleeping and playing, may occur more frequently in captive groups, but the range of different activities, and the context in which they occur, are generally appropriate. This is not the case when animals have been housed in barren environments or reared in social isolation, although even in these conditions animals still respond, so far as opportunity permits, in species-typical fashion (Bernstein 1971; Eisenberg & Kuehn 1966; Southwick 1967). Bourliere (1961) pointed out the fallacy in the widespread belief among ethologists that data are valuable in proportion to the difficulty of collecting them. Perhaps a greater fallacy is the belief that the behavior of captive animals, even those that are housed in environments approximating the wild, is in and of itself unworthy of study.

In 1984, the Santa Barbara Zoo acquired six members of a troop of lesser spot-nosed guenons (*Cercopithecus petaurista*) caught in the wild. The arrival of this essentially unstudied species of guenon and the zoo's determination to manage the troop as well as could be done prompted it to undertake a comprehensive and long-overdue study to document and analyze the behavior of this species. All possible resources were explored to ensure proper management and care of the species.

Field studies conducted by Struhsaker (1967), Bernstein (1970), Bourliere et al. (1970), and Kingdon (1971) included brief descriptions of the behavior of the lesser spot-nosed guenon, but this species was never the primary object of study. Observers have generally agreed that this species dwells in mid to lower canopy and moves through the trees by quadrupedal running, jumping, and aimed leaping. Their diet consists of leaves, fruit, bark, flowers, and insects. Their social system has not been investigated although they have been sighted in mixed-species groups with only one adult male in attendance. The species is found in the western plateau of the Gold Coast of Africa (Bourliere et al., 1970).

In 1970, Struhsaker recorded the vocalizations of several guenon species, including *C. petaurista*, in an attempt to clarify phylogenetic relationships. Each species had its own vocalizations and tended not to react to the social calls of other species; all, however, responded to contraspecific alarm calls. In pattern and specific elements, contraspecific vocalizations by lesser spot-nosed guenons were intermediate between those of *C. erythrotis* and *C. cephus*. Todt Bruser Hultsch and Lange (1982) have been the only laboratory workers to study the lesser spot-nosed guenon in the laboratory, describing nocturnal behavior and periodicity of behavior of infants.

Methods

Subjects and living conditions. The monkeys in this study (two males, four females, and one of undetermined gender) arrived at the Santa Barbara Zoo in 1984 as members of one troop, except for one infant conceived at the Zoo and born in September 1986. The wild-caught animals were conservatively estimated (based on size, subsequent growth, and dental eruption) to be between 1 and 4 years old at the time of arrival. They were immediately tested for tuberculosis, then caged in two groups of three animals each for 90 days of strict quarantine, during which time they were observed through a glass partition.

At the end of quarantine, each animal was captured again, weighed, measured anthropometrically, and had its teeth inspected. The original troop was then reconstituted and moved for 6 months into an interim holding facility consisting of two interconnecting cages each measuring 1.8 x 1.8 x 2.4 m. The top 1.2 m of each cage was chain-link fencing and the lower part concrete blocks. Each cage was equipped with a nest box, 0.6 m on each side, mounted on the rear of the cage. Tree branches ran diagonally from the nest box to the cage floor, and the cage was surrounded by a 15-cm wide plywood walkway at the juncture of the blocks and fencing.

On September 19, 1985, the animals were captured again, tested for tuberculosis, and moved into a display cage. This was a large aviary-type cage measuring 18.3 x 7.3 m and varying in height from 5.5 to 7.3 m. The cage contained a cliff 3.7 m long by 3.1 m high and 3.8 m deep. A stream flowed over the cliff creating a waterfall, a pond, and a stream at the bottom of the cliff. One live and two dead trees, assorted logs, boulders, stones, tall grass, and a network of flexible vines and ropes were also part of the exhibit. Concealed within the cliff, and accessible through holes in its face, was a series of interconnecting cages, each 1.2 x 1.2 x 2.4 m.

The animals were fed twice a day during quarantine and holding. After the troop was moved to the display cage, the feeding schedule was changed to one large meal at night and a generous supply of forage (rice, raisins, seeds, peanuts, grapes, etc.) scattered randomly about the exhibit every morning. The monkeys also foraged on the grass and a live tree (for as long as it lived) in the display cage.

Novel objects were often added to the display. When provided with freshly-pruned branches (browse), the animals would run their hands along the leaves, dive into the pile of branches, carry the branches around, and finally eat the smaller leaves. Such acts resemble components of the behavior of manufacturing tools for dipping honey (Westergaard & Fragaszy, 1987). Other novel organic items in the display included such things as coconuts, edible flowers, sticks, and by chance a few baby birds.

After the animals had been put on display, we began training them. The first action to be trained was to enter the off-display holding cages upon command. The animals were also trained to come to the fencing of the holding cages for close visual inspection. Standard operant-conditioning procedures were used with preferred food items as reinforcers.

Observational procedures. The classes of behavior to be monitored were selected from a comprehensive behavioral catalogue that had been developed for primates at the Santa Barbara Zoo. Descriptions of these behavior classes were derived from 50 hours of naturalistic observations under varying environmental conditions. A category of behavior was included in the catalogue only when it had been identified by two independent observers.

Thirteen general categories, comprising a mutually-exclusive and exhaustive list, were selected for measurement. The taxonomy is listed in Table 1. It includes three kinds of social behavior: specific, nonspecific, and communicative; four kinds of nonsocial behavior; self-maintenance; social contact with an individual outside the guenon group; and four actions typical of this species including various forms of kissing, handwiping, head twisting, and tail licking.

Table 1
Behavioral Taxonomy for the Lesser Spot-Nosed Guenon

SOCIAL BEHAVIOR

Nonspecific

Rest: dorsal-ventral, ventral-ventral, lateral

Specific

Groom: lick, chew, comb with fingers
Solicit: crouch, flank, lie on side
Mutual: lick, chew, kiss, comb with fingers
Play: leap, wrestle, push, pull, chase, pounce, swing, hang
Agonistic:
- *Direct*: push, pull, grab, bite, chase, jump, mount
- *Indirect*: displace, yawn, threat face, branch shake, head bob
- *Submit*: present, flee, crouch, look away, fear face
- *External*: threat, shake branch/cage

Reproductive: mount, copulate, present, smell genitalia
Communication: chirp, squeak, shriek, lip smack, loud call
Maternal:
- *Mother*: nurse, groom, pull, slap, pick up, carry, play, huddle, vocalize
- *Aunt*: groom, pull, slap, pick up, carry, play, huddle, steal

NONSOCIAL BEHAVIOR

Rest Alone: sit, lateral, huddle, head between knees
Travel: run, walk, jump, climb, aimed leaps, pace
Ingestion: eat, forage (pouch, search, beg), drink
Eliminate: urinate, defecate
Environmental Survey:
- **Visual**: internal object, external object, standing
- **Olfactory**
- **Tactile**

Self-maintenance: groom, tail lick, stretch, scratch, suck fingers

OUTSIDE CONTACT: the occasional interaction with an observer, visitor to the zoo, keeper, or peacock.

UNKNOWN FUNCTION: handwipe, head twist, tail lick

Data collection. Data were collected on a particular animal and on the entire troop (instantaneous scan) in every setting. One animal's behavior was recorded into a court stenograph or a tape recorder every 2 seconds for 10 minutes; the pace was set by a digital timer that sounded every 2 seconds. Each animal was observed in a randomized but balanced order 5 times during each of 4 two-hour periods: 0831-1030, 1031-1230, 1231-1430, and 1431-1630.

Instantaneous scanning data were collected on a check sheet once every 2 minutes. Data recorded included location of each animal, the presence of an animal in or on a prominent feature of the display, the distance of the animal from the floor of the display, and what each animal was doing.

Animals were observed from April to September of successive years, first in the interim holding facility, and then (continuously after being moved) in the display cage. This study developed a comprehensive behavioral taxonomy including descriptions of both species-specific and idiosyncratic behavior. Activity profiles were constructed that took account of the frequency and duration of 15 activities. Differences across settings in the duration or percentage of time spent in an activity were tested for significance with *t*-tests. Differences in the likelihood of occurrence of an activity in specific areas of each exhibit were tested for significance with the Mann-Whitney *U*-test for nonparametric data.

Results

Behavioral taxonomy. Fifty hours of observation during quarantine resulted in identification and description of 97 discrete activities that could be reliably identified (r = .90) by different observers. Eighty-three per cent of the categories emerged in the first 10 hours of observation. Maternal behavior was not included because the females were nulliparous while in quarantine. Both males displayed dominance mounting, but only the older adult male copulated with the oldest adult female. Head twisting was the only other action that was not recorded during the initial observations but was added to the taxonomy when it was observed later. Most acts could be classified under the broad and familiar headings of *Social Behavior* (groom another, play, communicate, mate, act aggressively) and *Nonsocial Behavior* (ingest, eliminate, travel, rest, self-groom, and survey the environment).

The exact form and pattern of many of the acts were unique either to the species or genus. For example, threat was typified by rapid vertical and lateral head movements followed by branch or cage shaking. This pattern is similar to that observed in other guenons including Green monkeys (*Cercopithecus sabaeus*) and Red-tailed monkeys (*Cercopithecus ascanius*), as well as other monkey genera such as owl monkeys (*Aotus trivirgatus*) (Kingdon 1971).

Four unique actions occurred consistently in the lesser spot-nosed guenons we observed at the Santa Barbara Zoo: mouth-to-mouth contact (*kiss*), rapid handwiping, head twisting, and tail licking. The *kiss* was typified by the approach of one animal toward another in a head-down, tail-up posture followed by brief (1-second) mouth-touches without vocalization. It was occasionally associated with overt eating; and since guenons have cheek pouches, the observer could not rule out the possibility that stored food or a nonfood item in the mouth of one or both participants was the stimulus for *kissing* at times when eating was not obvious. Usually this action occurred when two animals had been separated for some time.

Members of this species of guenon also frequently wipe their hands vigorously against a tree trunk, vine, bench, etc. It appears as though the animal were trying to remove something from its hand; but what, if anything, it might be was not always obvious to the observer.

Head twisting was not noticed until 4 months after the animals began living in the display cage. While looking through fencing, tree branches, or tall grass, the animal rotated its head in a figure eight pattern or in small circles in the vertical plane. Reports on the wild Red-tailed guenon also describe this action. Kingdon (1971) speculates that it has some communicative function.

Finally, five of the seven animals excessively groomed the end of their tails, resulting in a bright red knob on the tail's end. The adult male damaged his tail the most, to the point of exposing the bony end. This activity became more frequent toward the onset of the breeding season.

Activity profiles. Significant changes in activity profiles occurred when the guenons were moved from one habitat to another. Proportion of total time, frequency of initiation, and duration of rest, travel, ingestion, environmental survey, and social behavior differed significantly across settings. Figure 1 shows that activities were distributed more evenly in the display cage than in the interim holding facility.

One of the most obvious changes in behavior was less inactivity when the animals were moved from the interim holding facility to the display cage ($t = 68.87$, $p < .001$, df = 5). In the interim holding facility, the guenons rested 41% of the time -- 29% alone, 3% in a group, and 9% in one of the nesting areas. In contrast, when in the display cage the animals rested only 19% of the time ($t = 28.47$, $p < .001$, df = 5) – 13% alone, 3% in a group, and 3% in the nest area. Average duration of rests also differed significantly between habitats. Social resting had longer duration in the interim holding facility, averaging 26 sec in the interim holding facility but only 14 sec in the display cage (t = 8.24, p < .01, df = 5). On the other hand, resting in the nest area had longer duration in the display cage: an average of 15 sec in the interim holding facility and 29 sec in the display cage (t = 8.52, p < .01, df = 5). Throughout, the adult male spent the greatest time resting alone, while the oldest female spent the least.

Time spent moving around the habitat also decreased significantly when the animals were moved from the holding facility to the display cage (t = 48.90 p < .001, df = 5). In the interim holding facility, they moved unsystematically around the perimeter of the habitat in slow running or trotting. A bout of this activity lasted 10 sec on average. In the display cage this was replaced by more diverse forms of locomotor activity: the animals performed spectacular aimed leaps, bounding, running, and rebounding to and from various furnishings within the habitat. These actions were usually brief, lasting between two and three sec.

Several categories of social behavior were affected by change of habitat. About 13% of the guenons' time was spent in social behavior while they were in the holding facility (social rest made up about 3%, grooming 1.5%, playing 4%, agonistic behavior 5%, and reproductive behavior less than 1%). In the display cage, the animals engaged in social behavior 25% of the time. About 13% of this was social grooming, a significant increase from the holding facility (t = 68.87, p < .001, df = 5); 3% was resting socially, 6% was play, 2% was agonistic encounters, and 1% was reproductive behavior. The adult male was the principal recipient of grooming, with the three females often grooming him simultaneously. Although the percentage of time playing did not differ significantly between environments, the form of play did. Playing changed from wrestling and play-biting in the holding facility to chase-and-tag in the display cage. An average bout of playing decreased from 7 sec in the holding facility to 4 sec in the display cage. Although this might be an effect of maturation it is doubtful because four of the animals were already 4 years old at the beginning of the study. Bouts of direct aggression lasted about 7 sec on average in the holding facility and 2 sec in the display cage (t = 2.87, p < .05, df = 5); the average times for indirect aggression were 3.13 and 2.75 sec, respectively.

The category of behavior that showed the largest increase when the animals were moved from the interim holding facility to the display cage was ingestion (eating, foraging, drinking). The troop spent a total of 5.5% of its time ingesting while in the holding facility, but 27% in the display cage (t = 61.68, p < .001, df = 5). In both environments food was available nearly *ad libitum*, scattered about the enclosures. Items of food were larger and more visible in the holding facility, and this may be why less time was taken up with feeding. The average duration of individual bouts of foraging, eating, and drinking did not differ significantly between environments, nor did the proportional distribution among the three ingestive activities.

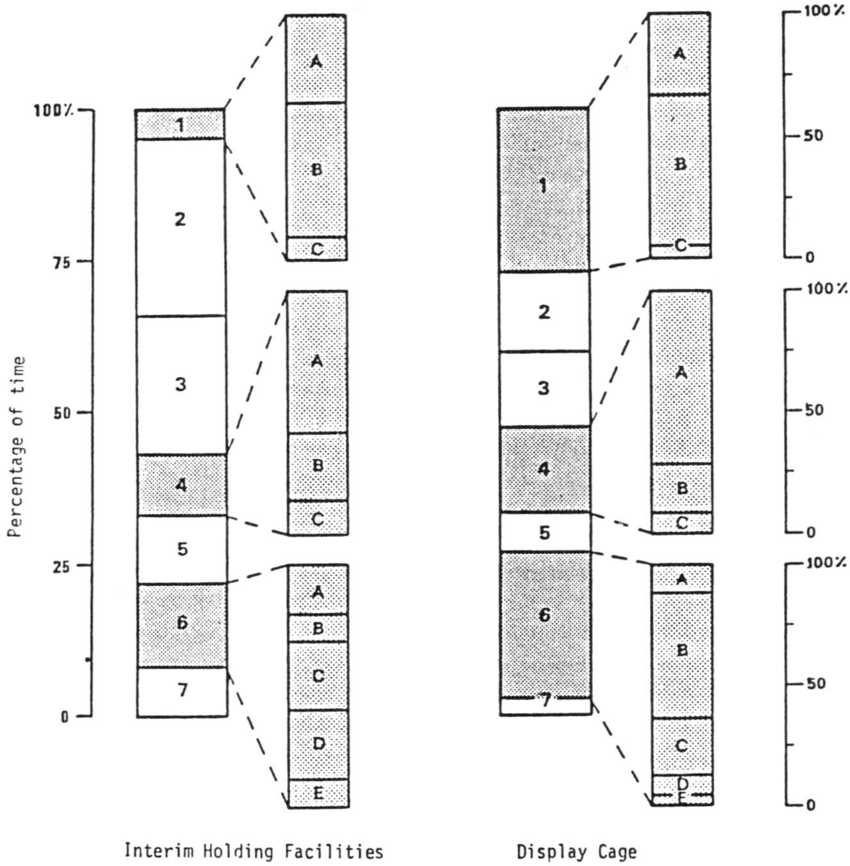

Fig. 1. Activity profiles for the guenons in holding facilities or display cage.
1-INGEST A:Eat B:Forage C:Drink 2-REST 3-TRAVEL
4-SURVEY ENVIRONMENT A:Visual B:Olfactory C:Tactile 5-SELF-MAINTENANCE
6-SOCIAL A:Rest B:Groom C:Play D:Agonistic E:Reproductive
7-OUTSIDE CONTACT

Use of vertical space. To compare the way the guenons used the space available in the two different habitats we monitored time spent at each of five vertical levels in each habitat. Only time spent at the lowest and highest levels differed significantly ($U = 2.84$, $p < .016$, $N = 6$). In the interim holding facility the animals spent 2% of their time on the ground, 83% in the mid levels, and 15% climbing across the roof or with their heads touching it. In the display cage, they spent 12% of their time on the ground, 84% in the mid levels, and 4% at the highest levels. Time on the ground could be accounted for by seven activities: 54% eating or foraging, 23% visual survey, 18% running or walking, 5% playing or agonistic encounters. In the display cage the guenons spent a higher proportion of time on the ground than might be expected in an arboreal species, but feeding strategies imposed by the caretakers account for this. The monkeys spent a large proportion of time on the ground filling their pouches before moving to higher areas to eat. This was true in both environments. In summary, moving animals from one setting to another profoundly affected activity profiles, but regardless of the environment, the animals occupied vertical space as one might expect for arboreal primates.

Since some of the changes we saw might have been maturational, it would have been ideal from a research point of view to follow an ABAB design: return the animals to the interim holding facility for a time to see if activity profiles would match those seen when the animals were in the holding facility before; then return the guenons permanently to the display cage. Such reversals are impractical in a zoo, however.

TOWARD OPTIMAL MANAGEMENT

Resolving the interests of institutions and captive primates

Despite differences between the two studies reported here, one dealing with aging in laboratory monkeys and the other with the activity profiles of monkeys in contrasting zoo habitats, the two studies taken together make the point that the same principles of optimal care hold for both settings. The first author has had experience in both settings and was struck by the common elements required for humane and sound care. Optimal conditions for working with any captive animal are by necessity a compromise among the specific requirements of the organism, the goals of the institution, and finances. Generally, the goals of research laboratories dictate a different set of conditions from those of zoos and aquaria. This does not mean that barren, nonsocial, or restricted laboratory habitats ever are necessary or even profitable, though it may mean that very large, complex social environments are impractical.

Insuring health and safety of captive nonhuman primates

The stated goals of the American Association of Zoological Parks and Aquaria (AAZPA) are conservation, education, research, and entertainment (in that order). Fulfillment of these goals requires housing and management that ensures the health and safety of the animals, minimizes stress and intervention, and allows a wide range of behavior to be displayed at the appropriate rate and in the appropriate context.

The *Guide for the Care and Use of Laboratory Animals* (NIH 1985) states that "Proper care and humane treatment of animals used in research, testing, and education...requires scientific and professional judgment, which is based on knowledge of the husbandry of each species and special requirements of research, testing, and educational programs." The goals of insuring the health and safety of laboratory primates can effectively be accomplished through regular, *noninvasive* health checks, screening new additions to the collection, providing clean quarters, providing appropriate diets, and a conservative course of preventive medicine. All this must be balanced, however, with efforts to minimize stress caused by capture and handling.

Capture and restraint of sick or injured animals may at the time seem unavoidable, but every effort should be made to avoid direct physical contact between caregivers and animals. Noninvasive procedures such as urine and fecal analysis can be effective diagnostic tools and should be used whenever possible. Injury and stress associated with restraint can be minimized further by training animals to assist in examination procedures. Most primates can be trained to move into restricted quarters on cue, climb into a small carrying cage for close inspection or transport, and even take medication voluntarily. Primates can also be trained to tolerate limited contact and will offer limbs for quick blood samples or present their rump for vaginal smears. Another possibility, as we suggested earlier, is to house an animal that is not in an experiment or display close by but not with the target troop, and monitor this animal to determine the presence of infectious diseases.

Removing and reintroducing an individual that is part of a social group impacts not only that individual but the entire group. The consequences can be aggression focused on the reintroduced individual resulting in its injury or even death, or, at the least, complete social rejection of the reintroduced animal. Erwin (1979) reported that removal of a high-ranking male from a troop often leads to elevated aggression within the troop as the remaining males compete for higher rank; still more aggression may occur when the alpha male is reintroduced. Should anesthesia or restraint be absolutely necessary for any animal in a group it may be advisable to restrain or separate all of the animals at the same time.

Importance of conspecific companionship

Optimal conditions also mean opportunities for animals to display as much of their normal behavior as possible. One of the most effective, safest, and probably cheapest ways to accomplish this is by allowing each animal to be social. Providing each animal with a conspecific partner can have a dramatic impact on behavior and on physical appearance. A conspecific partner is responsive in a fashion unique to the species. Partners can groom each other, learn from each other, and warm each other.

Almost any two animals of the same species will learn to tolerate one another, but putting two unfamiliar adult males together is usually inadvisable. A subadult male, however, may sometimes be paired with an adult male temporarily. Mitchell Maple and Erwin (1979) reported that pairing an infant rhesus monkey with an adult male that had been reared in social isolation presented surprisingly few problems and seemed to ameliorate some of the adult's behavior anomalies . (Also see O'Neill: this volume; and the recent work of Viktor Reinhardt on conspecific pairings, reported in several 1987 and 1988 issues of *Laboratory Primate Newsletter*. Ed.) Where possible, however, pairing an adult male with an adult female, two adult females, or a subadult with an adult female presents fewer problems.

Animal technicians tend to house animals singly in laboratories for reasons of disease control and ease of restraint. However, if animals are trained to enter a carrying cage voluntarily, either one of a pair can be handled. Also, if a colony is protected from infectious diseases, the danger of a monkey's transmitting infection to a cagemate is minimized, and the animals gain from social interaction.

Approximating naturally-occurring social systems allows for the expression of diverse behavior. Simulating natural social systems is not much of a problem with species that form nuclear family units such as the White-Handed Gibbon, Black and White ruffed lemur, or Goeldi's monkey; but since most primate groups do not consist of a bunch of strangers who have bumped into one another in the savanna, complex social systems can be quite difficult to simulate artificially. Although the exact composition of social units varies greatly among species, most exist as families or collections of extended families, all very familiar to one another, some with a single male and others with more than one.

The sudden appearance of a strange animal within a troop often results in an attack on that individual. Erwin (1979) investigated various ways primate groups are formed. The fewest injuries resulted when all animals were strangers to one another. More injuries resulted when two established groups were combined or when a stranger was introduced to two animals that were already acquainted. Apparently, acquaintances form coalitions and engage in prolonged, intense aggression against the intruder. Erwin (1979) also reports that when several animals are to be introduced to a group, simultaneous introduction of all of them results in fewer injuries than introducing them singly, one after another. Difficulties encountered in forming complex social groups of primates are increased by a lack of data about many species. Zoos may display their primates in pairs or harem groups simply because it is the safest thing to do and has the advantage of known parentage of offspring. Furthermore, if the troop is left intact, harem groups will often develop into more complex units typical of the species.

Design of housing appropriate to the species

Appropriate behavior is only possible if the caging is properly designed. It must be carefully tailored to the requirements of each species because housing has a profound effect on both the behavior and physical appearance of the animals. Primates generally use vertical as well as horizontal space. Thus, if objects are provided for climbing up, for crawling onto, and for sitting upon, usable space is increased and animals can escape visual contact with dominant animals. While adding dimension to an enclosure, a designer can capitalize on behavior typical of the species. Their full behavior capacity can be displayed only if the environment provides the opportunity. Patterns of locomotion are one common example. Inadequate resources prevent the expression of nest building, tool using, and swimming and, in arboreal primates, brachiating or running quadrupedally through trees on elevated, interconnected vines, trees, ropes, or branches.

Enclosures that are too small are unfortunately common in laboratories and have long been a subject of concern. Crowding leads to excessive aggression, decreased fecundity, and increased lethargy among most animals. Elton and Anderson (1977) and Erwin (1977 1979) have reported the effects of laboratory crowding on baboons and two species of *Macaca*. When the animals weren't fighting, they generally slept or rested in small groups. Many of the negative effects of overcrowding were alleviated by increasing the usable space within the enclosure.

An enclosure can also be too large, although this condition has received little attention. (But see Novak & Drewsen: this volume. Ed.) Individuals of some reclusive species that are naturally nocturnal and arboreal do not tolerate large open spaces and spend much of their time huddled into a corner or sitting quietly in a tree. From the researcher's perspective, an enclosure, whether in the laboratory or thezoo, may be too large if the animals are difficult to see.

Preparation of displays of captive primates requires a thorough assessment and analysis of data collected both in the wild and in the laboratory. Some of this information can be found in the published literature, but if this is unavailable, direct communication with persons who have experience with a particular species can be invaluable. Often, management must be based on an assessment of the behavior of a near relative, combined with understanding the environment the target species is known to inhabit.

Regardless of design, or its reasons, the adequacy of the habitat can only be measured by its impact on the animals and an assessment of how this meets the goals of the institution. It is not enough to put primates into an expensive and artistically-designed exhibit and just assume the best.

Management practices to broaden behavioral diversity

Specific management practices can also act to broaden behavioral diversity. Altered feeding routines can have a profound effect. Most laboratories and zoos provide a basic diet of commercial primate chow and some routine fruit and vegetable supplement once or twice daily. This precludes one of the most time-consuming, unpredictable, interactive social activities available to a primate, the search for and consumption of naturally-occurring food. Captive primates quickly deplete any forage found in a confined environment, so artificial supplementation of the diet is essential. The San Francisco Zoo attempted to solve this problem by overwhelming their colobus monkeys with vegetation. Waves of cheap but edible tree branches and plants were moved into the exhibit and removed when defoliated. Eventually the novelty wore off, and the rate of plant denuding and replacement reached an acceptable rate. It is a lot cheaper and easier to provide at least one meal in the form of forage than to grow natural vegetation in the enclosure. Items such as seeds, shelled nuts, coconuts, edible flowers, rice, freshly-cut branches, grasses, grubs, and crickets can be hidden around an exhibit.

Most primate enclosures are predictable and boring. It is easier to reduce boredom than to eliminate it. Housing animals in social groups reduces monotony. Novel items, such as electronic games, have been provided in several zoos. Most of these items cease to be novel as soon as they are introduced, and games without the give-and-take of a partner quickly become predictable. (See Markowitz & Line: this volume; and Rumbaugh Washburn & Savage-Rumbaugh: this volume.)

Behavioral training also provides variety. Training animals to enter a carrying cage or engaging them in a series of learning experiments provides the animal the opportunity to interact with its environment and control some of its outcomes. Periodically changing the content of training also provides novelty.

Unpredictable introduction and removal of unfamiliar items can also reduce boredom. Items that an animal can control, such as vines, or turn on and off, like a water valve, can be stimulating.

In summary, what is necessary for chronic studies with primates, both in the laboratory and in the zoo, are well designed housing, opportunities for social interaction, variety of experiences, conservative medication, and avoidance of restrictions on freedom of motion.

ACKNOWLEDGEMENTS

The first author thanks Heidi Hattenback, Steve Schiff, and Robin Sullivan for endless hours of dedicated observation and data collection. She also thanks senior keeper Fred Marrion and lead keeper Sioux Sanford for their assistance and willingness to work within the observation schedule.

The second author's part of the chapter is the result of support for behavioral research by USPHS between 1952 and 1972 and contract No. No1-AG-52854 from the National Institute of Aging. This part of the chapter is based on an invited paper by the first author that was given at the 1986 meeting of the Canadian Gerontological Society, Quebec, Canada.

Part of this research was conducted at the Santa Barbara Zoo and was supported in part by a 1986 Institute of Museum Services general operating support grant.

— 15 —
Solution to Psychological Enhancement of the Environment for the Nonhuman Primate

William M. Blackmore, D.V.M.

INTRODUCTION

In the early days of suborbital flight, when chimpanzees, not astronauts, were going into space, the University of Southern California's human centrifuge, its medical pioneers in space telemetry, and chimpanzees Amos, Rocky and Hamm, all demanded the best possible environment. A good part of the chimpanzees' lives was spent riding in the centrifuge and helping to develop the telemetry that was necessary to monitor all astronauts who would go into space in the future. In the spirit of creating a better environment for the nonhuman primate in research, an exercise cage was developed at the University of Southern California (USC).

EXERCISE FACILITIES AT USC

Several versions were developed over nearly 20 years before we arrived at the current design for our exercise cages. We wanted to have the exercise cages near the research labs and at the same time, although space in the research areas is limited, make them large, comfortable, and interesting. We believe the current exercise cages enhance the psychological environment of animals that are actively involved in research. Our monkeys willingly accept the opportunity to enjoy the exercise cage, and we are convinced that they are more content. The evidence is the noticeably low noise level and lack of aggression in our colony of mixed *macaca* species. Before the development of the exercise cage, any visitor to the colony room was greeted by loud calls and banging on the cages, making normal conversation difficult. Now all animals, both those in the special exercise cage and those housed in regular cages in the same room, quietly and attentively look at the visitor. The animals in the large exercise cage climb the cage wall to examine the visitor calmly, at eye level. Occasionally they utter soft calls. The animals in the regular cages follow the lead of those in the exercise cage. The contrast with the monkeys' behavior before introduction of the exercise cage is striking to anyone who has observed them before and after.

Room 1

Our exercise cages are paired, modular, and adaptable to the existing space in most large primate colony rooms. A flushing drain must be located near the bottom of the cage to facilitate cleaning. In our case it was not necessary to install new flushing drains as they were already in place. The first pair of cages at USC was built to conform to a room (Room 1) measuring 10 ft x 20 ft. Each cage of the pair occupies about a third of the floor area (5 ft x 12 ft x 7 ft high), leaving room for a work area with a sink, and space to wheel in our moveable, individual cage units or transfer cages. There are no other monkeys in Room 1 except the two being exercised.

The animals, one in each exercise cage, can see each other through the wall of wire mesh, but they are not in physical contact. With this arrangement, each animal helps stimulate the other to exercise, but with the screen barrier between the cages, biting and other injuries are non-existent. Although the evidence is subjective, we have preliminary data to support the facilitative effect on activity of keeping single animals in side-by-side exercise cages. Animal care personnel, who are in daily contact with the animals, report that a dominant monkey in one cage will stimulate a dominant monkey in the adjoining cage by challenging it, grooming itself in front of the mirror, or actively playing with soccer or soft balls. Further, my staff and I have noticed that most animals seem to grow stronger and more active with each successive exercise period. We believe this reflects improved muscular tone and comfortable familiarity with the unit.

Each cage has the following features:

1. **Gate.** This is a sliding gate with a transfer port.
2. **Squeeze-back.** This facilitates quick and easy transfer.
3. **Trapeze.** Admittedly this is not used much by ground dwellers but our macaques use it occasionally.
4. **Hammock.** This is made of freighter cargo netting. It extends the full depth of the cage, folding out of the way when the squeeze-back is brought forward. The animals lounge in it as well as running along it, as it bridges the cage from back to front. It is high enough from the floor that animals can also travel underneath it. Every monkey visiting the cage uses it. In fact, the hammock is an outstanding success and is in constant use. The smaller primates chew

extensively on the netting, but larger monkeys do not. The hammock is cleaned and sterilized periodically in the cage-and-rack washer and the autoclave.

5. **Mirror.** This is a large (6 ft x 3 ft) stainless-steel mirror that provides stimulation and pleasure for the animals. When the animals are first introduced to a mirror they treat the image as though it were an unknown monkey, first being aggressive, then cautious. Later they sit in front of the mirror and groom themselves.
6. **Cyclone fencing.** Each cage is made of cyclone fencing. The open-weave pattern provides hand and foot grips and so encourages the monkeys to climb to the highest region of the cage, yet they can be quickly brought to the transfer area by use of the squeeze-back.
7. **Radio.** Room 1 has a radio that plays during the day. According to the animal care personnel, the monkeys seem to be calmer when classical or other soft music is played.
8. **Television.** We are currently evaluating the effect of children's television programs. The response appears to be similar to their response to music, that is, they are soothed rather than aroused.
9. **Soccer balls and other diversionary items.** We supply soft balls and other soft children's toys. The monkeys eventually chew them up, but there seem to be no ill effects from this.

Room 2

The second room (Room 2) that we equipped with paired exercise cages was a large colony room housing macaques, both stumptail and rhesus. We made the new pair of exercise cages as large as possible (9 ft x 6 ft x 7 ft high). This still leaves room for individual squeeze-back cages that house 15-18 macaques.

From the very beginning of the installation of this new set of exercise cages, in December, 1986, noise from pounding on the home cages and screeching has been noticeably reduced. When anyone walks into the room, the two macaques in the exercise cages immediately stop what they are doing to observe or challenge the person, but they are not usually vocal. After a few minutes they either climb the cage wall nearest the visitor to get a better look, or resume play with the toys, or groom themselves. The other animals in the room are unusually quiet and observe the exercise cages, the occupants, and the visitors.

Our four exercise cages are occupied 24 hours a day, with one animal per cage, in shifts: 7-11 AM, 11 AM-3 PM, and 3 PM-7 AM. We are careful to schedule equal exercise opportunitiesfor all monkeys throughout the week.

Social Environment

Resocialization into groups, and maintaining nonhuman primates in groups, is not an easy task. It requires management by trained and experienced primatologists. Zoos that display large numbers of primates are experienced at resocialization techniques, and those of us with nonhuman primates in individual cages in research facilities must consult with zoo resources if a resocialization program is seriously to be considered.

The primatologists I have consulted have encouraged the continuance of the dual-cage concept so that monkeys with potentially similar dominance rank can see and interact with, but not injure each other in competition over rank.

The pair of animals occupying the dual exercise unit in Room 1 experience minimal interference from other animals or handlers. The exercise cages in Room 2, on the other hand, share space with singly-housed monkeys. Research cage tandems, one atop another on wheeled dollys, line the room, and all animals in these cages have a clear view of the dual exercise unit. Thus, animals in Room 2 are stimulated in several ways:

a. The two animals in adjacent exercise cages socially stimulate each other, as in Room 1.

b. They are also stimulated by the special features of the exercise cage, such as its large size and the cyclone fencing that allows easy exploration of all parts of the cage. The large stainless-steel mirror presents a reflection that is at first a social challenge, posing a threat as of another, equally excited monkey. Threat reactions are later replaced, however, by more relaxed self-grooming activities in proximity to the mirror. The hammocks have proven to be universally accepted, are used frequently, and enjoyed by all monkeys. Finally, the soccer balls and other diversions are the same as in Room 1.

c. The monkeys held in regular tandem cages are interested in the activities in the exercise cages, and their attentiveness in turn stimulates the animals in the exercise cages.

d. The caretakers are also a social stimulus for the animals. Since the exercise cage is contained within the colony room, cleaning, feeding and transferring of research animals continue to take place.

In comparing the different atmospheres of Rooms 1 and 2, I believe the animals make equally good use of the cages in both rooms, and benefit emotionally and physically by the atmosphere of either room. If resocialization in group caging were contemplated soon, I believe the more diverse social environment of Room 2 would be a better step in that direction. We plan to pursue this concept with consultant primatologists.

We hope soon to try a variety of exercise times, ranging from 2 to 16 hours. We will continue to consult with primatologists, principal investigators, and their respective staffs, to establish an optimal daily and weekly exercise schedule. With the current program, which divides the day into exercise shifts of either 4 or 16 hours, we are able to provide an overnight stay for up to 28 animals a week and to exercise every animal in the colony three times a week, which I believe is optimal.

HUMANE CONSIDERATIONS AND GOALS FOR THE FUTURE

If we are to continue the privilege of using nonhuman primates as animal models for important, potentially beneficial long-term research, we must enhance their environment. The question of length of time for long-term confinement of nonhuman primates must be defined by primatologists. As a beginning, I suggest long-term be defined as "over one month." Confinement of primates used for any purpose requires humane management, exercise, and large cage resocialization.

I suggest that we must accept the responsibility to improve the humane management of captive nonhuman primates by adopting an explicit scheme such as the following one, which takes account of the needs of the primates as well as of the research in which they are being used:

I. Close Confinement

This is the human equivalent of hospital bed confinement. It is required by research projects when the animals are actively engaged in research or are being closely observed. These animals have daily social interaction with the research team, the animal support technicians, and the veterinary staff. At the University of Southern California, it is these animals for which the dual exercise cages are intended.

II. Halfway House

This could be a two-story structure distinguished by its large exercise area, relaxed environment, and resocialization opportunities. Skylights, allowing trees and smaller plants to be grown indoors, would contribute to a more pleasant atmosphere. The two-story structure would allow the primates to move about both vertically and horizontally. The Halfway House could be an adaptation of the quarantine facility that has been for years an essential part of a well-managed colony. It could be built of pre-cast concrete at a modest cost per square foot.

The Halfway House should be a separate part of the program to enhance the environment of captive research primates. It should not substitute for the exercise cages intended for animals living in close confinement. Whereas the relaxed atmosphere of the large cages in the Halfway House does not lend itself to an intensive research program requiring frequent human interaction with the animals, placing exercise cages adjacent to research labs does allow a degree of controlled freedom even to animals that must be closely confined to permit frequent handling and close monitoring.

Animals in the Halfway House would be divided into categories depending on their treatment requirements:

a. **New animals.** These would be in quarantine and undergoing veterinary diagnostic procedures to establish their health status.
b. **Research animals in quarantine.** These would be under observation to determine the impact, if any, of prior research procedures.
c. **Research animals in isolation.** These would be long-term research monkeys held subsequent to a phase of frequent observation, in order to permit observation and testing at longer intervals, for example, to photograph the uvea every three to four months.

Unlike the exercise cage, which could serve all primates, the Halfway House would serve only a relatively few animals at a time.

III. Free-Roaming Environment

This could be accomplished in zoo-like facilities close to urban areas or in large compounds in rural areas. Primates in this environment would be headed for long-term observation and group resocialization. The Free-Roaming Environment, as with the Halfway House, is adaptable only to a small number of the total research colony. Although such an environment may augment the exercise cages, it cannot substitute for them. Closely-confined primates cannot be denied exercise until fulfillment of research needs permits them to be transferred to a Free-Roaming Environment.

I hope these concepts continue to grow, with acceptance and financial backing as primary requisites. I also hope that one day soon the total program of environmental enhancement for the nonhuman primate will include the three steps I have outlined, or some variant of them: Exercise Cage, Halfway House, and Free-Roaming Environment. The acceptance of the exercise areas by our monkeys has been a morale builder for all of us. With the exercise cages now an established success at USC, my next goal is the Halfway House.

Guided by our USC experience, and the guidelines that are being written by the USDA-APHIS committees who are studying the psychological enhancement of the environment of nonhuman primates, we hope to establish an optimal program when the new USDA rules are promulgated.

ACKNOWLEDGMENTS

I thank the following consulting primatologists for their help and advice: Roy Hendrickson, D.V.M., University of California Berkeley, formerly of the Primate Center of the University of California Davis; Deiter

Kramsch, M.D., a private investigator with eighteen years' experience with research primates; and Warren Thomas, D.V.M., Director of the Los Angeles Zoo.

I also wish to thank my friends and staff who have encouraged and contributed to the development of the concept of psychological enhancement of environment for nonhuman primates.

Figure 1. Monkey sitting on hammock in exercise cage and reaching out to touch the animal caretaker's hand.

Fig 2. Monkey taking a piece of chow from the animal caretaker. Note individual tandem cages in background, all facing the exercise cage. (Room 2)

Figure 3. A clear view of the sturdy hammock, with monkey standing on it.

Figure 4. Closeup view of hammock attachment. The monkey shows no distress at being photographed.

Figure 5. View of stainless-steel mirror, soccer ball, and floor drain, with individual cages beyond.

Figure 6. How monkey is transferred into and out of exercise cage.

— 16 —
Environmental Enrichment for Monkeys Used in Behavioral Toxicology Studies

Steven G. Gilbert and Ellen Wrenshall

ABSTRACT

The appropriate husbandry of laboratory monkeys has received increased attention not only by animal welfare groups but also by government research and testing agencies. Environmental enrichment can be a major factor in improving the laboratory monkey's psychological wellbeing. Enrichment is appropriate not only from the perspective of humane treatment but also from a scientific perspective. Long-term studies in behavioral toxicology require special management to ensure good scientific models. About 12 years ago, a colony of cynomolgus monkeys (*Macaca fascicularis*) was established for use in behavioral toxicology studies of neurotoxicants at the Health Protection Branch of Health and Welfare Canada in Ottawa. The monkey's natural instincts for socialization, manipulation of objects and exercise were considered in creating an enriched environment for them. Social groups of two to nine monkeys were developed according to age, sex, dose groups, and social compatibility. At the same time experimental requirements, including blood sampling, weighing and behavioral testing demanded priority. As our enrichment program developed, some colony rooms were made into exercise rooms by addition of stainless steel swings, climbing apparatus and portable swimming pools. Play objects such as metal cups, bowls, chains, hard nylon balls, and recently rubber objects encourage exercise and reduce boredom. Overall, this program helps to reduce self-mutilation and vices associated with caging and by considering the monkey's natural instincts along with experimental requirements, benefits both the science and the monkey.

INTRODUCTION

Public scientific and governmental concern about the appropriate care of confined animals, particularly nonhuman primates, has grown in recent years for several reasons. One is the growing appreciation of an animal's cognitive and communicative abilities. Numerous studies have examined the ability of apes to communicate with humans and amongst themselves (Savage-Rumbaugh, 1985; Patterson, 1981), while other studies have examined the nature of animal thinking (Terrace 1985), and even the nature of minds in other species (Gallup, 1985). A second reason for concern has to do with reproductive success of captive animals, especially endangered species. A third reason is the recognition that appropriate care of laboratory animals is essential for valid scientific results.

The monkey's physiological similarity to humans makes it the model of choice in a number of areas of research on human health and disease (McClure, 1984; King & Yarbrough, 1985; Newell-Morris & Fahrenbruch, 1985; Mitchel, 1988). For example, the monkey's long gestation period and complex central nervous system are well suited for long-term studies of the effects of chronic low-level exposure to neurotoxicants. The breeding and care of laboratory monkeys present special problems that grow in importance with the duration of a study. To solve some of these problems the Primate Behavioral Toxicology Program of the Health Protection Branch of Health and Welfare Canada has developed, over the past decade, a program for environmental enrichment of captive monkeys.

The Primate Behavioral Toxicology Program

The Primate Behavioral Toxicology Program was established to study the effects of low levels of food contaminants, such as lead or methylmercury, on central nervous system development (Rice, 1983). The monkeys are dosed with a toxicant at a level designed to mimic human exposure and produce subtle effects on the nervous system. The gross, overt behavior of treated and control monkeys is indistinguishable, thus, in part, providing the rationale for more detailed assessment of learning and memory function. The monkeys are bred specifically for this program; newborn infants are separated at birth from their mothers, moved to the primate nursery and reared according to standard practice (Willes, Kressler & Truelove, 1977). The oldest monkeys in these studies are now over twelve years of age and there are plans to maintain them to evaluate the effects of aging. The long-term physical and psychological health of these monkeys is extremely important for the study's successful conclusion.

This chapter reviews some of the practical reasons for providing a "good and appropriate" environment in primate research, and explores some of the problems and solutions for enrichment in a toxicology laboratory.

THE LABORATORY ENVIRONMENT

Advantages and Drawbacks of Laboratory Research with Captive Primates

The laboratory's primary advantage as a place for research is that it provides control over the environment, enabling standardization of conditions except the ones of experimental interest so that interpretation of results is not confounded. However, such standardization necessarily results in a predictable, monotonous environment for the experimental subjects. This can have adverse effects, especially on monkeys, as they are arguably (but not proven to be) more intelligent and therefore more sensitive to their surroundings than nonprimate species.

Toxicology Research: Special Needs and Husbandry Considerations

The most obvious aspects of laboratory husbandry are housing and nutrition. In the design of toxicology studies, a uniform living environment for all animals is a priority, so as to ensure that factors like temperature, humidity and housing do not differentially affect experimental outcome. In many laboratories, as in our facility, monkeys are housed singly in stainless-steel cages meeting the specifications of the Canadian Council of Animal Care. Although the monkeys can see and hear one another within individual colony rooms, the environment is very restrictive and does not change from day to day unless specifically provided for in the study design. The diet must be not only nutritionally adequate but also free from contamination by any agent that might confound the results. Although commercial monkey chow provides the monkey with all essential nutrients, fruits or vegetables are often given as a supplement to provide variety in an otherwise bland diet.

Manipulations required by a toxicology study may include obtaining blood or other biological samples, weighing, behavioural testing, and other handling. It is usually desirable to keep manipulations consistent and uniform across all individuals. A researcher must be concerned with the disadvantages as well as advantages of each manipulation and consider ways of reducing undesirable consequence of a procedure. For example, catching a monkey can raise hormone levels which may influence blood parameters (Kaplan, 1986); stress produced by various manipulations may be an important determinant of pregnancy outcome in reproductive studies. To mitigate such effects, handling should be performed according to a specific routine, so that individuals can anticipate what is coming. Small treats may also be offered to the monkeys as a reward following handling. Small forms of enrichment such as this are helpful in making the monkey more tractable.

When working with captive monkeys the laboratory researcher is obligated to consider the impact that the resultant impoverished environment may have on the experimental results. Lack of opportunity for constructive activities results in very destructive activities in primates, often selfdirected (Spinelli & Markowitz, 1985). The causes of such behavior are not presently fully understood, but it is best to avoid this situation both for health and testing reasons (Anderson & Chamove, 1981; Anderson & Chamove, 1985; Levitt, 1985). For example, a monkey prone to self-injury may not be suitable for studies involving the immune system because of increased risk of infection. Endocrine and cardiovascular, as well as other organ systems, also may be significantly affected by factors in the physical or social environment.

Consideration of the potential effects of environment on behavior is particularly important when behavior is a variable of interest. It is well documented that isolation-reared monkeys are less curious, more fearful, and less adaptable than their socially reared counterparts (Harlow, 1959, 1962). Even monkeys raised singly with their mothers but without the opportunity for peer interaction display behavioral anomalies (Suomi, 1986). Since such effects will certainly confound results of behavioral studies, they must be recognized and mitigated to the extent possible, by socialization, exercise, and opportunity for play. Furthermore, individuals reared under different environments should not be included in the same study, because their behavioural baselines may be very different. Even in adult monkeys, prolonged boredom and social isolation may lead to at least temporary intellectual deficits. The interpretation of results with subjects thus compromised would be at the least, questionable.

EXERCISE AND SOCIALIZATION

Adverse Effects of Isolate Caging

Exercise and socialization are two of the most important aspects of monkey husbandry, and they are closely linked. These two factors have a major impact on behavior and health, and thus influence a variety of dependent variables. In nature, monkeys are very active, exploring, testing and examining their environment (Morris, 1964). It is this ability to manipulate their environment and to learn from experience that makes them such attractive subjects for behavioral studies. In the laboratory, their environment is drastically restricted, and it becomes necessary to provide them not only with food, water and shelter, but also with the opportunity for activity and socializing and with other sources of stimulation. Considerable research has demonstrated the adverse effects of isolation and a restricted environment. Erwin and Deni (1979) described abnormal behavior commonly found in caged nonhuman primates, such as bizarre postures, self-mutilation, pacing, appetite disorders, and depression. Self-mutilation is often a result of boredom, inactivity, excessive self-grooming, and self-directed aggression, all apparently related to a lack of exercise and social isolation. Harlow and his colleagues (1959, 1962) extensively demonstrated that young monkeys reared in isolation with only mechanical mothers fail to adjust socially when mature, and fail to breed successfully. Even monkeys that remain with their mothers do not develop a full social repertoire unless, as juveniles, they are allowed to play with their peers.

Opportunity for Peer Play and Group Exercise

The design of behavioral studies at the Health Protection Branch requires that consistent breeding and rearing conditions be maintained. Female monkeys are singly-housed and time-mated to allow collection of pregnancy data. Infants are separated from their mothers at birth and reared according to standard practice (Willes et al., 1977). In this way each infant receives similar nutrition, housing and nurturing which may not occur, for example, if this is the mother's first infant or the mother does not have adequate milk. Early separation also facilitates some experimental procedures such as early dosing and behavioral assessment (Rice, 1979). Although hand-rearing of infants assures some consistency of conditions, it also presents complications, such as providing appropriate diet, housing, control of infectious diseases (Alford, 1985), and socialization.

The recognition that isolation-reared monkeys may be less curious and more fearful than their socially-reared counterparts prompted a number of strategies to circumvent these undesirable effects. Simple forms of enrichment such as placing infants within sight and sound of each other are incorporated into experimental design. Allowing monkeys to exercise in groups provides them with a change of scenery, greater freedom of movement, and the opportunity for social contact which increases activity. When infants are as young as 3 weeks of age, exercise groups are formed, primarily on the basis of treatment regimen; the groups include infants of both sexes. Younger monkeys are exercised in Plexiglas playpens furnished with shelves and grip holes to allow climbing and perching. For the very young, diapers are available for the security of something soft to cling to, at least until they learn to cling to each other. Playpens are typically used until the infants are about six months old when they are introduced to the stainless-steel runs.

Scheduling Exercise Groups: Coordinating with Experimental Design

Experimental requirements like dosing and treatment regimen are essential considerations in designing an exercise or socialization program. Treatment regimen often dictates which animals can exercise together and even in what order they are exercised. For example, a monkey exposed to lead will excrete lead in the feces (Truelove, Gilbert & Rice, 1985), which requires that only monkeys of similar treatment groups be put together. For the same reason exercise schedules are designed to ensure that untreated groups go out before treated groups or that exercise runs or play-rooms are washed before untreated monkeys are put out to exercise. Behavioral testing schedules are taken into consideration to ensure that monkeys have an opportunity for socialization after testing.

Groups are carefully constructed and monitored. Schedules are created to allow as many monkeys as possible to be exercised as often as possible, while respecting study requirements. Groups are developed on the basis of sex, age, treatment regimen, compatibility and group size. This is more easily accomplished if socialization starts early. Monkeys need time to establish a hierarchy, and some aggression is to be expected during initial group formation. Once groups are established, as few changes as possible are made, because close bonds of grooming and play usually form.

The order in which the monkeys are put out for exercise can minimize fighting and is important for group stability. The optimum strategy appears to be working from the bottom up, that is, the least dominant monkeys out first and the most dominant last, always in the same order. Care is taken that one animal does not remain away from a group too long (e.g. due to injury or illness) as this may change the hierarchy, causing aggression. In one case, when the dominant female was removed from her group for only two days, then reintroduced, she began to fight excessively with a second female who had established dominance during her absence. Peace was finally restored by removing the second female, allowing the first female to regain dominance, then reintroducing the second female.

As the monkeys mature, the sexes are separated to prevent breeding. This has detrimental effects on some groups. Close bonds are broken, and intense fighting occurs when monkeys from other groups are introduced into existing groups. New bonds prove much harder to develop. Eventually fighting subsides, but often very little social interaction, grooming or play develops in these groups. In contrast, monkeys in groups that require little or no change can be seen running after each other in play or huddling in a group, grooming each other.

Exercise Runs

To exercise and socialize the monkeys requires appropriate equipment. Space and group size are two important and interrelated factors to consider. Group size is dependent on space but available space rapidly decreases as the monkeys grow larger. Groups of adult males or females range from 2-3 monkeys. Juvenile and subadult groups range from 4-8 monkeys. Large exercise runs of stainless steel are designed to be practical and efficient while providing the maximum area for exercise. These runs include swings, multi-leveled perches, feed buckets, water-bottle holders, and transfer-box holders (Figure 1). Because of their dimensions and the fact that they have wheels, the runs can easily be put through the cage

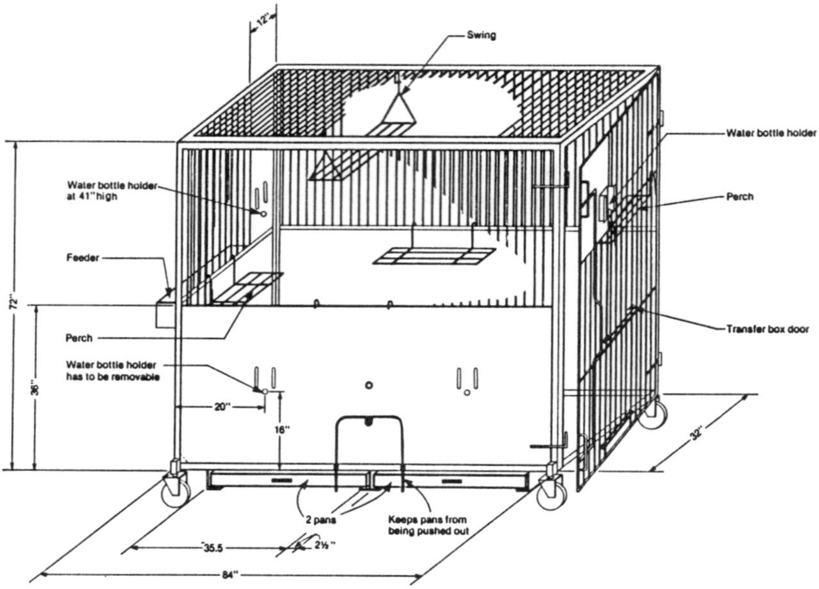

Figure 1. Easily-cleaned stainless-steel cage designed to be used for either exercise or gang housing of a group of monkeys.

Figure 2. Swimming pool: Stainless-steel frame with 3/8" Plexiglas sides (121 x 80 x 78 cm). Spigot on bottom allows drainage but can be opened by inquisitive monkeys. The monkey in the water is diving for raisins.

washer for cleaning, and they are readily adaptable for gang housing, or general exercising. For the safety of the staff and the monkeys, older animals are trained to enter a portable transfer-box that can be attached to either their home cage or the exercise cage. If there are any training problems, a reward of a prefered piece of food is used to encourage the monkey into the transfer box.

Playrooms

A fairly new feature of the exercise program is the use of playrooms, which offer greater exercise areas and accommodate larger groups. Colony rooms are modified by addition of perches, shelves and swings; with the addition of a jungle gym the room becomes very versatile. All additions are stainless steel to prevent their destruction by the monkeys and to allow proper sanitation. The increased space reduces intergroup conflict and is used to introduce a new member to a group. The extra room and play objects within the room allow monkeys to get away from an attacking monkey. During the period of introduction, the extra room helps to reduce injuries and facilitate the adjustment of the entire group.

A portable swimming pool in the playroom. A Plexiglas pool on wheels is used in the playrooms (Figure 2). The pool has been a tremendous success with the younger monkeys, who adapt easily to water and are instinctively good swimmers. The pool encourages play, increases activity and helps clean the monkeys' fur. Monkeys who are not introduced to the pool until they are older take longer to adapt to swimming. Swimming is encouraged by placing raisins at the bottom of the pool. They will swim under water with their eyes open looking for the raisins and playing with each other. Pool use is coordinated with room cleaning schedules to minimize extra work of mopping up spilled water.

A Plexiglas wall and door separate the playroom and anteroom. The anteroom prevents monkeys from escaping down the halls when the room is entered. The Plexiglas wall also made it possible to install a door for a transfer box, so that larger or more aggressive animals can benefit from the playroom. The same procedure that is used to transfer monkeys to and from the exercise cages is also used for transfers to the room. Monkeys first learn to transfer between home and exercise cages before they are introduced to the playrooms.

Exercise, social contact and swimming provide the monkey with diversions from the laboratory routine. However, not all studies lend themselves to such elaborate programs. The design of any enrichment program must take into account the anticipated impact on research. For example, regular group socialization leads to a dominance hierarchy, and this may influence hormone levels or produce undesirable stress in certain monkeys. It is one of the obligations of the researcher to weigh the advantages and the disadvantages of captivity as well as enrichment programs when using monkeys in research.

PLAY OBJECTS

"If we can't mimic the animal's natural environment and we insist on 'civilizing' the animal, why not give it some 'civilized' activities?" (Spinelli & Markowitz, 1985). Monkeys are constantly active; when not performing tasks for the basic necessities, they like to play and to investigate anything within reach (Bernstein, 1984). These proclivities have many advantages in nature, which provides a vast resource of things to play with. However, in the very restricted environment of a research laboratory, there are few or no stimulating objects unless explicitly provided by the caretakers. One of the advantages of early introduction of novel play objects is the promotion of more adaptive behavior (Mineka, Gunnar & Champoux, 1986; Westergaard & Lindquist, 1987).

Some facilities have developed sophisticated devices that challenge monkeys to learn skills to operate food-vending machines and thus attain some control over their daily schedule (Markowitz, 1979; Bloomstrand, Riddle, Alford, & Maple, 1986). Some of these monkeys became so adept that they were able to accumulate a surplus of food. This would not suit many laboratories where food and water consumption are closely monitored, especially for animals on test.

Finding indestructible play objects was a trial-and-error matter. Golf balls were used initially and fairly successfully. Although they were good for the juvenile monkeys and the perfect size and weight to carry around, we feared that they might get stuck in the mouths or throats of adults (although this never actually happened). Also, a golf ball could eventually be cracked or chewed open, exposing the inner material, which in some cases may be hazardous. Another drawback to the golf ball was that it was too small to leave in most cages permanently as it could be easily thrown out between the bars. As monkeys enjoy chewing on things, rawhide dog bones were tried. These satisfied the monkeys' urge to chew but became unsanitary very quickly and were difficult to clean. Another toy that was tried, with the idea of giving the young monkeys some sort of visual stimulation, was the Fisher-Price Butterfly Ball. This is a large clear plastic ball for toddlers that, when rolled, spins around the colourful butterfly enclosed within it. Initially it sparked great interest, but somehow this large ball (originally thought too big to carry) was dropped from a height of about 6 ft. and broke.

Most successful is a hard nylon ball designed for dogs, and available in 3 sizes, ("Nylaball" by Nylabone Corp., Box 27, Neptune, N.J., 07753). The largest or "wolf" size Nylaball is now a permanent fixture of all monkey cages, being large enough that the monkeys can't throw it out between the bars. The ball remains in the cage as it goes through the cage washer - it can also be autoclaved. The "wolf" size is given to the larger monkeys with no fear of it getting stuck in their mouths. The smallest or "regular" size is quite close to the size of a golf ball and much easier for the smaller monkeys to use - and they do use them! Nylaballs get carried, chewed on, thrown, dropped, rolled and don't appear any the worse for the wear. Certain animals have been seen poised in positions where they would normally bite themselves but instead they bite the Nylaball. For variety, which is considered to be the spice of life, we have experimented with rubber toys specifically made for monkeys ("Kong Toys" by The Kong Company, 300 S. Lamar Ct, Lakewood, CO, 80226). These toys come in various sizes and in red or black rubber. Raisins or other food can be place inside of them and provide the monkey with a lively new challenge.

A very easy and inexpensive item that the monkeys also enjoy is stainless-steel chain, in segments of about one foot. They provide the monkeys with something to drag and swing around, play tug-o-war with and simply chew on. Stainless-steel cat feeding bowls have also proved very successful. Monkeys invent endless games with bowls and chains including walking around with the bowl on their head. We have never had any injuries as a result of these objects.

A little imagination and variety are important keys to finding successful play objects. Due consideration must be given to possible disadvantages of a play object. For example, can it increase the possibility of disease, change absorption of the agent under study or otherwise interact with the study design? The toxicologist must consider any potential toxic substances that may be associated with an object. For example, woodchip litter fortified with food has been successfully used to reduce intergroup conflict and help maintain a more hygienic environment (Chamove, Anderson, Morgan-Jones & Jones, 1982), but monkeys will consume these chips, and they can contain pesticide residues that may have an undesirable impact on a toxicology study. Monkeys are notorious pickers and chewers, so consideration must be given to wall covering, particularly in older facilities that may have walls painted with lead-based paints.

CONCLUSIONS

Environmental enrichment has been an integral aspect of experimental design since the inception of the Primate Behavioral Toxicology Program at the Canadian Health Protection Branch. Over the past decade we have gradually expanded and improved our program to include large exercise cages, exercise rooms, swimming pools and a variety of play objects. From a scientific perspective the primary benefits are a better animal model for assessing the behavioral effects of toxic compounds to which humans are exposed and more reliable and reproducible data.

Although the enrichment program complicates experimental design and costs money, the benefits from all perspectives clearly exceed the costs.

These efforts were undertaken to provide an environment in which the laboratory monkey could grow and develop as normally as possible given laboratory restrictions. We believe this program has provided for the physiological wellbeing of the monkey which is in every way compatible with the research objectives. However, the merits of any enrichment program must be compared to any anticipated impact on the outcome of the research. The researcher using monkeys in an experimental design must consider the influence captivity has on the monkey and in turn the influences enrichment may have on both the monkey and the experimental results. It is the researcher's obligation first to insure that monkeys are truly necessary for a particular study and second to insure a well-balanced study design that includes environmental enrichment.

— 17 —
How Not to Enlarge a Stable Group of Stumptailed Macaques *(Macaca arctoides)*

Ramon J. Rhine and Roberta L. Cox

INTRODUCTION

A major problem in the management of captive macaque colonies is the need to constitute and reconstitute groups for research or management purposes. Available experience shows that for some macaque species, notably *Macaca nemestrina* (Dazey & Erwin, 1976; Erwin, 1977, 1979) and *M. mulatta* (Bernstein, 1964; Bernstein & Gordon, 1974; Southwick, 1967; Southwick, Siddigi, Farooqui, & Pal, 1974), the introduction of unfamiliar individuals into established groups triggers xenophobic aggression, sometimes with fatal results. Severe aggression is especially likely with sequential introduction of individuals, but introduction of several animals at once is also risky.

In 1972, two groups of stumptailed macaques (*M. arctoides*) were combined after removing the adult males from one of the groups. There was an initial conflict between females of the two groups, but in due time roles were sorted out. The females of both groups remained healthy and subsequently reproduced.

Eleven years later, all but one of these females were still together and quite old. It was decided to study the social life of the aged females, for which a comparison group of younger females was needed. Encouraged by the successful 1972 experience, four young adult females were designated for inclusion into the group. The 1972 experience suggested that the four would be well integrated in a year or so, after which comparison with older females in the same group and environment could be made.

This paper is the story of what happened following the introduction of the young females. Their introduction was closely monitored to determine

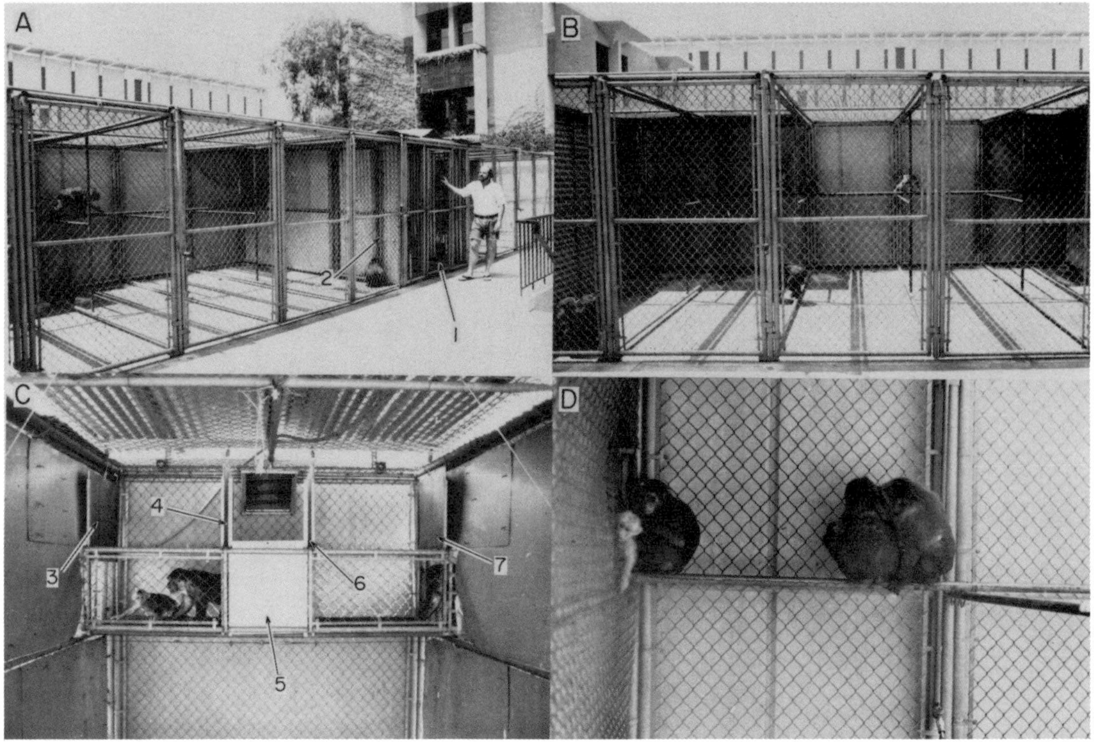

Figure 1. Panel A: Door (1) to a safety area, and door (2) from safety area to an enclosure. Panel (B): One of two connected enclosures, with an animal on rear walking rails. Panel C: Inside the safety area, with five numbered guillotine doors of the tunnel connecting the two enclosures. Panel D: Rear rails of enclosure at entrance to tunnel. After Bunyak et al., 1982.

how they were treated, how their behavior differed from that of others, if at all, and what social behavior was prominent after introduction, or changed in frequency over time.

METHOD

Animals and Environment

The monkeys were housed in a pair of enclosures connected by a tunnel running through the back of a safety area (Figure 1). The guillotine doors shown in part C of Figure 1 could be closed to seal off one enclosure from the other. Each enclosure was 3.7 x 6.1 x 2.1 m high. Further details of colony housing appear in Rhine, Hopper, Harvey, and Bunyak (1981) and in Bunyak, Harvey, Rhine and Wilson (1982). The enclosures had housed two separate groups until 1972 when a single group was established by opening the connecting doors. Members of the merged groups were then able to move freely between the enclosures, giving them a chance to leave the space and immediate sight of others with whom they might have come into conflict.

At the time of the present study, the adult *residents* of the colony were seven females and four males, described previously (e.g., Harvey & Rhine, 1983; Hendy-Neely & Rhine, 1977; Rhine & Hendy-Neely, 1978a, 1978b). They had lived together in a single social group for eleven years. There was also one subadult female who had lived with the group all her life, and four immatures. The four subjects to be introduced into this colony were one subadult and three young adult females whose ages in months were 42, 56, 68, and 69. They had already been together for several months before the present study began, but they were *strangers* to the colony residents. Throughout this paper, for purposes of analysis and exposition, the subadults are lumped together with the adults of their respective groups, giving twelve adult resident subjects and four adult strangers.

Procedure

Introduction of strangers into the enclosures: The four female strangers were introduced to the colony residents on February 7, 1983. In the morning, the colony residents were herded into the left enclosure, and the guillotine doors to the tunnel were closed (Figure 1). The strangers were all placed in the right enclosure at 1000 hours. An attempt was made to let the dominant resident male of the colony join them, but he could not be enticed to their side. At 1500 hours, the doors separating the two enclosures were opened and observations began.

Activities recorded: Activities recorded during this study were defined by Rhine and Kronenwetter (1972), and included *eat, drink, flee, fear-grimace, cringe-crouch, fight, charge, chase, threat, grunt-bark, fear-scream, present, mount, grooms another,* and *is groomed by another.* Additional activities whose definitions may be less self-evident are as follows:

Travel : Any change in body location of a full body length or more.

Auto: Self-directed activity including scratching, self-grooming, etc.

Manipulate: Carrying, pushing, or pulling objects with the hands, mouth, or feet. (Objects include mostly food pellets, but also bits of paper, etc.)

Rest: No movement for an entire time unit (30 sec.) with the eyes closed or partially closed.

Passive: Remaining in one location throughout a time unit without engaging in any other activity and with the eyes open.

Social investigate: Examinating the body orifices or wounds of another.

Huddle: Extensive body contact with one individual's weight partially supported by another. One or both may hold the other and one animal's head may rest on the back of the other's neck.

Contact: Any contact not otherwise indicated.

Still : Remaining inactive within one ft of another for at least 5 sec without touching.

Sampling procedures: Sixteen observers worked simultaneously, one to each of the 12 adult residents and 4 strangers. In addition, RJR spoke a running commentary into a tape recorder throughout all observations, attempting to characterize the overall behavior of the group and to record noteworthy activities of specific individuals. The four most experienced of the sixteen observers were assigned to the four strangers as their focal animals. These four observers, like RJR, recognized all animals individually, and so were able to specify for social behavior the identity of the animals with whom their focal animal interacted. The observers of the 12 resident monkeys indicated their focal animal's activity, but could not identify interactants. Except for *is groomed,* records of social behavior indicated what the focal animal was doing to others.

Data were collected using the one-zero sampling method with 30-sec observation periods (Martin & Bateson, 1986; Rhine & Ender, 1983; Rhine & Linville, 1980). A checksheet listed activities as columns and 30-sec intervals as rows. For each 30-sec interval, a tally mark was entered in the appropriate cell for any activity occurring at least once. (Thus, tallies could be marked for more than one activity during a 30-sec interval.) If an activity did not occur at all during the interval, that cell of the checksheet was left blank. *Frequency* in this paper means the total number of cells bearing a tally mark: the number of 30-sec intervals during which an activity occurred at least once. The *one-zero percentage* of an activity is its frequency divided by the total number of recorded 30-sec intervals. If for any reason a 30-sec observation could not be recorded, the corresponding interval on the data sheet was crossed out (and did not enter into the denominator for percentage calculations), but this rarely happened. The mean number of recorded 30-sec intervals per focal animal was 311.6 out of a maximum possible of 320.

One method of recording social behavior was used with the adult residents and a different method with the strangers. Since observers of residents did not recognize all group members, they recorded the occurrence of social activities but not information about the identity or number of interactants. Hence, the meaning of their one-zero frequencies and percentages was the same for social behavior as for individual behavior. For the social behavior of strangers, on the other hand, the more experienced observers recorded the identity of each animal who interacted with the focal animal (a method used previously with the study group -- Rhine, 1972, 1973). These data allow separate determination of the frequency and percentage of the focal animal's interactions with each of the fifteen other adults in the group.

The sixteen observers and RJR made half of their observations on *Day One*, the day the strangers were introduced into the group, and half on *Day Two*. RJR made additional observations during these two days as well as periodic observations over the subsequent 60 days.

RESULTS

Focal Animal Observations

Residents' reactions to strangers: A set of analyses was done in on the activity percentage scores (defined above) of residents. A two-factor analysis of variance for each activity had the resident's sex as one factor and Day One vs Day Two as the other.

Statistically-significant differences were mostly between Day One and Day Two (Table 1). Grooming increased dramatically on the second day, while several agonistic activities decreased. The amount of threatening done by residents substantially decreased. It is especially noteworthy that fleeing by residents decreased by several orders of magnitude from Day One to Day Two. Consistent with decreased fleeing, there were changes in two individual activities. There was more self-attention and less moving about the cage from Day One to Two.

Table 1
Mean Percentage Scores for Statistically-Significant Differences in the Behavior of Residents Between the Day When Strangers Were Introduced into the Group (Day 1) and the Next Day (Day 2)[a]

Behavior Category	Day One	Day Two	p
Affiliative			
Groom	.3	20.2	< .001
Is groomed	.1	10.7	< .01
Sexual			
Present	1.1	2.6	< .025
Agonistic			
Flee	4.1	0.6	.01
Fight	9.1	2.6	< .025
Charge	2.2	.2	< .05
Chase	1.6	.2	.05
Threat	15.6	.7	.001
Individual			
Auto	7.0	23.4	.001
Travel	49.7	36.2	< .01

[a]Males socially investigated and mounted more often than females. There were two interactions: females were more often still on Day One and males on Day Two, and female travel decreased from Day One to Day Two. These effects were significant at the .05 level.
N's: 12 residents (8 females, 4 males), 4 female strangers (all adults).

There were two sex differences. Social investigation, which most often was examination of the female genital area, was more a male than a female activity ($p < .05$). Males also did a considerable amount of mounting, whereas females did none. On Day One, male mounting was recorded in a mean of 14% of the 30-sec intervals; this dropped to only 1% on Day Two. Male mounting and female fleeing are central to a major dynamic of group events that is discussed below.

Differences between residents and strangers: A second set of two-factor analyses on activity percentage scores again had days as one factor; the second factor was residents vs strangers. In the previous

Table 2
Mean Percentage Scores for Statistically-Significant Main Effects in the Behavior Exhibited by Residents and Strangers[a]

Behavior Category	Group Status: Residents	Strangers	*p*
Affiliative			
Groom	10.2	2.6	<.05
Huddle	9.0	60.0	<.001
Agonistic			
Fight	5.9	.1	<.05
Threat	8.1	.3	<.025
Cringe-crouch	.3	39.2	<.001
Fear-grimace	1.6	2.7	<.001
Individual			
Travel	43.0	8.7	<.001
Eat	11.7	1.7	.01

[a] Statistical tests were not feasible for mount, charge, and chase because they were never performed by the strangers. The respective mean percentages for residents are 2.6, 1.2, and .9.

Table 3

Mean Percentage Scores for Statistically-Significant Interactions in Analyses of Variance of Table 2[a]

Behavior Category	Day One		Day Two		
	Residents	Strangers	Residents	Strangers	*p*
Groom	.3	.6	20.2	4.5	<.025
Contact	12.5	12.4	14.4	35.5	<.025
Present	1.0	4.3	2.6	.8	<.025
Threaten	15.6	.2	.7	.5	<.025
Cringe-Crouch	.1	55.8	.3	22.7	<.001
Fear-Grimace	.4	3.9	.7	1.6	<.01

[a]Day One vs Day Two was one factor and Residents vs Strangers was the second.

analyses, differences between days were for the 12 residents only, but now the 4 strangers' scores were included as well. Residents' scores in these analyses were the same as in the previous analyses, so differences between days would be expected to be similar in the two sets of analyses, and they were.

The statistically- significant differences between residents and strangers are shown in Table 2. Residents groomed more often than strangers, while strangers more often huddled together. Whereas residents were responsible for almost all of the aggressive behavior, strangers exceeded residents in fear-grimacing and especially in cringe-crouching. Consistent with differences in cringe-crouching, which is an immobile posture, residents did far more traveling about the enclosures and ate much more than strangers.

Table 3 gives the mean percentage scores for the six significant interactions. In four cases there was a significant main effect for residents vs strangers, so interactions in these cases are refinements of the information in Table 2. The reasons for the interactions are self-evident from the mean scores. For example, grooming was primarily by residents on Day Two; the significant interaction in cringe-crouching was due to a big drop in cringe-crouching by strangers from Day One to Day Two. Although main effects for presenting and contacting were not significant, two of the interactions were due primarily to increased contact and decreased presenting by strangers on Day Two.

Social interactions of strangers: The previous analyses concerned social activities without regard to who interacted with whom. Since the strangers' records included interactants' identities, it was possible to determine the mean percentage of observation intervals during which the four strangers interacted with resident males, resident females, or with each other ("other strangers"). Analyses of variance were done using interaction percentages for each social activity except fear-grimace and cringe-crouch, which are not directional enough to specify with confidence the particular animal being responded to. (Transcription of RJR's tape-recorded commentary indicate that these activities were performed primarily in response to female residents.) The analyses had days as one factor and strangers' responses to resident males, resident females, and other strangers as a second factor.

Statistically-significant results from these analyses are shown in Table 4. Significant differences in hostile activities, prominent in previous tables, are missing because the only focal animals here are strangers, from whom hostile acts were rarely recorded. Differences in behavior directed toward others were in three degrees of social closeness, increasing from still, to contact, to huddle (Rhine, 1973).

Table 4

Mean Percentage of Strangers' Interactions with Residents and Among Themselves on Days One and Two[a]

Behavior Category	Day One	Day Two	p	Interactants: Resident Female	Resident Male	Other Strangers	p
Still	.9	1.7	< .05	.8	1.2	2.5	ns
Contact	1.0	3.9	.001	.5	1.1	7.8	< .001[b]
Huddle	7.4	11.9	< .05	0.0	0.0	38.7	< .01[c]

[a]Lists only categories with one or more statistically-significant effects.
[b]There was also a significant interaction ($p<.001$): contact was primarily between strangers on Day Two.
[c]ANOVA was not appropriate with two means of zero; however, even with an *n* of only four strangers, a mean percentage of 38.7 is still significantly different from zero ($t = 6.10$, $p<.01$).

Table 5

Adult Composition of Two Groups Before They Were Merged in 1972

Group 1.5 Male	Group 1.5 Female	Group 2.5 Male	Group 2.5 Female
Abe	Maria	Yuri[a]	Joan
Ute	Heather	Karl[a]	Thelma
			Doris[a]
			Gail[b]
			Lois[b]

[a] Permanently removed prior to combining the groups.
[b] Had previously lived with the adults of Group 1.5.

Huddling is the nearest thing in adults to ventral-ventral cradling between mothers and their infants (Rhine & Hendy-Neely, 1978b). Huddling occurred very often between strangers, but never between strangers and residents. The mean percentage of all three social behavior categories increased from Day One to Day Two. Contact and huddle were primarily between strangers on Day Two. Being close to another without contact did not occur significantly more frequently with other strangers than with residents.

Overall Group Observations

Transcripts of the tape-recorded observations of the group as a whole revealed several facets of group dynamics not reflected or poorly reflected in the checksheet data. Four points stand out.

Timing: As soon as the tunnel between the enclosures was opened, several of the residents entered the strangers' side and chased them. Aside from initial fleeing and submissive gestures, the strangers offered almost no defense to attacks. They sat hunched over or huddled in corners, and allowed the residents to abuse them.

Before half an hour had passed, the level of aggression had noticeably subsided. Thereafter, aggression came and went in erratic bursts, depending upon how aroused the resident group became. The transcript contains numerous comments such as, "Everything seems to be remarkably quiet" or "It has been pretty quiet during the entire hour," followed by descriptions of further flare-ups. The canines of the resident adult males had been removed years previously to control slash and deep puncture wounds. At the end of the first day, the closing comment of the transcript indicates that the strangers "are in pretty decent shape, and they all look lively." There were no severe wounds, though all strangers had abrasions and some blood had flowed.

During Day Two, affiliative interactions between residents and strangers notably increased. The group was quieter. Although the strangers spent a great deal of time among themselves, as time passed they moved more and more on their own. Residents gave considerable time to picking over scabs on the strangers' wounds, which was a main reason why their grooming scores were elevated on Day Two. There were affiliative touching interactions between residents and strangers, and overall the strangers seemed, after only one day, to be much better accepted by the residents. Nevertheless, flare-ups still occurred, involving further attacks on the already-weakened strangers. At the end of Day Two, it appeared that the worst was over and that the strangers would become integrated into the group.

Resident female aggressors and male protectors: From the start, the primary aggressors were the resident adult females, led by Maria, the matriarch of the dominant matriline. All eight resident females attacked the strangers. Sometimes two or three simultaneously attacked a single unresisting stranger.

In contrast, the two most dominant males were less disposed toward aggression than toward repeated full sexual mounts. These two males were unequivocally dominant over the other two adult males and all females in the group. The alpha male, Sam, occasionally chased the second male, Paul, away from a stranger and then mounted her, but for the most part, each was able to force his sexual attentions upon the newcomers without interference from the other. These males seemed intent upon an orgy of repeated sexual mounting and thrusting, which they must have performed dozens of times during Day One, though much less often thereafter.

During the height of aggression by female residents, Sam and Paul protected the strangers. One of many relevant descriptions from the transcript is paraphrased as follows:

> Paul left a stranger who is now getting it from two residents; Sam came over and stood next to the stranger, and the two residents backed off; as soon as he left, the two started bothering her again.

Similar descriptions indicate that time and again Sam or Paul chased or threatened away hostile resident females. The strangers learned to move toward the males to avoid female aggression, even though repeated active mounting of their bruised bodies must have been painful and tiring. The coming and going of the males seemed to contribute to the sporadic nature of flare-ups.

Arousal: The worst flare-ups occurred when many residents became aroused at once, sometimes for reasons undiscerned by the observer. A scream or the chasing of a resident by one of the dominant males would send a wave of arousal through the group, and the outlet would be threats and attacks against the helpless strangers. Stimuli which earlier elicited arousal did so less often on Day Two.

One important arousing stimulus was any sign of greater initiative by the strangers. To paraphrase the transcript, the strangers were often okay until they started moving, which aroused the whole group, including Sam and Paul. For example, when one of the strangers started to move out of a corner, a low-ranking female resident saw her and screamed, and the stranger went right back into the corner again. Also, if a stranger looked up, it might start an arousing action from one or more residents.

Immatures and low-ranking adult males: Resident immatures and the two low-ranking resident adult males were outside the group action during much of the first day. Thus, the transcript reads, "Ned [subordinate adult male] has hardly been in the action at all," and "Ivan [the other subordinate adult male] is joining the threatening, but not much." Occasionally, a subordinate male would be threatened or chased by another resident, especially during times of group arousal. During Day Two, the dominant males were less sexually active and less supportive of the strangers, possibly because there was less overall arousal in the group and less aggression by resident females. With the dominant males more in the background, some aggressive interactions and sexual contacts were attempted by the subordinate males.

Withdrawing the Strangers

By the end of the second day, the group had quieted down and the introduction of the strangers appeared to be successful. In fact, it ended in failure. On Day Three, one of the strangers was so obviously in a weakened physical condition that she had to be removed from the group. Subsequently, the remaining three seemed increasingly integrated into the group, though still clearly less-well accepted than residents. On Day Eight, a second female became noticeably very weak, and she too had to be removed.

The remaining two were not fully integrated at that time, but they seemed to have made good progress. As the days passed, they continued to spend much of their time either together or alone. Although the frequency and intensity of the initial hostile reactions toward them were much decreased, they continued to have hair pulled out and from time to time were bitten or chased. They ate and moved actively, but appeared depressed. Two months after being placed with the group, one of the remaining two strangers was found dead. The fourth stranger remained with the group, surviving adequately, but never becoming well- integrated.

DISCUSSION

Reactions to Introduction of Strangers

Resident females' reactions to female strangers were predictable from research summarized by Erwin (1979) in which residents' levels of threatening and other antagonistic activities were raised. The strangers' reactions were less predictable. They might have fought back, as occurred during a previous combining of groups, but in fact there was no contest from the start. On the first day, the strangers did relatively little traveling or eating, and a great deal of frightened cringe-crouching.

Perhaps the dominant males' reactions were also predictable since the presence of males is known to inhibit hostility among females (Dazey, Kuyk, Oswald, Mortensen, & Erwin, 1977; Oswald & Erwin, 1976; Sackett, Oswald, & Erwin, 1975). Elevated frequency of fleeing by residents on Day One was in response to males' chasing resident females away from the strangers. Without the males, who afforded some protection, the strangers' plight on Day One might have been much worse. As it was, following about 20 min of intense hostility, the attacks upon the strangers settled into a pattern of sporadic, fairly serious flare-ups, with the cumulative effect of depressing and weakening the strangers.

Selection for Female Xenophobia

Resident stumptail females gain an advantage from xenophobic reactions to strange females. Xenophobia is a conservative mechanism, tending to maintain the status quo of group social structure (Southwick, et al., 1974). Female xenophobia is expected to be especially strong in macaques and other species in which it is primarily males who transfer out to another troop. Intense sexual interest in the newcomers was evident in the two dominant males. Among stumptails, dominant females hold their position in the social structure in part by association with dominant males (Rhine, 1972). Thus, competition for association with males by sexually mature outsiders poses a threat to the resident females. Maria, the dominant matriarch, who was in the thick of the fighting with the strangers, on an earlier occasion had been protective of two immature yearlings, a male and a female, when they were introduced into her group. In that case, the dominant male, with whom Maria had a close relationship, showed no particular attraction to the yearlings. Interestingly, the subadult of the four strangers was the only one who was able to remain with the group. Xenophobia would be subject to selection if females who tended in that direction were more likely than others to enjoy increased and more-nearly exclusive association with males, and thereby greater success in production and rearing of offspring. Females who are able to limit competition will have more options in choosing mates who can benefit their offspring (Tilford, 1982).

How Not to Enlarge a Group

Perhaps something useful can be learned by considering a puzzling question: Why did two stumptail groups combine successfully in 1972, but not in 1983? Possible answers may be useful to those who find themselves faced with the need to combine or enlarge macaque groups. Clues to the answers can be sought in the differences between conditions of the 1972 and 1983 groups.

Number of females: In 1972, two groups, Groups 1.5 and 2.5, were merged by opening the tunnel between their separate enclosures (Figure 1). The adult composition of these groups is shown in Table 5. Prior to the merger, both adult males and one adult female were removed from Group 2.5, leaving four adults in each group, including two adult males in Group 1.5 but none in Group 2.5. The females from Group 1.5 had the occasional backing of Abe, the dominant male of Group 1.5 and also of the merged group. When the dust settled, the adult females of his group were dominant over the adult females of Group 2.5. During the transitional time, the chief conflict pitted the two females of Group 1.5 against Joan and Thelma from Group 2.5. Each pair often stood shoulder to shoulder facing and threatening their opponents. The other females of Group 2.5, Lois and Gail, had previous familiarity with the members of Group 1.5, and they mostly steered clear of the intergroup conflict.

In comparison with the 1972 situation, the four female strangers introduced in 1983 found themselves facing 12 adult residents including eight females. Against such odds, it may well have been adaptive to give up without a challenge, but the outcome was to place each of the strange females at the mercy of eight relatively-confident females who persisted in treating them as intruders.

Age and experience: All the females in 1972 were prime adults in the age range of approximately 9-12 yr (Harvey & Rhine, 1983), whereas the 1983 strangers were younger, in the range of 3.5-5.8 yr. Prime females may be socially more mature and better able to judge how assertive they dare become. In 1972, the females of neither group capitulated easily; those who eventually won were probably unable, even then, to push the others too far without a response. The younger animals in 1983 might have been better off if they had not signalled a total unwillingness to resist, even under fierce duress.

Cohesiveness: Each of the two 1972 groups was comprised of animals who had known each other since 1968. The members of each group formed a more cohesive unit than the four strangers of 1983, who had been together a much shorter time. There was a greater willingness of pairs of females in both 1972 groups to stand together against members of the opposing group. In the long run, such mutual support may help keep aggression from escalating too far. Mutual support may also help eventual losers avoid the depression that seemed to set in before the 1983 strangers' condition deteriorated to a level necessitating removal.

Enclosure experience: Another important difference between the 1972 and 1983 situations was familiarity with the enclosures. The strangers in 1983 were placed in an enclosure that for the 11 years previous had been part of the residents' home. The strangers spent only five hours alone in that (right) enclosure before the doors separating them from the residents in the left enclosure were opened, and they remained there after the doors were opened. In contrast, when the doors of the tunnel were opened in 1972, the females from Group 1.5 crossed over from the right enclosure to the left. As a result, much of the female conflict in 1972 took place in the left enclosure where the eventual losers had been living for over 1.5 years.

In a sense, the 1983 strangers were trapped in a situation worse than the predicament of a small wild group, without adult males, that inadvertently invades the core of a larger group's range. The strangers' predicament was worse because they could not flee back to their own range. Being in unfamiliar surroundings may add to the psychological misery of an inescapable plight.

A Final Word

Probably all four of these differences between the 1972 and 1983 situations contributed to the problem experienced in 1983. However, it would be dangerous to assume that taking just these differences into account would assure successful group mergers. The differences described seemed to be important factors, but something equally or even more important may have been missed. Furthermore, no two groups of macaques are identical because of individual differences among these intelligent monkeys. Severe discomfort, even mortality, is risked when one or both of the groups that are being merged have been living cohesively with little disturbance of social structure. The best advice, based on our experience with established groups of stumptails, is to combine groups, or introduce adult animals, only as a very last resort, and then with great care and assiduous monitoring.

ACKNOWLEDGEMENTS

This research was supported by intramural grants to the first author from the University of California, Riverside. We thank Doris Rhine for editorial assistance, Michael Costello for critical comments, and Todd Little for assistance with computer analyses.

— 18 —
Influence of People on the Behavior of Display Primates

John E. Fa

INTRODUCTION

Zoos and wildlife parks have made considerable efforts in recent years to improve husbandry so as to encourage the propagation of captive primates (Markowitz 1982). The study of the interrelation between behavior and the built environment, which Maple and Finlay (1986) call "applied primatology," is attracting increasing attention. Investigations into ways of improving the appeal of exhibits to the public (Rhoads & Goldsworthy 1979) are also emerging. These approaches have primarily concentrated on the environment within the display (the intrinsic environment); little, if any, attention has been given to factors outside the immediate surroundings that may influence the animals' behavior (the extrinsic environment). By display primates I mean animals that are kept, whether in conventional zoos or in semi-natural (free-ranging or enclosed) conditions, as intact social groups, or groups that promote social behavior characteristic of the species. Solitary individuals on display in zoos are not considered.

The conditions under which monkeys and apes are displayed in zoos and semi-natural environments vary widely. Variations occur in (i) absolute and relative space, (ii) quality of the intrinsic space (furnishings of enclosures, variety of biota in semi-natural habitats), (iii) distribution and availability of food, and (iv) composition and relationships of the primate group(s). Cage environments can reach moderate complexity (Hinde 1971), whereas semi-natural habitats (free-ranging or enclosed) can more nearly approximate the wild condition (typology after Drickamer 1973).

Semi-natural free-ranging bands include colonies of monkeys kept on islands for study (for example, Cayo Santiago: Rawlins & Kessler 1986) and groups that are partially provisioned and visited by humans, such as

Japanese macaques (*Macaca fuscata*) in Japan (Koyama, Fujii & Yonekama 1981), other primates in Southeast Asia, and Barbary macaques (*M. sylvanus*) in Gibraltar (Fa 1984). Groups in very large, fenced habitats, such as the Barbary macaque park enclosures in France and Germany (de Turckheim & Merz 1984), are defined as semi-natural enclosed groups.

Both zoo and park environments may fall within the same continuum of interacting social and ecological factors. There is no need to perpetuate a central axiom of captive primate research, that a "confined" or "altered" habitat works against exhibiting "whole" animals (Gledhill 1982) - animals that show the full range of their species' behavior. Although a substantial body of evidence links certain forms of captivity with serious pathological behavior (self-aggression and stereotypy, for example; see Chamove and Anderson 1981), most captive primates studied show no significant qualitative or quantitative differences from wild or non-captive groups (Bernstein 1970 1972; Drickamer 1973; Hall 1969; Hornshaw 1981; Rowell 1967). The behavioral solutions adopted by primates in captivity or in semi-natural conditions reflect the constellation of environmental and social features of their situation, and represent a greater or lesser degree of psychological fitness.

Wellbeing can depend on factors beyond the basic elements that make up the intrinsic living environment. Whether the animals are in cages, in large enclosures or provisioned, people's teasing, offering food or merely passing by provides extra variables that can modify the behavior and ultimately the reproductive potential and survival of the animals. Although the display of animals to the public is the primary objective of zoos and wildlife parks, human visitors must also be considered the major element in the animals' extrinsic environment, and taken into account in evaluating the entire array of environmental parameters important for rational maintenance of display primates. Knowing the effect of human visitors will not only allow us to better the animals' welfare, but will also enrich the interpretative potential of zoos and parks.

This paper aims to present some general ideas on the impact of humans on the behavior of display primates by referring to examples of monkeys in zoos and in semi-natural (free-ranging and enclosed) systems. I will draw particularly on two studies my associates and I conducted on the influence of visitors on (i) a group of green monkeys (*Cercopithecus aethiops sabaeus*) at the Mexico City Zoo and (ii) Barbary macaques in Gibraltar (Fa 1984 1986a in press).

VISITOR BEHAVIOR IN ZOOS AND SEMI-NATURAL DISPLAYS

Little interest has been shown in the behavior of visitors to zoos and wildlife parks and their interaction with the animals that are on display. Literature on the subject is scant but cursory remarks on the importance of taking the relationship between human and exhibit animal into account are sometimes made. For example, Markowitz and Spinelli (1986) comment in passing that people's throwing food and objects into the gorilla (*Gorilla gorilla*) exhibit at the San Francisco Zoo works against satisfactory display of the animals, because it encourages the gorillas to concentrate in an unattractive part of the exhibit that is devoid of natural foliage. Markowitz and Spinelli propose as a partial solution that interesting foods be placed in the trees to induce the gorillas to forage rather than beg. Although this idea follows from Neuringer's (1969) report that in some circumstances animals will work for food even when food is freely available, placing food in trees may not compensate for the attractiveness of the types of food offered by visitors, and other aspects of visitor behavior. Fa (in press) has shown that the Gibraltar Barbary macaques, which are substantially provisioned and not deprived nutritionally, still seek out the sweets and other "junk" foods offered to them by passing tourists. Segal (1987) has in fact demonstrated that when a test Barbary macaque was given choice trials with a number of different foods, a preference for the sweeter and saltier items prevailed, consistent with the inherent tendency of animals to select sweet and salty foods (Moskowitz 1971). Promoting changes in the behavior of the humans who visit the zoo may be as expedient as providing enriched primate habitats.

Hediger's (1964) principal rule for zoo management is that the animal biologist's duty "not only lies in taking appropriate measures to protect the public from the animals but also, conversely, to protect the animals from the public." Hediger attempted to characterize the kinds of people who visit zoos in order to decide the degree of attention to be placed on each by zoo staff. Quite apart from exceptionally abnormal zoo visitors (Hediger mentioned psychopaths, petty criminals, the weakminded, idiots, and the homicidal and suicidal), the general public can present three important problems for the animal exhibits. 1) People can be a source of infection, particularly for apes and monkeys, 2) unsuitable foods can cause bouts of serious indigestion and other gastro-intestinal ailments, and 3) there is a possibility of direct harm or indirect disturbance to the animals by worrying, teasing, frightening or exciting them. Murer and Dobberstein (cited by Hediger 1964) noted that after days when the number of visitors at the Basle Zoo was large, more (largely non-lethal) digestive disorders occurred than after days when visitors were fewer. Although the frequency of visitor interactions with animals will vary between zoos as a function of the establishment's size, focus on public education, vigilance, and just how animals are displayed (open exhibits or glass-fronted ones), visitor interactions are still likely to be an important occurrence in most zoos. Markowitz and Spinelli's (1986) observations of people feeding and teasing animals at the San Francisco Zoo indicates that even the wealthiest zoos are not exempt from the problem. Only monkeys and apes exhibited behind glass are protected from problems arising from direct contact with the public.

The magnitude of the problem of people's disturbing, teasing, throwing objects and feeding display animals can probably be gauged from a zoo or park's entrance statistics. Although there are no published data for this, high visitor turnover in a zoo or wildlife park should be an indicator of greater disturbance (area of the zoo and density of people around exhibits affect this relationship). The Mexico City Zoo is the most visited of the world's major zoos (International Zoo Yearbook 1987); it is in the most populated city in the world and no entry charges are levied. Twelve and a half million people are reported to visit it every year. Vigilance is virtually impossible with such a volume of people and interactions between animals and visitors are significant. At Gibraltar as an example of a semi-natural situation current visitors to the macaque troop that is accessible to tourists (Queen's Gate) may be up to half a million a year. This is a conservative estimate based on unpublished data I obtained during summer 1987; it represents a quarter of the two million tourists who visit Gibraltar annually (Fa 1987). Although there are notices posted prohibiting visitors from feeding the monkeys, there is no vigilance, and interactions between animals and people go unchecked.

It is reasonable to expect a positive correlation between the number of visitors to a zoo and the incidence of human-animal interactions. Data obtained for the green monkey exhibit at the Mexico City Zoo and for the Gibraltar macaques (Figure 1) confirm this and show that the probability of interactions increases, perhaps exponentially, as the number of visitors to the monkeys increases. Despite the lower number of visitors, there are more interactions between people and monkeys at Gibraltar, owing to the greater opportunities for direct contact that is typical of semi-natural displays. In both situations, even though variability in the data may indicate heterogeneity of visitor attitudes, the proportion of people violating the anti-feeding rules increases as the number of visitors rises, and not necessarily linearly; social facilitation may become important when visitors reach a certain number. When there are large crowds, it is probably easier for visitors to feed or otherwise interact with the monkeys without being noticed. This is seen better in the Gibraltar data, presumably because of the unusual situation, which permits people to crowd around the macaques (Figure 2).

In the Mexico City Zoo, and at Gibraltar, interactions between people and display primates show a temporal pattern. Interactions are not isolated events but occur frequently throughout the times when animals are exposed to the public. In both places, feeding interactions were recorded at all hours of the day, peaking at the main visiting times (Figure 3). A mean of 2.7 +/- 1.2 visitors per hour interacted with the monkeys in the Mexico City Zoo while there were 57.4 +/- 46.3 per hour at Gibraltar. Throughout the day there was a positive correlation between number of visitors and interactions per hour in both study locales (Spearman rank-order correlation, Mexico

City *rho*=0.76, P<0.05; Gibraltar *rho*=0.68, P<0.05). Recorded quantities of food given ranged from a few grams to almost a kg at Mexico City, and from a few grams up to 3 kgs in Gibraltar (unpublished data). Types of interactions between people and monkeys varied from direct contact in Gibralter (where monkeys are enticed by the offer of food to sit on tourists' shoulders for photographs), to the provocation of novel reactions by throwing children's toys into the monkeys' enclosure at the Mexico City Zoo. Observations at Mexico City indicated that, of 146 recorded incidents of visitors' providing tangible stimuli, animals failed to react in only 4% of the cases. Reactions were mostly to food (17.1 +/- 8.6 events per day) but also to objects (3.3 +/- 0.9 events per day). At Gibraltar, 98% of the recorded interactions were feeding incidents and monkeys' responses were linked to the type of food offered. There was never any attempt to take staple foods such as bread, but the monkeys would violently snatch sweets from tourists.

THE IMPACT OF EXTRINSIC ON INTRINSIC ENVIRONMENT.

Any attempt to describe the effect of extrinsic factors on the intrinsic environment of an exhibit necessarily forces the investigator to describe the latter's strongly interrelated network of social and ecological variables. Because no two captive or semi-natural environments are the same, generalizations are not often possible. In fact, captive environments can be as heterogeneous as field situations and can represent "ecological variants" for any particular species (after Rowell 1972). However, we can approximate the degree to which the exhibit's environment, both intrinsic and extrinsic, influences the behavior and ecology of the display species if we can tease apart ecological and social variables contained within a display or affecting it from outside. Such an exercise is not far from attempts usually made by ecologists in the field to understand the surrounding variables of an animal population. In theory this should be easier for display situations.

In captive environments, ecological variables consist of the space available, elements within the space, the nature and distribution of food, and gross environmental features such as microclimate. Social variables consist of the patterns of social interaction within the group, which are largely determined by group composition and history (length of time together, etc.). These variables may interact. A basic element influencing the expression of adequate psychological standing, reproductive potential, and even survival of the animals in display environments is the absolute space available. This ultimately affects inter-animal contact, albeit through the mediation of patterns of resource distribution and abundance. Even in semi-natural free-ranging conditions, resource distribution and abundance will necessarilly influence the absolute space available to the animals. Thus, if overcrowding forces animals to violate one another's individual distance (Hediger 1964), behavioral (Calhoun 1962) and physiological (Christian 1963; Clarke 1977) changes may occur. In turn, these features can modify aspects of the animals' reproductive capacity and survival. People's feeding the animals affects food distribution and abundance, and impacts individuals' use of space. There is the added problem of monkeys' eating unsuitable foods and dedicating attention and time to obtaining food (usually highly palatable and rich in calories) from people, activities that directly affect the animals' activity budgets, reproduction and survival. People themselves are semi-novel stimuli, and as such, influence exhibit animals. This network of interactions is a tentative interpretation that can serve to guide research on the effect of different extrinsic stimuli. The level of disturbance (whether auditory, visual or gustatory) that is permissible for an exhibit will only emerge from examining empirically the relation between intrinsic and extrinsic environmental components. Abnormal behavior typical of apes kept in inadequate cages, such as spitting and throwing dung, could be explained by the relation between impoverished intrinsic environments and attractive extrinsic ones.

Even though the suggested interactions are inferred from a strong empirical base, there is still disagreement as to how universal these principles are and which factors contribute more to a group's wellbeing. For instance, Hornshaw (1981) pointed out that, contrary to expectation, observed association patterns and social relationships within primate groups are not always symmetrical nor transitive.

Zoos

It is obvious that the absolute amount of available space per animal is less in zoos than in semi-natural free-ranging or enclosed environments and this in itself has implications for the kind of behavior typical in each. The problem is a function not only of the forced proximity of the animals and their greater visibility but, as mentioned above, of the feeding procedures (distribution and abundance). Behavior of zoo primates kept in cages is thought to follow three main trends: 1) interindividual spaces are expected to be perforce smaller than in free-ranging animals; 2) tension is expected to be greater because food resources are localized or concentrated, not widely dispersed (Chalmers 1972; Rowell 1972; Southwick Beg & Siddiqi 1965) and 3) aggression in general is also reported to increase as social density rises but mediated by the composition of the group. However, not all conditions of captivity promote a response in the predicted directions.

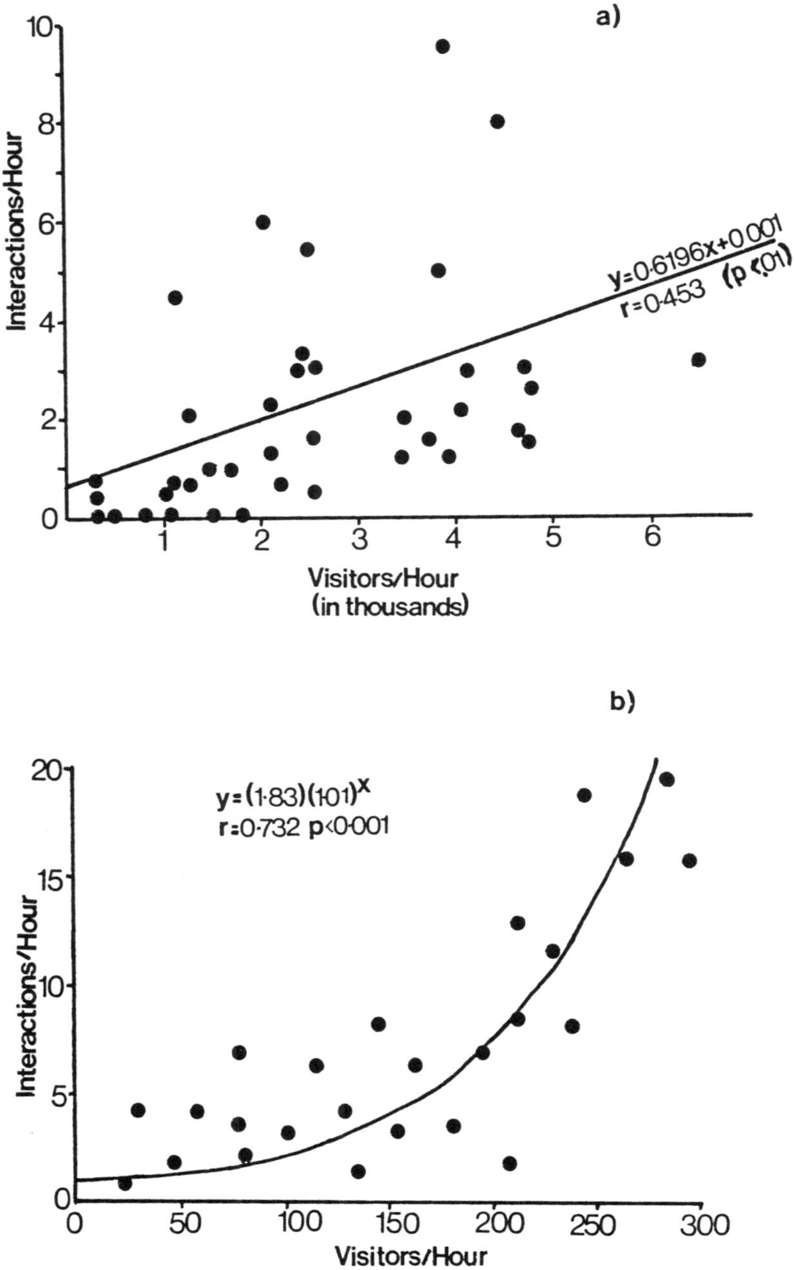

Figure 1. Interactions (feeding, touching, throwing objects) between people and monkeys related to visitors per hour. a) Mexico City Zoo. b) Gibraltar. From hourly observations March-July 1987, Mexico; 23 months, Gibraltar.

Figure 2. Regular visitor pressure on the monkeys of Queen's Gate troop disrupts their normal activities. (Photographs by JEF)

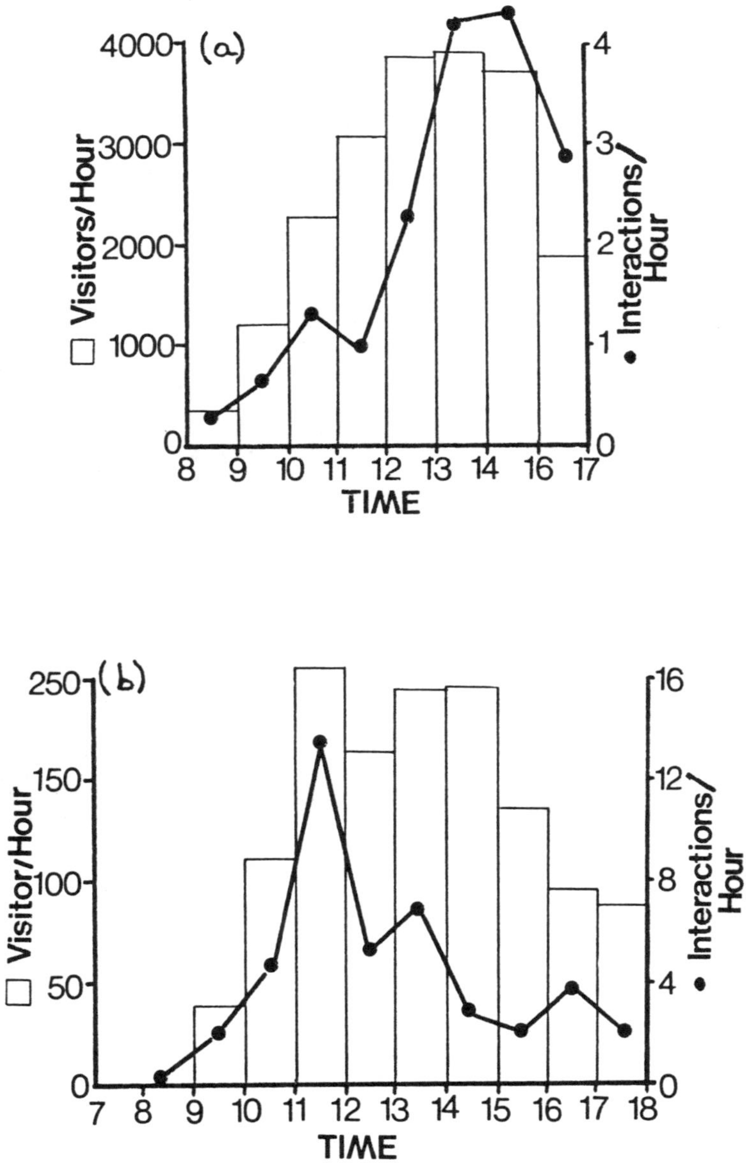

Figure 3. Diurnal changes in number of people visiting and interacting with a) green monkey exhibit, Mexico City Zoo; b) Queen's Gate Barbary macaques, Gibraltar.

Captive groups may express significantly larger interindividual distances than non-captive ones despite similar distribution of resources in both and the more reduced space of the captives (Ehardt-Steward & Bramblett 1980; Hornshaw 1981 1984; Weigel 1980). Kummer and Kurt (1963) and Rondinelli and Klein (1976) also observed, in their comparisons of wild and captive primate groups, that ecological differences alone were inadequate to explain observed differences and similarities. They propose that social principles operating within groups, in particular genealogical associations influencing group cohesion, are more important. Nonetheless, there could well be a preferred range of "mean distances" (Innis Balph & Balph 1984) between nearest neighbors in a non-competitive context, reflecting a compromise of attraction and intragroup competition.

Carefully controlled studies of 14 groups of 207 pigtail macaques (*M. nemestrina*) by Erwin (1978) revealed that considerably less aggression resulted when groups had access to one room than when they had access to two (also see Novak & Drewsen, this volume). Erwin showed that the frequency of aggression was reduced by the mere presence of an adult male, and that the provision of shelter where subordinates could hide had a similar effect. Chamove and Anderson (1979) also saw, unexpectedly, that aggression in a group of macaques increased when food was scattered on the cage floor. Aggression decreased when food was amassed into two piles, decreased further when the food was buried in woodchips, and was least when monkeys could not see many other animals because of an opaque partition. (They have also observed that aggression may increase by a factor of four when there is just a single food source - Chamove & Bowman 1978.) Chamove and Anderson (1979) suggest that there may be a critical distance, still undetermined, beyond which scattering of food abates conflict, less than which the same characteristics may increase aggression because dominants attempt to control access to all the food that is visible. These observations suggest that resolving the impact of the extrinsic environment on display zoo primates requires full consideration of the intrinsic environment, both ecological and social, in each study situation.

Although how variables interact within the intrinsic cage environment has a major effect on display animals' behavior, the presence of people is indisputably an important extrinsic factor that can cause changes in the nexus of social and ecological parameters. Research on this topic is still lacking because most attempts to discern the quality of captive environments have been based on simplistic assumptions about the intrinsic environment - i.e., that all ecological factors within cages were similar. For example, Bernstein (1976) postulates that the behavior of captive primates reflects "genetic heritage of the species," even though he did not examine the role of ecological or social variables within the study cages.

The influence that people have on the behavior of animals in exhibits can be directly estimated by comparing the behavior of an exhibit group subjected to alternate days of presence and absence of visitors. The green monkey group at the Mexico City Zoo fulfilled this criterion as the zoo is closed to the public two days a week. Comparisons of a number of parameters - feeding times, food quantities offered by the zoo staff, diurnal temperature ranges - were made between weekend days (which had a mean of 20,242 people visiting the green monkeys) and Mondays and Tuesdays, when the zoo was closed to the public. For the five months of the study, activity patterns and budgets as well as agonism rates and dispersion within the enclosure were noted for all troop members. There were no significant differences in the intrinsic environment in terms of food quantity and distribution between weekends and non-visitor days, so that behavioral differences between blocks of days had to be a function of human presence.

Activity budgets for the entire troop demonstrated that monkeys spent more time resting and socializing during non-visitor days (Figure 4). On weekends, time spent feeding on food provided by visitors comprised about 4% for females and 3% for males, in contrast to 12.5% and 10.5% of time spent feeding on provisioned food, respectively. Animals spent about 9% of their day time feeding on provisioned foods during non-visitor days. There was a significant positive correlation between numbers of people visiting the monkey exhibit and percentage of time spent by the troop feeding on visitor food (Fig. 5). No differences in agonism rates were observed between visitor and non-visitor days, as competition for visitors' titbits was not observed. A clearly-defined hierarchical system of priority of access to food could have been responsible for the lowering of tension (Fa 1986; Singh 1969; Southwick 1967). But a clear change in the use of space and, therefore, in the contact values between monkeys was detected. On weekends, monkeys spent more time around the edges of the enclosure, where food was offered by visitors, and tolerated smaller interindividual distances. The ultimate consequences of this elevated exposure to people five days every week has still to be examined but the group's failure to breed may be attributed to this.

Semi-Natural Free-Ranging & Enclosed Primate Displays

Adequately designed naturalistic environments that include forested areas no doubt help promote the fuller expression of a primate's behavioral repertoire (Markowitz 1982). The relatively new concept of allowing the visiting public to enter large enclosures (usually fenced-in patches of forest) where the monkeys roam freely, or of allowing controlled access to areas where wild monkeys are consistently provisioned, engenders a unique form of contact between people and the primates on display. In the former (e.g., Barbary macaque parks in France and Germany), the display has a direct commercial benefit, whereas in the latter (e.g., public access to the Queen's Gate troop of Barbary macaques on Gibraltar), visitor interest in the monkeys may have indirect economic consequences by promoting tourism.

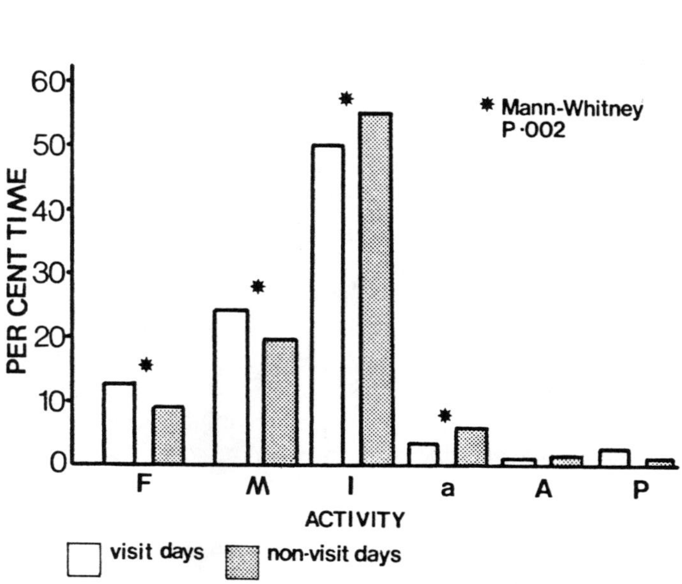

Figure 4. Activity budgets of green monkey group at Mexico City on visiting and non-visitor days. From 15-min. scan samples of the group (160 observation hours). F:Feed M:Move I:Inactive a:Allogroom A:Autogroom P:Play

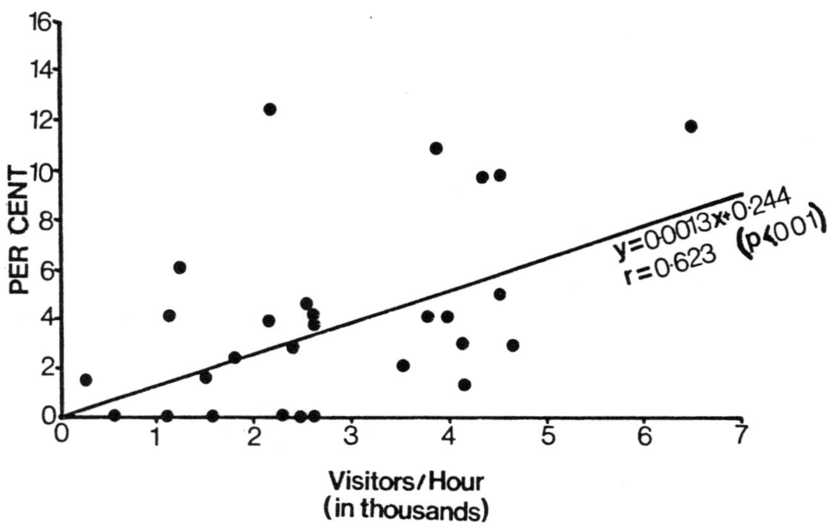

Figure 5. Number of hourly visitors around the green monkey exhibit and time spent by the group feeding on supplemental food offered by people.

Table 1
Known Semi-Natural (Free-Ranging and Enclosed) Primate Displays

Location/Species	Numbers	Source
FREE-RANGING		
Provisioned, non-commercial, tourist-visited		
Japan		
M. fuscata	variable 60-431	Koyama et al. (1981)
India		
M. mulatta	variable	Southwick et al. (1965)
M. radiata	variable	Singh et al. (1984)
Presbytis entellus	variable	Winkler (1984)
Nepal		
M. mulatta	variable	Teas et al. (1981)
Sri Lanka		
M. sinica	variable	Dittus (1980)
Indonesia		
M. fascicularis	variable	Southwick (personal comm.)
Gibraltar		
M. sylvanus	30-40	Fa (1984)
ENCLOSED		
Provisioned, commercial		
Goulds Monkey Jungle USA		
Cacajao calvus	150+/island groups	DuMond (personal commun.)
M. fascicularis		
Alouatta seniculus		
Kintzheim France		
M. sylvanus	211 on 12.5 ha.	de Turckheim & Merz (1984)
Rocamadour France		
M. sylvanus	137 on 10 ha.	de Turckheim & Merz (1984)
Salem W. Germany		
M. sylvanus	382 on 14.5 ha	de Turckheim & Merz (1984)

There are still few examples of commercial animal parks around the world, but the probable impact of encouraging such close contact between people and primates is worth exploring with reference to free-ranging provisioned groups. A summary of the main characteristics of known semi-natural display situations is presented in Table 1. All these parks are able to exhibit large numbers of various species to the public.

The composition of semi-natural free-ranging primate groups differs fundamentally from that of enclosed groups. The free-ranging groups generally although not always consist of wild monkeys, living in their native habitat, that have been conditioned to gather at provisioning points through the consistent offer of food there. These groups are well-integrated socially and able to utilize the habitat's natural food supply. Provisioning affects the animals' use of the habitat and promotes a substantial increase in numbers due to greater infant survival or breeding potential or both (Chalmers 1979, qp. 33; Koyama Norikoshi & Mano 1975; Loy in press). On the other hand, semi-natural enclosed primates are always groups of monkeys or apes introduced to a habitat that is not native to them. Such groups may not consist of intact social units but rather of individuals brought together from diverse locales, not necessarily of the same geographic origin.

These two types of semi-natural primate displays have a common ecological feature - the animals are provisioned and space is relatively unlimited. Provisioning has an important impact on the animals' behavior because food is more abundant, more reliably to be found (in time and space), and usually higher in quality than natural sources. The animals' different response to such a food supply may be linked to how energy availability changes relative to the number of monkeys competing for the resource, but competition may be a direct function of the composition and history of the groups (see above). Provisioned foods can influence the time an animal needs to devote to feeding. Compared to wild primates, food-enhanced primates [a term coined by Fa & Southwick (in press) to identify groups of monkeys and apes that receive all or part of their diet from humans] can spend proportionately less time searching for food and more time resting and socializing. As a corollary, provisioned primates have considerably smaller home ranges.

In certain situations, the relation between food and activity profiles of provisioned groups can be affected by the persistent offer of small quantities of very attractive items outside the normal provisioning regime. People's offering monkeys supplemental foods can provide another level of response for the animals even when they are nutritionally satisfied. Some foods, such as sweets or cooked foods, are superoptimal or extranormal releasers of positive gustatory reactions, even stronger than those evoked by natural items (Fa in press). I have suggested that the magnitude of this response is comparable to a supernormal stimulus effect. The reinforcement received from the pleasantness of sweet items can lead to an "addiction" to them, with people becoming the search image.

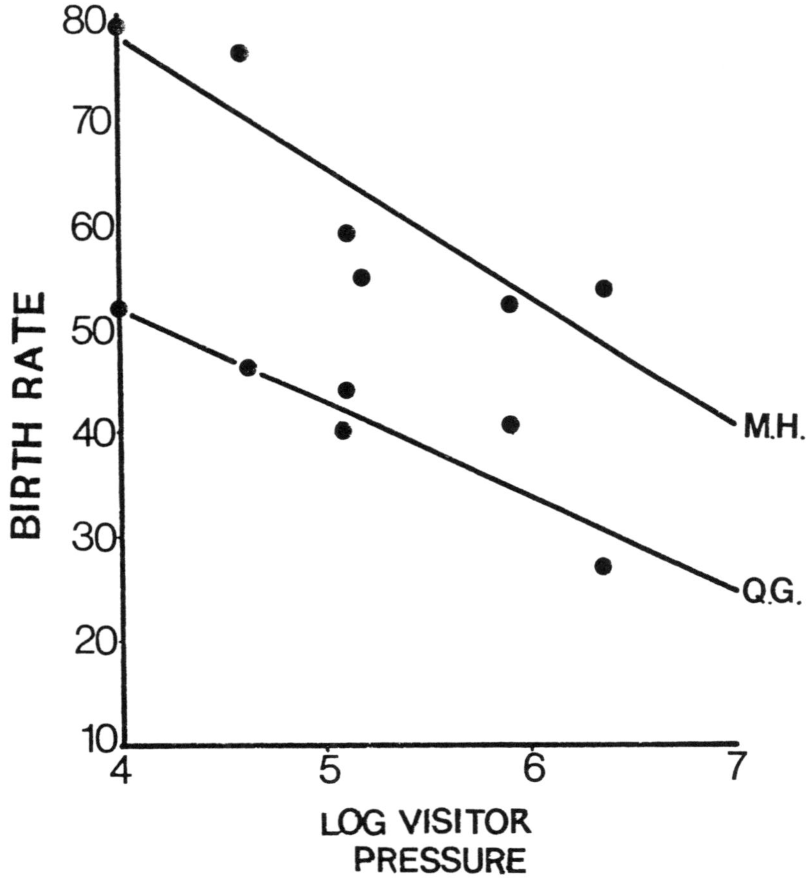

Figure 6. Changes in birth rates at the Middle Hill and Queen's Gate Barbary macaque troops in Gibraltar in relation to tourist numbers per half-decade periods (modified from Fa 1986b). Visitor pressure is taken as index resulting from recorded tourist numbers visiting the Rock each year. The supposition is that there is a positive relation between overall tourist numbers at Gibraltar and level of interference with the monkeys.

The constant presence of people among monkeys that know the appeal of the foods they offer can have a significant impact on the animals' behavior and ecology. From an analysis of 43 and 34 years, respectively, of demographic data from the Queen's Gate and Middle Hill troops of the Gibraltar monkeys (Fa 1984), I observed temporal fluctuations in birth rates of both troops during certain periods and a significant difference in mean annual birth rates between troops (Queen's Gate = 48.5%; Middle Hill = 63.2%). Both troops are provisioned, but Queen's Gate has always been a tourist attraction, wheareas Middle Hill has been isolated in Ministry of Defense territory since 1972. The discrepancy in breeding performance between troops was linked to the level and manner of feeding and the food types given by passersby. The feeding of Queen's Gate troop by tourists appeared to disrupt the animals' behavior by affecting time spent on other activities, and this resulted in dampened reproductive potential. Across nearly five decades of data for the Gibraltar macaques, a significant fall in birth rates in both groups appeared when there was an increase in tourism during a boom period from 1960-1968. In fact, a significant negative correlation (Pearson's r, Queen's Gate r=-0.85, $0.02 < p < 0.01$; Middle Hill r =-0.89, $0.01 < p < 0.001$) appeared between half-decade birth rates and tourist numbers in Gibraltar; Middle Hill showed a steeper drop in birth rates in relation to tourist numbers (Figure 6). From observations of how humans interact with the monkeys, I suggested that people affected breeding adversely. Other hypotheses, such as a common climatic event affecting breeding in both troops simultaneously, changes in group structure, density, or inbreeding depression promoted by the high relatedness of the troops, do not satisfactorily explain such a drop in fertility (see Fa, in press).

The mechanism of disruption I propose lies in the manner and type of supplemental food offered to the monkeys by people. Firstly, high-caloric and nutritionally-attractive foods are given, and through constant exposure to them, monkeys know their palatability and actively seek them. Secondly, there is a continuous flow of visitors throughout the day who feed the animals or at least present the possibility of doing so. Thus, animals will remain in the same area, in constant view of each other, waiting to be fed by the visitors. Supplemental feeding occurs when a monkey is singled out for feeding by a visitor. Younger (more subordinate) animals are more appealing to the public and this leads to a whole series of dominance displacements accompanied by high levels of aggression. All offers of food to single individuals inevitably depress the ratio of food to monkeys in a manner unknown in *ad libitum* provisioning. Competition for the more attractive foods offered by people intensifies as the number of visitors increases (see section above). Because the macaques in Gibraltar breed for a period of no more than 5 months, with mating concentrated in one month, and since the socionomic sex ratio of the troops has usually been 1-2 males for up to 10 females, any increase in distractions from the strictly-seasonal mating activities would have the effect of depressing reproduction. At

Gibraltar, adult males will abandon estrus females to feed on items offered by people. Malik and Southwick (in press) have also observed this for rhesus macaques in India.

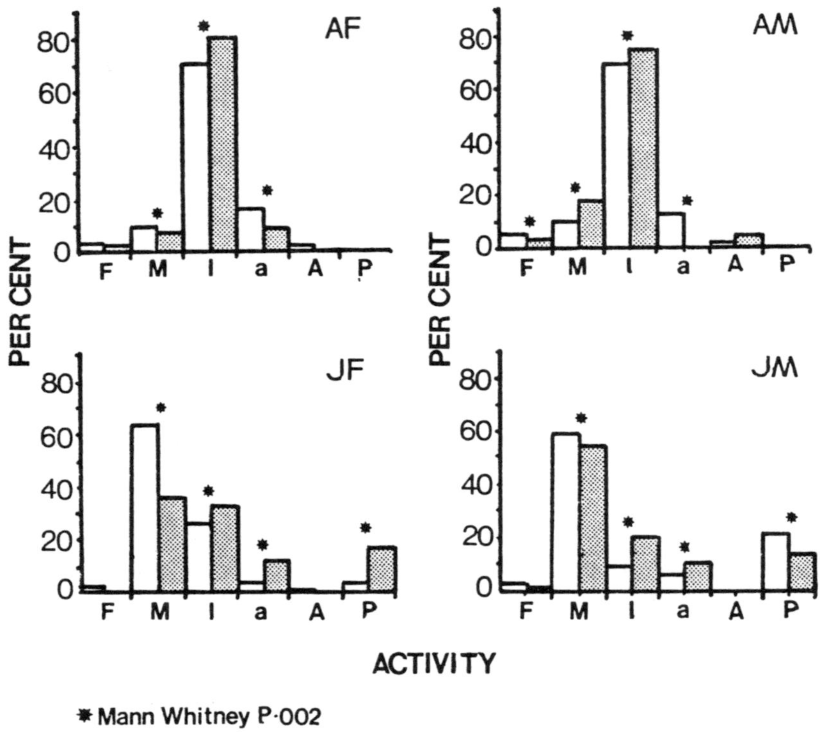

Figure 7. Activity budgets of main age-sex classes in Queen's Gate troop in a low tourist year (1979) (open bars) & a high tourist year (1987) (shaded bars). Troop composition troop varied between study periods but means for each age-sex class are given. Data taken in summer high tourist season (July-August). AF:Adult Female AM:Adult Male JF:Juvenile Female JM:Juvenile Male

At a more proximate level, human interaction with the monkeys can alter their overall behavior to such an extent that the monkeys may no longer be "adequate representatives" of the species (Markowitz & Spinelli 1986). At Gibraltar, the group of monkeys that is exposed to people differs from the more protected troop in overall activity budgets, use of space (Queen's Gate roams in 1 ha. compared to 21 ha. for Middle Hill), and synchrony of activities within groups (Fa, in press). These differences are attributed to disruptions caused by people's passing by the group throughout the day and throughout the year. Differences in troop activity were not related to provisioning level (as Middle Hill was provisioned more than Queen's Gate) nor to ecological differences (the troops are separated by less than a kilometer).

Overall differences in activity budget between the troops correlated with the older age structure of the Queen's Gate troop. Because there was a negative correlation between age and activity, the lower Queen's Gate activity could have been related to the presence of older monkeys. The rates of human interaction with monkeys during the time (1979) of my study were low (15 +/- 6 visitors per hour) compared to levels during high tourist years (Fa 1984). Despite this, the effect of people on synchronicity in troop behavior was pronounced. Confirmation that people do have a significant effect on the monkeys' behavior comes from observations undertaken in the summer of 1987, when visitors averaged 120 +/- 82 per hour. The later observations clearly showed that at a higher visiting level, the monkeys were more affected; they spent more time in the area where tourists congregate, and showed greater inactivity and lower socializing levels (Figures 7 & 8) than during my 1979 (Fa, in press) study. The age structure of Queen's Gate was younger in 1987 than in 1979; thus, older age cannot be the factor responsible for lower activity in the Queen's Gate troop.

CONCLUSIONS

Displaying primates in zoos or parks (semi-natural free-ranging conditions requires an understanding not only of the intrinsic but also the extrinsic environmental factors that can affect the animals. The extrinsic environment consists primarily of the behavior of human visitors toward the animals. In most situations where visitors have access to monkeys or apes on exhibition there are exchanges between primates and people that range from passive interference by people, through noise and visual stimulation, to active interference through the offer of foods or throwing objects. The study of the effect of these factors on the animals' wellbeing and ultimately on their reproductive potential is still incipient. It is inevitable that people will interact with display primates in zoos or parks but their impact on the animals must be better determined in order to create a better quality of life

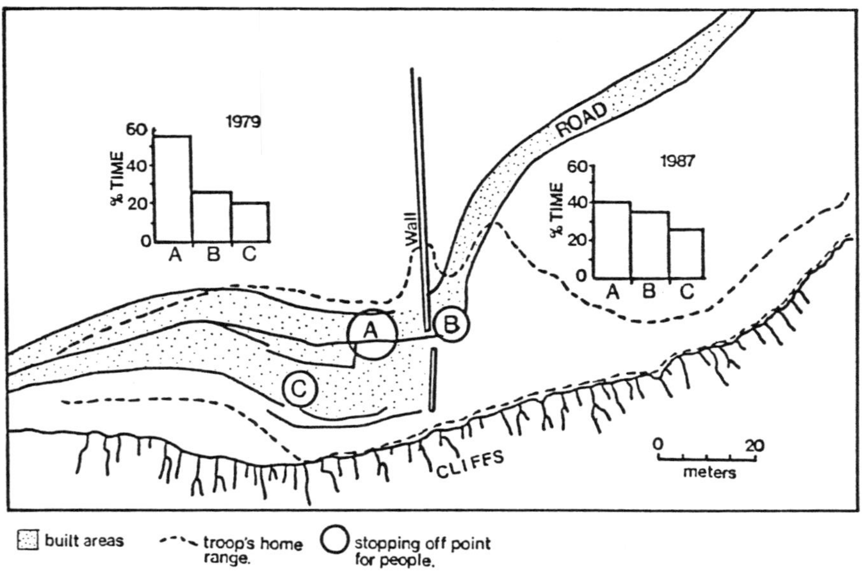

Figure 8. Percentage of time Queen's Gate troop spent in main points where people fed monkeys during 1979 & 1987. A, B, & C: areas where people fed monkeys on the road. Frequency diagrams (insets) show percentage of daylight hours troop spent in each site. In 1987 they spent more time (M=87.5% of daytime) in A & B, areas more used by people; in 1979 they spent more time in site A (M=60.7% of daytime).

for the animals through the techniques of environmental engineering and applied primatology. This is difficult, time-consuming, and above all expensive. Not all zoos or parks are likely to attain this goal.

For the time being, more practical measures can be taken to educate the public about the natural behavior and ecology of animals on exhibit and the harmful effects of interfering with their social relations and normal activities. In theory, education should or could do away with the deleterious interactions between people and display primates but stricter vigilance to make it difficult for people to feed or tease animals can be enforced. There is no doubt that the impact on the animals of exchanges between them and visitors is not simple or unidirectional; rather it depends on the quality of the animals' intrinsic environment as well as the quality of outside stimuli presented by human visitors.

An intermediate option between allowing "free for all" interactions between people and monkeys, as at Gibraltar, and totally restricting contact, is that developed successfully in the Turckheim parks (see Table 1). These allow, in fact encourage, interactions between visitors and monkeys by providing the public with bags of popcorn to feed the animals. Popcorn is of low caloric content and not likely to cause digestive problems. The offer of this food satisfies the public's desire to interact without upsetting the animals' diet. Yet it still carries the cost of some behavioral disturbance and the possibility of cross-infections through close contact. Gastrointestinal diseases and human viral pneumonias in the Gibraltar monkeys are linked to increases in tourist pressure (Fa 1984). As recently as August 1987 (peak tourist season), about 25% of the Queen' Gate troop died as a result of an outbreak of pneumonia (*Gibraltar Chronicle,* 1987).

Neither is the problem of the disruption of activity patterns and the subsequent "taming" of monkeys overcome by "managed interaction." In an effort to enrich the environment of a group of caged zoo mandrills (*Mandrillus sphinx*), Markowitz Stevens Mellen & Barrow (1982) provided them with a competitive speed game to play with the public. Although the mandrills became proficient at the game, the question still arises, to what extent these learnt activities are likely to curtail breeding activities.

It is inevitable that wherever there are primates on display, there will be interactions that will affect the monkeys' behavior and consequently their reproductive potential and survival. The aim of captive primate establishments must be to continue to explore ways to minimize the adverse consequences of primate-human contact through environmental engineering, and through taking into account the extrinsic as well as the intrinsic environments of exhibits.

ACKNOWLEDGEMENTS

I am most grateful to Carolina Valdespino and Jaime Ramirez for the collection of the Mexico data and for help in their analyses. To Debra Forthman Quick, Don Lindburg and to my wife I extend thanks for a critical reading of the manuscript. Permission to work in the Mexico City Zoo, and assistance, were afforded by the Director, Ms Marielena Hoyo. The Gibraltar Regiment, and especially Sgt. E. Asquez, gave their help and permission to work with the Gibraltar monkeys. I thank Peter Olney for providing me with the zoo entrance data.

— 19 —
Langur Monkeys *(Presbytis entellus)* in Captivity

Mark A. Taff and Phyllis Dolhinow

INTRODUCTION

If nonhuman primates are to be kept in captivity and, in addition, are to become research subjects, it is our responsibility to maintain them in a manner that minimizes the stresses of captivity and assures them the least compromised life style possible. These conditions will vary with the species of primate and can only be known after the free-ranging behavior of that species has been studied carefully. There is no shortcut to understanding the physical, social, and emotional or psychological needs of monkeys or apes; it requires knowing their life style and life history patterns in the environment of origin. This, then, is a baseline of knowledge to evaluate the effects of compromises made necessary by captivity.

It would be artificial and misleading to consider the components of wellbeing individually since the physical, social, and emotional environments of life interact in a continuous, multiply-contingent process. Some authors have suggested that the only defensible measure of wellbeing is to determine whether the animal is suffering from stress (Moberg 1985). Others (Curtis 1985) see wellbeing as a dynamic state varied in its manifestations and enormously complex. It can differ from individual to individual and from time to time. It is, further, unrealistic to expect that wellbeing will be the case at all times for an individual; stressful events occur in the life of all individuals, whether captive or free-ranging. We know from observation and experimentation that life events, including stresses, are quickly translated into an animal's physiology with tremendous immediate, as well as oftentimes long-term, effects (Levine Johnson & Gonzales 1985; Moberg 1985; Harbeck & Hoffman 1981; Reite Short Kaufman Stynes & Pauley 1978; Reite Seiler Crowley Hydinger-MacDonald & Short 1982; Reite & Short 1978; Reite & Field 1985; Hennessy Mendoza &

Kaplan 1982).

A captive primate must have adequate and appropriate companions, space, diet, and care if it is to flourish (Erwin 1986; Maple & Finlay 1986; Markowitz & Spinelli 1986; Ullrey 1986). Simple survival may require much less, but at a tremendous price in terms of the quality of life and the quality of research. Behavioral or biomedical research on compromised animals can only produce compromised results.

We present here a case study, the maintenance of a colony of Indian langur monkeys (*Presbytis entellus*), an Old World Colobine monkey. This colony was established in 1972, mainly with wild-captured animals, and is now a multigenerational set of social groups living in large outdoor-indoor enclosures.

LANGURS IN INDIA

The gray langur is the most widely distributed of all the colobine monkeys and is found in Nepal, throughout the Indian subcontinent, and Sri Lanka (Wolfheim 1983). Langurs have been studied in a number of different sites by many observers (for example, Jay 1963; Sugiyama 1964 1965a 1965b 1966 1967; Sugiyama Yoshiba & Parthasarathy 1965; Vogel 1971 1976; Dolhinow 1972; Hrdy 1977; Curtin 1975; Bishop 1975; Boggess 1976; Ripley 1965 1967; Yoshiba 1967 1968; Mohnot 1971a 1971b). Habitats are varied: from sea level to more than 3,600 m (Bishop 1975 1979; Boggess 1976; Curtin 1975), under conditions from snow to arid drylands where the animals may not drink for 4 to 5 months (Sugiyama 1964).

Langur diet is almost entirely vegetarian (Jay 1963). They are able to eat and digest quantities of relatively unnutritious mature leaves by means of a special diverticular form of the stomach allowing a ruminant-like partial fermentation involving anaerobic and cellulose-digesting bacteria (Bauchop & Martucci 1968; Bauchop 1971 1978). Food species vary according to area and season (Roonwal & Mohnot 1977). Langurs are one of the largest of the colobines, with adult males weighing from 9 to 23 kg and females 7.5 to 20 kg (Roonwal & Mohnot 1977). They may spend as much as 80% of the day on the ground (Jay 1963).

Large monkeys, langurs are strong jumpers and often make spectacular leaps from tree to tree. One stride of a running adult male can easily cover ten feet. Their build is long and lean in comparison with the compact rhesus macaque with which langurs share the Indian countryside and villages. With their large size, height, and long arms and legs, langurs need more square meters of living space than macaques (Cercopithecine). No area of confinement should ever be so small that the enclosed primate cannot stretch to its fullest length.

Free-ranging langur groups exhibit a range of size and social structures. These include mixed-sex groups with a single adult male (one-male groups), mixed-sex groups with multiple adult males (multi-male groups), and groups composed of only males (all-male groups). Males aged 4-5 year and older commonly leave their natal groups and live for a time in the all-male groups. Each kind of group may be found in every type of habitat. Groups range in size from 3 to more than 100 individuals. Population densities range from 2.7 animals per square kilometer (Jay 1963) to more than 135 (Sugiyama 1964).

THE LANGUR COLONY

The Berkeley langur colony is located on the grounds of the Animal Behavior Research Station, University of California, Berkeley. This is secluded acreage in the Berkeley hills, with access restricted to researchers, animal technicians, veterinarians, and students undertaking supervised research projects. The climate is temperate with a mild rainy season during the winter months. Three outdoor enclosures (6m x 23m x 4.6m high), each with attached heated sleeping quarters, provide living space for three langur groups. Each enclosure has a cement base, cyclone mesh on the front and top, and corrugated metal siding on the back and far end. Each enclosure contains a variety of platforms, suspended branches, tires and other structures on which the animals rest, move, and interact. We maintain three groups, each structured as a typical free-ranging langur group would be in India. Two of the groups contain one adult male, several adult females, and immatures; the third group consists of adult males ranging in age from 6 to 12 years.

RESEARCH EMPHASIS

The langurs have been observed on a regular basis in a long-term study of social development since the colony was formed in 1972. Animals born into the colony have grown to adulthood and together they provide data spanning four generations. Because our research emphasis has been on normative patterns of development, manipulations of the groups have been kept to a minimum. The one exception to this policy is when tension develops between a maturing adolescent or young adult male and the dominant adult male. Because of the potential for injuries and group disruption, the younger animal is transferred from the mixed-sex group into the all-male group.

No invasive physiological data-gathering has been attempted, and while this restriction has lost us some information, it has also made it possible to observe development in an exceedingly stable social environment (Dolhinow & Murphy 1982; Dolhinow & Rickenbach 1988). An important aspect of our investigations has been the animals' freedom to establish their social relations and maintain them according to patterns typical of the species. A brief investigation of the nature of early relations included the removal of the mothers of a small group of 6- to 8-month old immatures, once each, for from 2 days to two weeks. Each mother was then returned to the home cage (Dolhinow 1980; Dolhinow & Murphy 1983 1988). Many other aspects of behavior have been and are being investigated, including social development (Dolhinow & Murphy 1982), adoption (Dolhinow & DeMay 1982), weaning (Dolhinow & Rickenbach 1988, Rickenbach 1988), play (Taff & Geary 1988), proximity patterns (Pearson Williams & Allison 1988), female behavior, reproduction and aging (Dolhinow & Krusko 1984; Dolhinow McKenna & Von der Hass Laws 1979; Harley 1983 1985), and male behavior (Dolhinow & Taff 1988; Moorhouse 1980; Curtin 1981).

Teaching is a second major use of the langur colony. Both undergraduate laboratories and graduate seminars are held at the research facility. Small laboratory classes train students in all aspects of observational research: project design, selecting the behavior repertoire, making and recording observations, data analysis, and writing formal reports. This training has prepared many students for fieldwork.

Using our facility as a model, we discuss below four major areas of concern in the maintenance of a healthy captive colony: (1) group composition, (2) space, (3) diet, and (4) caretaking.

GROUP COMPOSITION

Although free-ranging langur groups are highly variable in size and structure, the space limitation of captivity has prevented maintaining mixed-sex groups with more than one fully adult male. As they mature, young males invariably come into conflict with the adult male of the group and the ensuing tensions create a situation in which fighting could occur. This, in turn, could lead to injuries and most certainly would be extremely disruptive to the group. The maintenance of group stability requires constant monitoring of all subadult males so that when the typical patterns of increasing subadult peripheralization and/or tension with the group adults appear, transfer of the adolescent is considered.

When an adolescent male is removed from his natal group, he is introduced into the all-male group. The composition of this group is identical to many such groups in India, and the membership of our all-male group has been relatively constant over the years. The social dynamics of the group have been studied and occasional removal of individuals, for example to replace an adult male in a mixed-sex group, has stimulated shifts in the group dominance patterns and alliances. Other internal structural changes take place as members mature (Taff 1988).

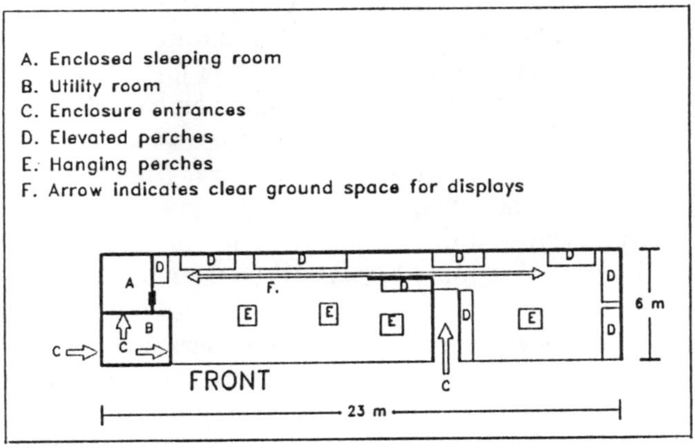

Figure 1. Top view of enclosure. Indicated is the clear space for running displays.

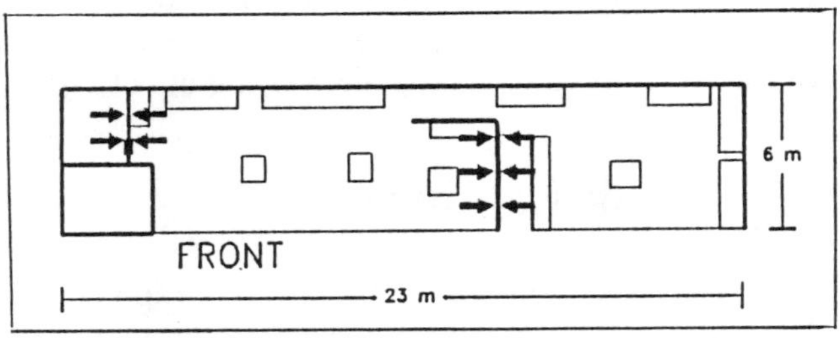

Figure 2. Visual separation. Arrows indicate barriers to visual contact.

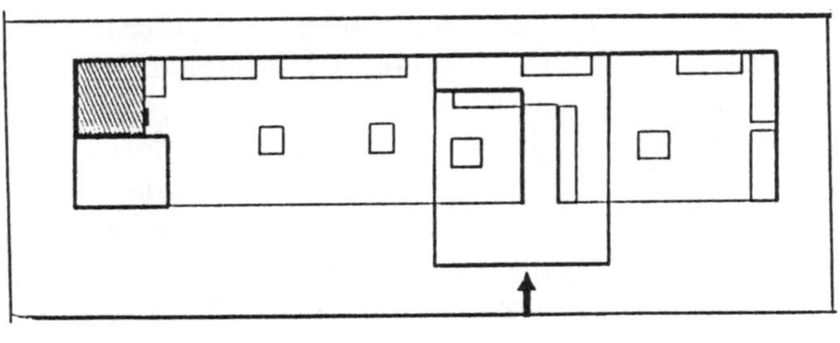

Figure 3. Sheltered areas. Arrow: sheltered part of outdoor enclosure – a metal-and-fiberglas roof covering sufficient perch space to allow many animals to sit protected. Cross-hatching: inside heated sleeping quarters.

Because the all-male group is composed of very active, large monkeys, it presents special problems of maintenance and care. The greater activity of this group requires constant checking of the physical structure of the cage and prompt repairs of damage. An adequate number of perches, visual barriers, and remote safety areas is a necessity. Boisterous high-velocity displays require running space away from the perches, where subordinate animals may remain well out of the way (Figure 1).

We have discovered that optimally there should be an even number of males in the all-male group because the males tend to form affiliative pairs. Each animal has a partner for grooming, huddling, and, at times, for support in an aggressive interaction. The membership of each pair remains stable for long periods of time if there are no changes in group membership. As with any social group, it is necessary to judge the compatibility of the members and should there be an individual whose temperament or personality makes him unacceptable as a group member, he must be housed elsewhere. Because of such incompatibilities, there have been occasions when it has been necessary to remove an individual from a group. It should be emphasized that animals forced by humans to live closely together must be compatible. It is our responsibility to assure that no group member terrorizes or injures others. After all, in the wild animals have the option of fleeing, or, at least, of getting and staying out of sight.

SPACE AND THE CAPTIVE MONKEY

Captivity is by necessity a condition of compromises. For example, the amount of movement in the typical activities of daily life ought to provide the amount of exercise required by each animal. There is no way that captive animals can do what they would in the field. Foraging for food, for example, does not entail a two-mile walk when food is dispensed in only a few locations within the enclosure. Theoretically, feeding practices could be developed to increase the amount of energy animals must expend to obtain their food, but in most colonies such strategies have not been attempted. At our facility, for example, it might be possible to structure multiple feedings during the day, placing small amounts of food throughout the enclosure, but this would increase activity levels only slightly, at a cost in personnel time that would be prohibitive. Further, the constant interruption of group activity would disrupt ongoing research, and dominant animals could restrict access to food piles. Developing feeding procedures that replicate free-ranging feeding conditions is extremely difficult.

This is only one aspect of daily free-ranging life that cannot easily be replicated in the cage; the conclusion is inescapable that even in a large cage, it is unlikely that the levels of exercise will be optimal for the animals. This is a significant factor in estimating the bioenergetic needs of captive animals. Immature animals are more likely than adults to exercise adequately provided they have peers to play with. Young adults seem to maintain a higher level of activity than do older adults, as is true in the field. A major difference between captive and free-ranging adults is that the former have so much to eat that with only a moderate decrease in activity level, weight gain may be a significant problem.

Cage design can accommodate and even encourage some types of active movement. Providing a corridor allows running displays to occur away from nonparticipating animals (Figure 1). For younger animals it is important to provide stimulating areas of play apparatus where they may develop and test skills of climbing, jumping, and fast movement. Play objects such as swings, hanging milk cartons, and large plastic barrels interest all animals in the group but they hold a special attraction for immatures. The latter show sustained interest in and will manipulate, investigate and play on these objects long after the adults lose interest and ignore them.

It should be mentioned here that, for langur immatures, what appears to be important is not the object or toy *per se*, but rather its ***novelty***. Periodically changing the locations and form of play structures provides novelty and the opportunity to increase the range of their familiar movements. We are presently testing the hypothesis that it is the object's novelty and not its specific qualities that is important by investigating the behavioral effects of altering first the quality and content of objects and then their location within the enclosure.

The structure and location of sitting areas and perches, should take into consideration the size of animals and the numbers that tend to sit or rest together. Langurs typically groom in small groups and sleep huddled together. Ledges that are high above the ground may be difficult for the older and somewhat physically compromised group members to reach. Easy access to all levels is important. Accordingly we offer perches that are staggered in height.

Privacy is an important consideration in the spatial arrangement of a colony. However difficult this may be to achieve in some facilities, it is vital. Limited space in most captive situations restricts the ability of animals to resolve disputes by moving away from the source of disturbance, the most common way of settling frequent spats in the field. Even when a subordinate animal in the cage moves away, there may still be constant visual contact. Providing the animals with visual barriers provides a chance to get out of sight. There should also be areas for resting and for a few animals to interact that are not in view of the rest, while making certain no part of the enclosure is out of sight of the human observer. It is possible to design visually isolated areas even in a small enclosure. Figure 2 illustrates how our enclosures provide visual separation.

Figure 4. Photograph of langur enclosure, showing some of the structures used to vary the interior.

Figure 5. Four animals huddle on suspended branch.

Figure 6. An adult male performs an energetic display, using the open corridor. Non-participants observe from out-of-the-way perches.

Figure 7. Mother grooms her infant while a second female watches. The animals sit on a branch elevated 6 ft above the floor. Above them two infants are playing.

Adequate protection from rain in portions of the outdoor enclosures allows our groups to remain outside at all times if they choose to do so, even during storms (Figure 3). Given the choice, most animals remain in sheltered areas outdoors rather than in the heated sleeping quarters during inclement weather, both day and night. During cold winter rains, observer complaints are far more frequent than any apparent increase in the level of monkey discomfort.

All primates are inquisitive, intelligent animals that are easily bored. Ours are no exception. This means that major efforts are required to enrich enclosures and to provide a degree of regular change in furnishings. This can be done in part by varying the spatial arrangement of the standard perches, ladders, swings, and the like. Novel items provide much stimulation and are often the focus for many hours of play. Although less manipulative (and destructive) than most macaques, langurs pay a great deal of attention to new items in the enclosures. For example, we have used children's swingsets, branches, bowling balls, tires, rust-resistant chain-link swings, skateboards, tin cups, and bales of alfalfa (which they also eat). PVC tubing, connected with rust-proof chain, is easily assembled into a variety of shapes, is nontoxic, easily cleaned, and durable. PVC piping is relatively inexpensive and can be obtained at most building supply centers. Many construction companies will happily donate unusable lengths of PVC tubing that they have on hand. .ixenrichment devices, durability of

Do not underestimate the strength and perseverance of primates. A small group of large primates with time on their hands can wreak impressive destruction on even the toughest of enrichment items. Displays, investigatory behavior, and normal use can destroy the structural integrity of all manner of furnishings. This does not reduce our responsibility to replenish what becomes damaged or destroyed. Frequent replacement of novel furnishings is important; they are disposable items , like food, and should be budgeted accordingly.

Periodic capture and/or restraint is a necessary part of life in captivity. Law requires regular TB testing in addition to daily visual inspection to detect any medical problems. When the need arises, immediate veterinary attention may require that an individual animal be caught and perhaps sedated. Enclosures must provide an easy method for capture or restraint. Although the stress produced by catching a group member may be substantial, it can be minimized with some thoughtful cage design. For example, the height of the cage should not exceed the distance a human can reach who is attempting to grab and remove an animal from the ceiling. Experience demonstrates that our enclosures are higher than necessary. Primates naturally will try to climb away from perceived dangers. Prior consideration of this problem in cage design would have eliminated many potential problems. One modification was the construction of a capture tube. If each member of the group is trained to run through a capture tube built into the enclosure, the monkey can be isolated and easily moved into a transfer cage for attention. It is necessary to train the animals to move freely through a capture shoot or tube by requiring them to move through the structure on a regular basis such as before feeding. We have tried this with great success.

DIET

Free-ranging langurs begin to feed at sunrise and, depending upon the season of the year, may spend upwards of 5 hours eating (P.D. personal observation). We cannot, of course, hope to duplicate the variety of plants and plant parts consumed in the field, and poor diet in captivity has been noted as a source of mortality (Hill 1964). Suggestions have been made as to appropriate foods including green vegetables and alfalfa for langurs (Crandall 1964; Napier & Napier 1967), combined with exercise and a vitamin supplement.

The daily diet fed to the colony consists of carrots, yams, apples, broccoli, bananas, alfalfa pellets, and monkey chow. Some foods that produce diarrhea are avoided such as lettuce and cabbage (Urbain & Nouvel 1954 recommend avoiding these for colobus) and citrus fruits. Hill (1964) suggested that food type and exercise are importantly related. It has been our impression that active monkeys evidence less gastric distress than, for example. relatively inactive old individuals. An adequate diet is most certainly a central factor in longevity.

The langur colony is fed early each morning after the enclosures have been cleaned. Since early morning foraging is typical of free-ranging populations, every attempt is made to feed our animals prior to 0930 hours. Feeding later than 1000 hours can result in restlessness and tension in the groups. The food is cut into small pieces and dispersed in several areas. By not concentrating food in a few locations and by selecting areas from which the animals cannot see other food piles, all animals have free access to food, and competition and fighting are reduced. Since free-ranging populations tend to forage again in the late afternoon, sufficient monkey chow is provided each morning so that there is food available later in the day. As a measure of how much chow is placed in the enclosure, enough is left so that only a few pieces remain at the time of cleaning the next day. Quantities of food are increased during cold weather since the animals choose to remain out of doors most of the time.

Novel food items are offered as treats and for variety. These foods are not used as rewards in training, are given on an irregular basis, and never immediately before, during, or immediately after observations. Treats include grapes and other seasonal fruits and vegetables not included in their standard diet, unsalted peanuts in the shell, sunflower seeds, and bird seed. Some food items act as environmental enrichers by creating novel situations. Coconuts in the husk will keep a group of langurs busy for hours first grooming off the hair and then trying to break them open. Scattered bird seed gives the entire group an activity that will keep their attention for hours. Corn in the husk also provides both busy work and a relished unusual food item. Langurs dispatching whole watermelons amuse human observers but are rather messy. Care is taken that any unusual new food is first given in small quantities to avoid the possible upsetting of langur digestion. Once it has been established that eating moderate quantities of an item does not cause any perceptible gastric problems, the items may be used freely. In at least one instance a food, raw onions, appealed greatly to the monkeys and caused them no visible distress, but provided an absolutely unbearable flatulent atmosphere that extended for a radius of 25 feet from the animals.

CARETAKING

Personnel responsible for cleaning and feeding the colony are among the most important people who have contact with the monkeys. The animals react quickly and strongly to different people, accepting some with no notice but threatening and chasing others. Since caretakers must sometimes be in the enclosures with the animals it is necessary to train all personnel carefully to avoid confrontation. Playful interaction and other kinds of friendly physical contact with the animals, including hand-feeding, are strictly forbidden. The quality of animal care must be noninteractive. It only creates potential problems when people make pets of large, untamed monkeys. The cute infant quickly grows into a strong sensitive adult that will not take kindly to human interference. Both human and nonhuman can easily misinterpret the other's cues.

The relation between researcher and subject should be based on mutual respect, but the form of the relation depends on the research. If the objective is to observe and record behavior with a minimum of human interference, then to establish friendly interactive relations would make noninterventive investigation impossible. If, on the other hand, research involves handling the animals, perhaps to obtain specimens such as urine or blood, then to be able to do so in a friendly, nonintimidating and nonstressful manner may be facilitated by taming and getting the animals used to being approached and handled. Rewards are valuable in this training (see Turkkan, this volume).

In addition to not being afraid of the animals, animal technicians must be interested in them for best care to take place. Because daily checks are made on the health of each animal, every person responsible for these checks must know each individual monkey well. Knowing the moods and tolerance of each adult is important in being able to work with them with a minimum of effort, stress and potential danger to the personnel. In addition, knowing individuals is invaluable in assessing the health and wellbeing of each animal. Because it is not always possible to hire experienced technicians, training is vitally important, and it should stress such general information as the species' basic characteristics, including behavior and temperament. Staff must be fully aware of the kinds of unusual and emergency situations that might arise and how to deal with them. Special emphasis should be placed on how to anticipate and identify potential problems before they happen. For example, variation in an animal's daily routine may indicate a change in social relationships or the onset of illness. New staff must be monitored by an experienced supervisor until both the animals and the worker begin to feel comfortable. It is as much an introduction of a strange human to the monkeys as a new species to a human. The presence of the more experienced and familiar person reassures both animals and new staff alike.

Every primate facility should have a basic manual in the facility for each kind of primate. The manual should include basic information on the species, how personnel should conduct themselves toward the animals, and information about individual animals. This gives everyone associated with the animals an opportunity to get to know what is typical. The manual should also include the basic goals of the research *project*. Ideally there ought to be videotapes showing how each species behaves in the wild to give an appreciation of *its* habitat and the richness of behavior patterns under normal conditions. Without this glimpse into the quality of life under free conditions, there is no way for most people to appreciate how much we have removed from the animals and how much we must strive to replace.

— 20 —
Beyond Chronic Catheterization in Laboratory Primates

Jaylan S. Turkkan, Nancy A. Ator,
Joseph V. Brady and Karen A. Craven

INTRODUCTION

While an empirical definition of "psychological wellbeing" remains elusive for nonhuman as well as human primates, many of us who work with laboratory animals in the same phylogenetic family as ourselves have seen a need to be especially sensitive to the conditions under which scientific data are obtained in laboratory settings. Over the years, much progress has been made in the improvement of housing conditions and experimental procedures used with nonhuman primates (e.g., Spinelli & Markowitz, 1985). In cardiovascular psychophysiology and behavioral pharmacology, for example, a number of procedures have been developed and implemented that not only widen the range of methodological options for conducting experiments with these animals, but also generally decrease the need for surgical procedures or restriction of the animals' movement. These improvements include, for example, increased freedom of movement at all times as well as new techniques that obviate, in many cases, chronic surgical implantation of catheters. In this chapter we will describe behavioral techniques for the indirect measurement of blood pressure and for oral delivery of drugs, both of which can be implemented in the primate laboratory as alternatives to chronic catheterization.

THE MEASUREMENT OF SYSTOLIC AND DIASTOLIC BLOOD PRESSURE

We have previously reported the development of a swivel-tether-harness system for the controlled intravenous delivery of drugs to baboons (*Papio cynocephalus*) (Lukas, Griffiths, Bradford, Brady, Daley, & DeLorenzo, 1982) and for the measurement of arterial blood pressure with use of a miniature transducer inserted into a backpack (Report of Task Group on Restrained Animal Considerations, 1984). Despite these improvements in freedom of movement, repeated and unpredictable loss of catheter patency or infections at surgical exit sites have led to interruptions of chronic blood pressure experiments. Catheter removal, recovery, surgical re-catheterization, and recovery of blood pressure baselines often involved months of experimental delay. As a result, fewer subjects were examined than advisable for statistical power or experimental control.

Figure 1 illustrates an alternative method for obtaining blood pressure measurements from individually-caged baboons. We have used behavioral procedures of shaping, fading, and reinforcement to train the baboon to extend an arm on command onto a shelf that is attached, at heart level, to the outside of the animal's living cage; and to accept repeated cuff inflations for manual auscultatory blood pressure measurements. Each measurement includes from two to three blood pressure determinations which are then averaged to give the final systolic and diastolic values for that measurement period. Blood pressure is measured three times a day (at 0900, 1200, and 1600 hrs).

The duration of training has varied; two animals accepted this procedure reliably after 2-3 months of daily training while other animals required 4-5 months of training. We have determined, however, that many brief training sessions distributed over the day are preferable to one long daily training session.

We recommend that the trainer be comfortable and experienced with auscultatory blood pressure measurement, which we have ensured by enrolling in a one-day training course offered by the American Heart Association. The animal training procedure is facilitated by habituating the animals both to the laboratory and to the trainer.

Other broad prescriptions for shaping behavior (applicable to other behavior as well as the present task) are that the act being trained at each training stage should initially be rewarded, but later not rewarded, in order to encourage the appearance of further acts that will be rewarded and practiced in the next stage; and that training stages should progress quickly to avoid overlearning and premature fixation of behavior at an early stage. As with any training procedure, the trainer must be sensitive to the need to fall back to earlier training stages when progress has halted, lest acts qalready trained deteriorate.

a

b

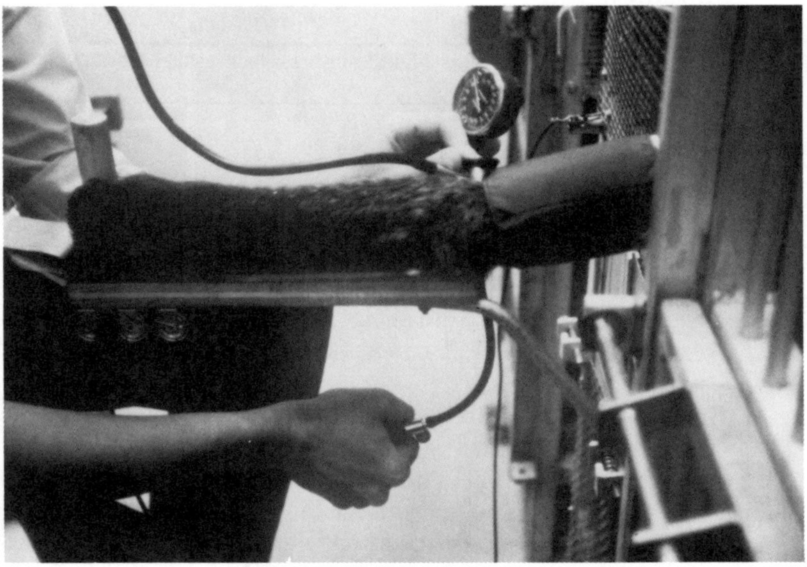

Figure 1. Manual auscultatory blood pressure measurement procedure. A: Animal extends arm onto shelf. Cuff is placed on arm. Stethoscope is placed on brachial artery. B: Cuff is inflated.

C

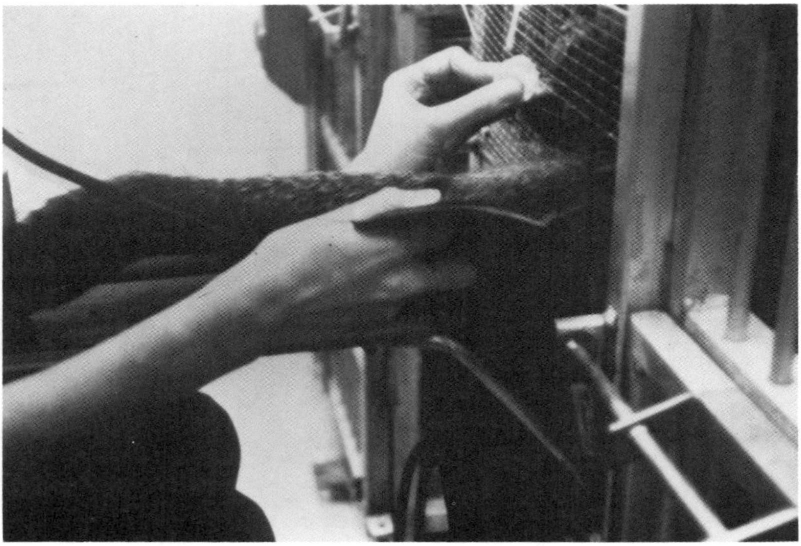

Figure 1C: Fresh fruit reward after 2-3 blood-pressure measurement cycles.

In our training program for baboons, an arm shelf, with a 4-in vertical post at one end, is attached to the animal's cage (Figure 1) and the animal is rewarded with a small food pellet for the following actions in progression:

* Extending an arm onto the shelf.
* Next, extending the left arm as far as the post; later touching the post.
* Grasping the post with the left hand, and next, holding the post for increasing durations.
* Allowing the arm that is holding the post to be touched and next, allowing it to be stroked with the blood pressure cuff.
* Allowing the cuff to be placed briefly around the arm. At this stage, the cuff is opened and closed repeatedly so that the animal will habituate to the sound of the velcro fastening and unfastening. At each step, food rewards are given freely.
* With the cuff in place, allowing the stethoscope to touch the arm

* At this next crucial step, with the cuff, stethoscope, and aneroid manometer in place, the trainer slowly inflates the cuff while delivering frequent food pellets. It is important to keep the training session short so that aversion to the cuff inflation does not have time to develop. Most animals begin to pull on the blood pressure apparatus at this stage, and the trainer needs a quick hand to rescue all the paraphernalia before the animal can pull them into its cage.

  Also at this stage, training is facilitated by improving the quality of reinforcer by switching from food pellets to fresh fruit chunks, or apple sauce dispensed to a food nozzle by means of an infusion pump.
* The trainer then increases the rate and degree of cuff inflation while decreasing the frequency of reinforcement.
* The trainer must deflate the cuff slowly and carefully to be able to note the appearance and disappearance of Korotkoff sounds. It is therefore necessary to reward the animal at this stage for sitting extremely still and waiting quietly through a period of nonreward while the trainer attends to the blood pressure measurement.
* A piece of fresh fruit follows the final measurement.

We want to emphasize that an episode of uncooperativeness by the animal must never be allowed to end a training session, because then the animal quickly learns to avoid discomfort by acting aggressively. When aggressive acts such as scratching the trainer have occurred, we have withheld reinforcement and imposed a few minutes delay, and then resumed training for several more minutes before ending the session.

Once animals are trained, they readily present the arm for blood pressure measurement even if they are actively engaged in an interaction with an animal in another cage. Blood pressure training took place between 0900 and 1600 hrs; animals were not fed their daily ration until 5 pm, which ensured that the food reward during blood pressure training would be a salient reinforcer.

Sample Results. Frequency histograms of individual blood pressure determinations obtained with the manual auscultatory procedure from three normotensive baboons and one hypertensive baboon are shown in Figure 2. The histograms cover blood pressure determinations over 37 (Baboon DES), 77 (Baboon PW) and 134 days (Baboons DI and DU).

These baboons are currently engaged in a project investigating the effects of standard orally administered antihypertensive medications on visual (color) sensory function and psychomotor performance. Mean arterial pressure of animal DES is experimentally elevated by approximately 20-25 mm Hg over levels of the other three animals. This animal and additional experimental baboons simulating clinical hypertension are used

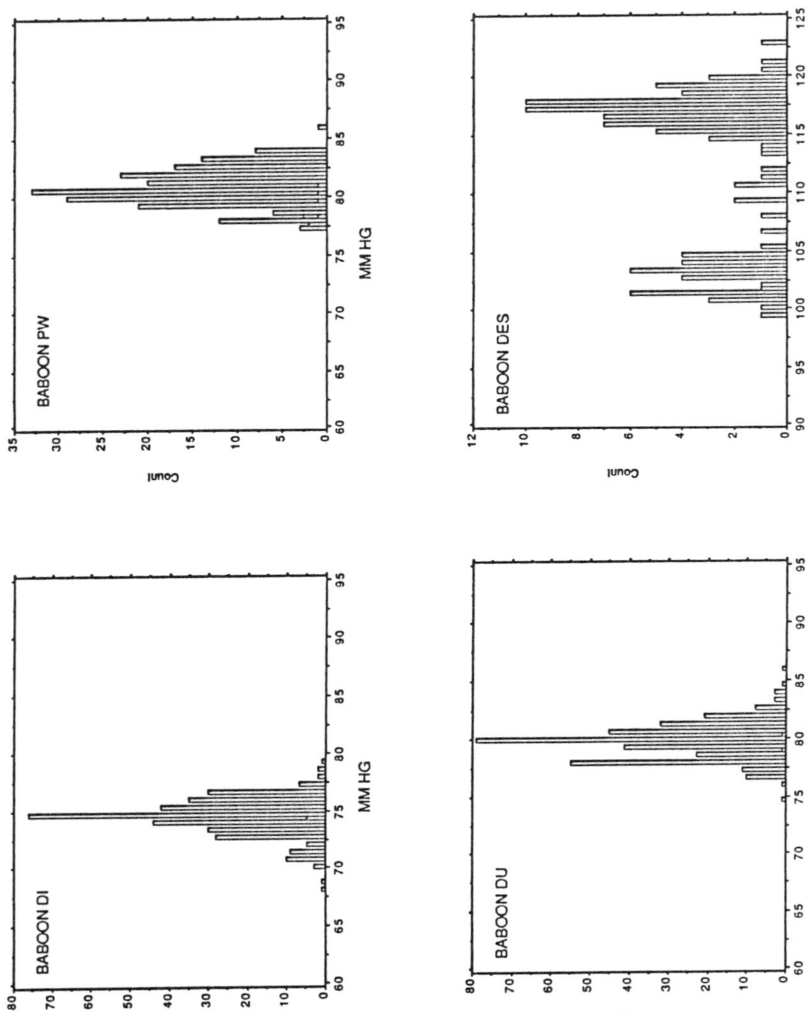

Figure 2. Frequency distributions of mean arterial pressures from four baboon subjects.

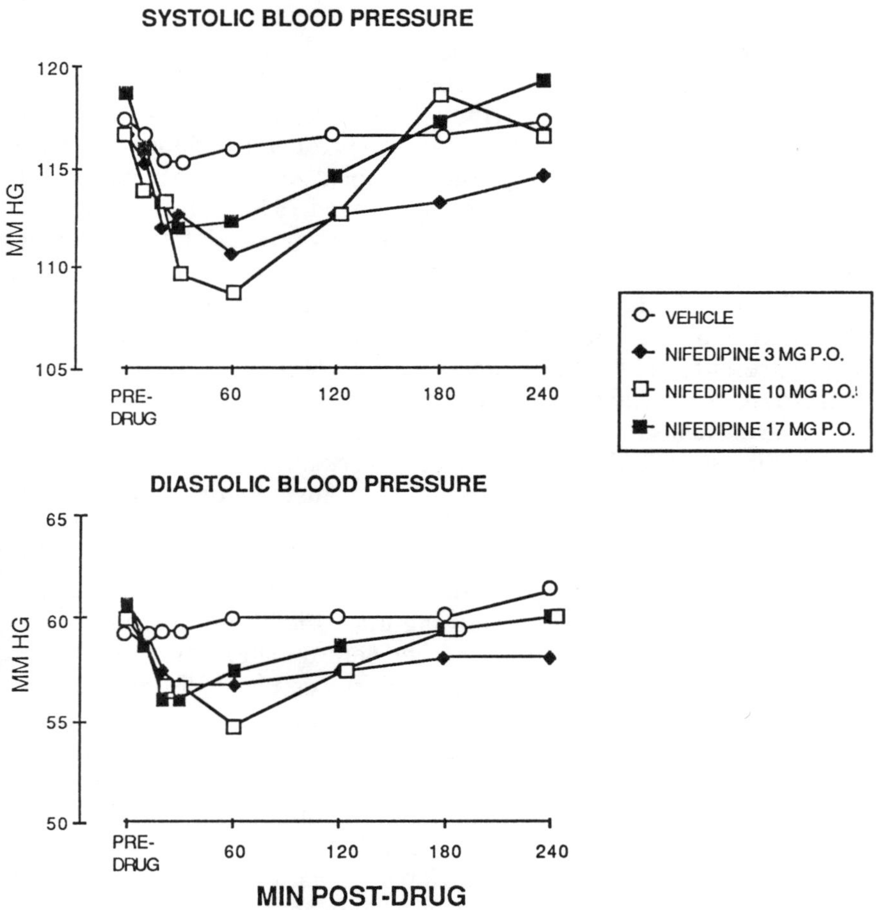

Figure 3. Mean systolic and diastolic blood pressure as a function of time; drug dose is shown as a parameter.

to compare behavioral impairments produced by antihypertensive drugs across normotensive and hypertensive subjects.

The hypertension was surgically induced by unilateral renal artery stenosis (two-kidney, 1 clamp renovascular hypertension) at the Southwest Foundation for Biomedical Research. The baboons were shipped to our laboratory approximately three months after the surgical procedure. During the same weeks that the baboons were being trained to present their arms for blood pressure measurement, they also were being trained in the combination reaction-time and color-discrimination task, which we use to assess behavioral impairment.

Mean arterial pressures in Figure 2 are derived from systolic and diastolic blood pressures at each determination {DBP + [(SBP-DBP)/3)]} and appear to be normally distributed, as were pressures we obtained previously from intra-arterial catheters during resting conditions (Turkkan & Harris, 1981). Mean arterial pressure, pooled across time of day, of the normotensive subjects DU, PW, and DI averaged 79.8(1.9, S.D.), 80.8(1.7), and 74.4(2.5) mmHg respectively (range of blood pressure from highest to lowest value in each distribution: 24.6, 8.7, and 37.3 mmHg respectively for 338, 193, and 332 individual mean arterial pressure values). The mean arterial pressures are in close agreement with an overall average mean arterial pressure of 78 mm Hg derived from systolic and diastolic blood pressures of chronically-catheterized baboons (range of individual means among seven subjects: 77-82 mmHg); these values also were obtained in a basal, resting state (Goldstein, Harris, Izzo, Turkkan, & Keiser, 1981; Harris & Turkkan 1981; Turkkan & Harris, 1984). Auscultatory mean arterial pressure of DES, the hypertensive baboon (Fig. 2) averaged 112.3(7.2) mmHg (top to bottom range: 28.7 mmHg for 102 determinations). The bimodal shape of the frequency histogram of Baboon DES is due to a change in the method of determining diastolic blood pressure after approximately 40 determinations; earlier determinations were based on Phase V (disappearance of K sounds), later ones on Phase IV (muffling).

The sensitivity of the manual auscultatory cuff procedure and its feasibility in acute dose-effect studies is illustrated in Figure 3. The first point in each panel is a pre-drug value, after which the drug was delivered orally as a powder mixed with banana. Systolic and diastolic blood pressures are averaged over three normotensive baboon subjects as a function of time before and after ingestion of vehicle or 3, 10 or 17 mg p.o. of the antihypertensive drug nifedipine (a calcium channel blocker) inserted into vehicle. Measurements every 10 min for the first half hour yielded expected immediate decreases in mean arterial pressure. Hourly measurements continued until 4 hr after drug administration, when blood pressure had returned to pre-drug baselines. These data demonstrate that acute dose-effect determinations can be obtained from the baboon by means of repeated indirect cuff measurements.

Further Developments

Our indirect measurement procedure is being improved currently in two ways. First, we are testing an automatic oscillometric blood pressure monitor (Lifestat model #200, Physio-Control) that measures blood pressures at pre-set intervals and provides a printout of time of day, systolic blood pressure, diastolic blood pressure, mean arterial pressure, and heart rate. In the oscillometric method, an air-filled bladder connected to a pressure transducer serves both as an occlusion cuff and a sensor to detect pressure pulsations transmitted from the underlying artery. Systolic blood pressure, diastolic blood pressure and mean arterial pressure are associated with changes in cuff pulsation amplitude as the cuff is gradually deflated. The advantage of this method over auscultatory machines is that placement of transducer elements on the baboon's arm need not be precise. Also, transducers in auscultatory machines are susceptible to acoustic and mechanical noise.

The Southwest Foundation for Biomedical Research in San Antonio has compared the indirect oscillometric method with direct measurement of radial artery pressure in eight baboons (Mitchell, Herbert, Widogsky, & Morris, 1980). Automatic oscillometric measurement pressure compared very favorably with intra-arterial catheter measurements: average difference between the two methods of determining mean arterial pressure in anesthetized animals was -1.01 (5.84) mmHg. Moreover, the oscillometric readings accurately tracked rapid changes in pressure after i.v. infusion of pressor and depressor drugs.

Indirect blood pressure determinations have also been improved in a second way, by use of a random-zero sphygmomanometer (generously donated for our use by the Hawksley Company in Sussex, England). This device reduces bias: It randomly varies the amount of mercury in the manometer system before each measurement (range of zero level: 20 mm Hg); the true blood pressure levels are calculated after the reading is taken. Random-zero syphygmomanometers are the standard for judging accuracy of oscillometric machines in human epidemiological studies (e.g., Whelton, Thompson, Barnes, & Miall, 1983). A systematic schedule of blood pressure measurement with the random-zero device provides for checks against bias.

DRUG ADMINISTRATION PROCEDURES

For many years, chronic catheterization has been the most widely used method for studying drug self-administration in nonhuman primates. Intravenous methods, originally developed for rhesus monkeys (Yanagita, Deneau, & Seevers, 1965), were extended to squirrel monkeys (Goldberg, 1973) and baboons (Griffiths, Wurster & Brady, 1975). More recently, improved methods have been reported for maintaining intravenous catheters in the baboon, as well as for implanting and maintaining chronically-indwelling intragastric catheters (Lukas Griffiths Bradford Brady Daley & Delorenzo 1982).

The advantages of chronic intravenous catheterization are that specific and precisely-timed drug doses can be reliably and automatically delivered to the animal, and that the effects are virtually immediate, a critical variable in the study of reinforcement processes. The disadvantages of chronic catheterization, whether intravenous or intragastric, are that the method is extremely labor- and material-intensive and requires a high level of training of all involved: surgery must be done under aseptic conditions; postoperative care must be rigorous to prevent infection; and catheter maintenance involves elaborate restraint systems, particularly with large primates, to protect the catheter.

Theoretical and practical considerations have led to greater interest in developing reliable methods for studying self-administration via the oral route. Studies using intravenous drug self-administration have shown that drug reinforcement fits well within general reinforcement theory (Johanson, 1978; Morse & Kelleher, 1977). The intravenous model is useful also because of society's persistent problems with abuse of drugs self-administered intravenously or by routes with comparably rapid onset of effect (e.g, intranasal cocaine). But several clinically useful drugs, particularly sedative/anxiolytic and several stimulant drugs, are prescribed to be taken orally and are most likely to be abused by this route. Similarly, laboratory models of alcohol abuse require a reliable way to generate chronic "volitional" oral self-administration (cf. Lester & Freed, 1973). Furthermore, researchers may wish to administer a drug chronically in order to study the physical or behavioral sequelae of such a regimen. Methods developed to study the reinforcing efficacy of a drug taken orally can provide a useful means of engendering chronic drug intake, even to levels of physical dependence.

The methods to be described below obviate forced drug-taking, such as making the drug solution the only fluid available or adulturating food. We will first describe a procedure for oral drug self-administration that induces baboons to consume drug solutions at behaviorally-active levels. The major focus of this procedure is typically on level of intake as the dependent variable. The second oral-dosing procedure we will describe induces the baboon to reliably ingest a fixed amount of drug in a brief period (less than 15 min), so that acute effects of the drug can be assessed. The latter procedure is useful when the focus is on drug administration as the independent variable.

Oral Drug Self-Administration

Methods of oral drug self-administration developed with rats and rhesus monkeys and reported by Meisch and his colleagues at the University of Minnesota (Meisch, 1977) have recently been successfully adapted for use with baboons to study oral self-administration of ethanol (Henningfield, Ator, & Griffiths, 1981), the short-acting barbiturate methohexital (Ator & Griffiths, 1983b), and the benzodiazepines triazolam and diazepam (Ator & Griffiths, 1987; Griffiths, Lamb, Ator, Roache & Brady, 1985).

In the oral self-administration procedure, three-hour experimental sessions are conducted at the same time each day. During the session, fluid is available only through a specially-constructed drinkometer (Dan Gieske Co., Minneapolis, MN, similar to the one described in Henningfield & Meisch, 1976) that is mounted inside the cage (Figure 4). Lights on the faceplate of the drinkometer signal fluid availability. Contact with the spout completes a circuit and permits fluid to flow through a solenoid valve for either a programmed interval or until the baboon's mouth is removed, whichever occurs first. Water-insoluble drugs (e.g., diazepam and triazolam) have been suspended in a matrix made with 1-4 gm/liter of Bio-Serv Suspending Agent K (Frenchtown, NJ) in distilled water. The amount of suspending agent used depends upon the maximum drug concentration to be studied.

The essential elements of the procedure for engendering self-administration of behaviorally-active amounts of a drug are two: (1) induce ingestion of a large volume of fluid in a short period of time and (2) very gradually increase the concentration of the compound to permit habituation to the taste. Initially water is available through the spout and the daily ration of dry food is delivered all at once 30 min after the session begins. Under these conditions, virtually no drinking occurs in the half hour before food delivery, but the baboon drinks a large quantity of fluid soon after eating the ration of monkey chow (food-induced drinking).

When the subject is reliably consuming a large volume of water (>500 ml for the baboon), a small amount of drug can be added. (If the drug to be studied is water-insoluble, then first the baboon is given habituation sessions with the suspending-agent vehicle alone.) Once drug has been added, the drug concentration is gradually increased (e.g., every 4 days, given that total volume consumed remains stable or increases).

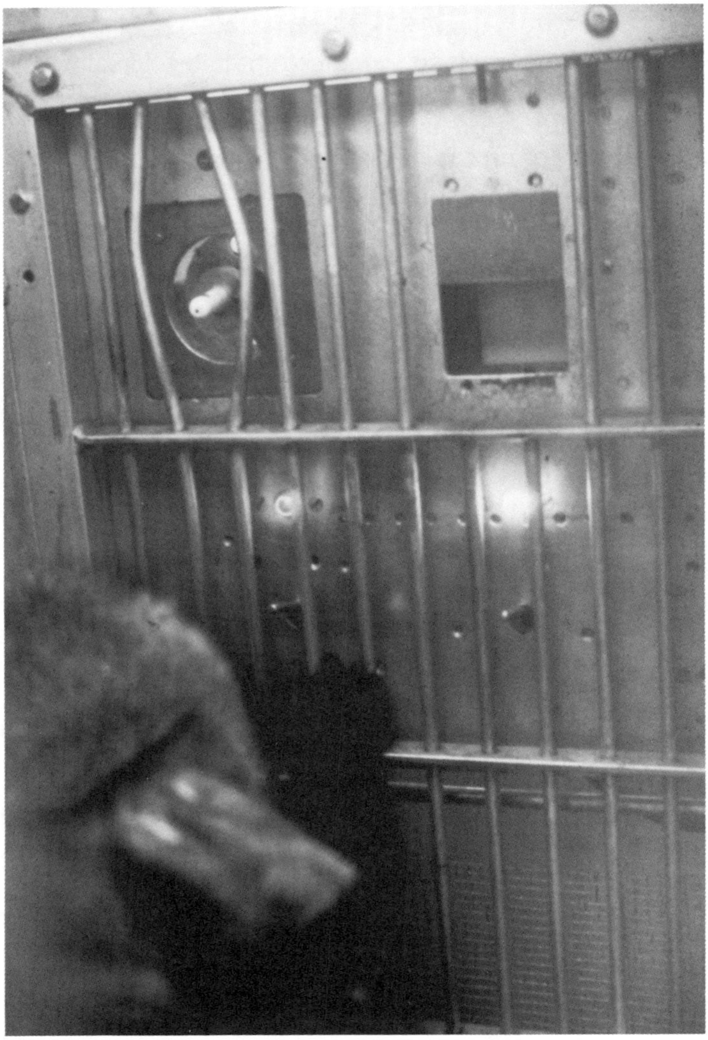

Figure 4. Intelligence panel containing drinkometer (upper left) used for oral self-administration and oral dosing procedures.

At some concentration, the baboon will then be ingesting a behaviorally-active drug dose during the drinking bout that *follows* food ingestion. With drugs that are reinforcing via the oral route, the baboon now begins to consume increasing amounts of fluid in the 30 min *before* food delivery. When the percentage of fluid consumed in the 30 min before food delivery reaches 25 to 30% of the total volume consumed in the 3-hour session, food delivery during the session typically can be eliminated, and oral drug self-administration will be maintained by the reinforcing properties of the drug itself. At this point drug concentration can be manipulated and an oral self-administration dose-effect relationship can be determined without resort to further food-induced drinking. All feeding then occurs 30-60 min after the experimental session ends.

Carroll and Meisch (1984) found it easier to induce oral drug self-administration in animals that had been food-deprived to a specified percentage of free-feeding body weight rather than allowed continuous access to food. While we have not systematically investigated this point with baboons, we find that delivering a fixed ration of monkey chow (as well as a piece of fresh fruit) once a day is sufficient to induce oral self-administration. The baboons typically consume the entire ration within 30 min and get no more food until the next day. During the food-induced drinking phase a ration of 150 g of monkey chow is provided during the session and a piece of fresh fruit later in the day. After drinking has been established and feeding has been shifted to after the session, the ration is gradually increased to an amount calculated to meet the caloric needs of baboons of different sizes. All baboons we have studied have slowly gained weight on these rations, and without drinking less. Note that the baboons have continuous access to water through a separate spout except during the three-hour experimental session and for 30-60 min after the session. Water deprivation plays no part in either inducing or maintaining oral drug self-administration.

Sample Results. We have compared volume of drug consumed at various concentraions to volume of vehicle-alone consumed, both when only one or the other fluid was available in the session and in choice tests with both drug and vehicle concurrently available. These tests revealed clear preference for ethanol (Ator & Griffiths, unpublished data; Henningfield Ator & Griffiths 1981) and methohexital (Ator & Griffiths, 1983b) over vehicle alone. Figure 5 presents the amount of drug (mg/kg) ingested by baboons in studies of oral ethanol and methohexital self-administration and indicates concentrations at which drug was preferred in choice tests. Recent work has also demonstrated that baboons would consume sufficient quantities of both triazolam and diazepam (e.g., 11-18 mg/kg triazolam and 9-15 mg/kg diazepam) to reliably produce physical dependence (Ator & Griffiths, 1987 and unpublished data; Griffiths Lamb Ator Roache & Brady 1985).

Oral Drug Dosing

Lessons learned from developing the method for inducing oral drug self-administration enabled us to develop next a method for reliable oral drug-dosing of these large primates. Oral drug-dosing is useful when an experiment requires delivery at a predetermined time of a specific amount of

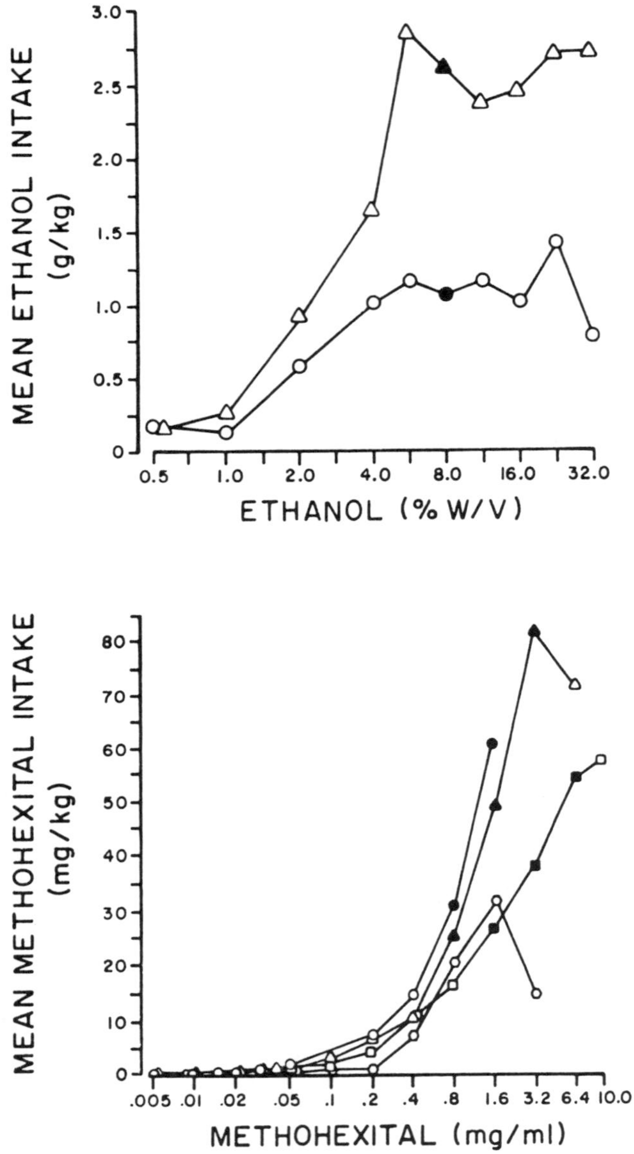

Figure 5. Oral drug self-administration: Mean intake of ethanol (g/kg) and methohexital (mg/kg) by each baboon in last 4 sessions at each concentration. Filled points: concentrations yielding clear preference over vehicle in choice tests (run at most methohexital concentrations and at 8% ethanol). Ethanol data adapted from Henningfield et al., 1981. Methohexital data adapted from Ator and Griffiths, 1983b. (Reprinted with permission)

a compound. For example, in preclinical studies of a drug believed to be potentially useful in treating depression, anxiety, or schizophrenia, the relation between drug administration and a number of specific activities (e.g., locomotor activity or eating) must be determined. This requires delivering a range of doses at specific times before testing (see recent behavioral pharmacology texts, e.g., McKim, 1986; Seiden & Dykstra, 1977).

Many such studies have relied on intramuscular injections. Although this method yields reliable and orderly dose-effect relationships for a number of drugs and types of behavior, some drugs are poorly absorbed via the intramuscular route (e.g., chlordiazepoxide in the baboon). The types of solvents needed for water-insoluble drugs may cause tissue damage with repeated use, leading to poor drug absorption at the injection site (e.g., those requiring high proportions of propylene glycol, such as lorazepam and diazepam). Oral drug delivery via gavage has been routinely employed with rats, pigeons, and smaller primates, but this method is prohibitive with larger primates such as the baboon. Although there are methods for implanting chronically-indwelling intragastric catheters in large primates (Lukas, et. al., 1982), doing so merely for the purpose of delivering a single oral drug dose once a day or every few days seems unwarranted.

It is sometimes possible to induce an animal to accept a drug dose several times by putting it in a flavored drink or other food, but the number of times this method can be used is usually limited because, without special training, the animal often comes to reject the adulturated food or drink or to slow the rate of eating or drinking sufficiently to introduce unacceptable variability in dosing time. Rejection of adulturated food or drink also severely limits the range of oral doses that can be investigated. Individual animals differ in the probability of dose rejection, thus introducing an uncontrolled "selection" factor on subjects that can be included in the study of any given dose. The problem with trying to conceal the taste of a drug in a flavored drink is that we generally overestimate how successful masking is. Experiences with the oral self-administration procedure described above suggested that it might be possible to habituate the baboons to a bitter taste and thus train them to accept such solutions.

The oral dosing procedure involves training baboons to drink a concentrated solution of quinine. Drugs (and the suspending agent if required) can then be added to the solution in the desired amounts. In well-trained baboons, the quinine can be omitted when the drug is given, but in many cases dosing is more easily accomplished by simply adding the drug to the familiar quinine solution. Although quinine may mask the taste of the drug, it is just as likely that the familiar taste of the quinine makes the fluid acceptable despite the addition of an unfamiliar ingredient.

To train baboons to ingest a fixed amount of drug rapidly, all fluid is made available through drinkometers (Figure 4) installed inside the cage. Access to water is controlled through a timer that turns off the drinkometer at a set time each day (e.g., midnight). The next morning, between 0830 and 0930 hrs, lights on the drinkometer panel are illuminated, signalling that fluid is available through the spout. Initially, 100 ml of an orange-flavored drink is offered (e.g., Tang, General Foods Corp., White Plains, NY). Baboons drink this avidly. After all the orange drink is consumed, the water supply is immediately reconnected to the spout and water is continuously available for the rest of the day. Over daily sessions, increasing amounts of quinine sulfate are added to the orange drink up to a concentration of 0.325 mg/ml.

As a standard protocol, drug doses are delivered in 60 ml of orange-flavored quinine vehicle and followed by 40 ml of orange drink alone to assure that the baboon takes all drug in the lines before water is made available. The immediate availability of pure orange drink may or may not be necessary; the essential factor in the success of this method clearly is gradual habituation to the bitter taste of the quinine. If drinking slows significantly or stops as the quinine concentration is increased during training, it is important to decrease the concentration until the baboon drinks readily and then increase the concentration again more gradually over days. Long periods of exposure to a concentration that is not accepted readily only slows the process of training the baboon to accept higher concentrations. Once the baboon is readily drinking the 0.325 quinine solution, it is very probable that it will tolerate the addition of drugs.

Using volumes that do not generally exceed 100 ml is important to completing the dosing in approximately 5 min. Typically, an upper limit of 15 min is set on dosing. If the total dose has not been ingested by then, the drug is removed and water is made available. Whether a test session is conducted after only partial dosing depends on whether the investigator is interested in the effects of the actual dose the baboon received. This dose can be calculated from the amount remaining when the 15-min dosing interval ends. Partial dosing may occur at extremely high doses of some drugs (e.g., 100 mg/kg which, in the baboon, might involve a total dose of 2 to 3 g). Complete delivery of these high doses usually can be accomplished by increasing the fluid volume slightly or by increasing the proportion of orange drink powder. Partial dosing may also occur with some drugs (e.g., caffeine, pentylenetetrazole) after the animal has experienced intermediate to high doses of those drugs. The behavioral effects of those drugs may be sufficiently aversive to decrease their later ingestion. We should point out, however, that merely experiencing gastric distress (e.g., vomiting) after consuming a dose of a drug has not resulted in later rejection of the drug.

Sample Results. For drug-discrimination studies, 11 baboons have been trained to accept oral drug dosing, making it possible to obtain complete dose-effect curves in individual subjects with a number of benzodiazepines, barbiturates, and other sedative/anxiolytic drugs as well as compounds from other classes (e.g., Ator & Griffiths, 1983a, 1985, 1986,

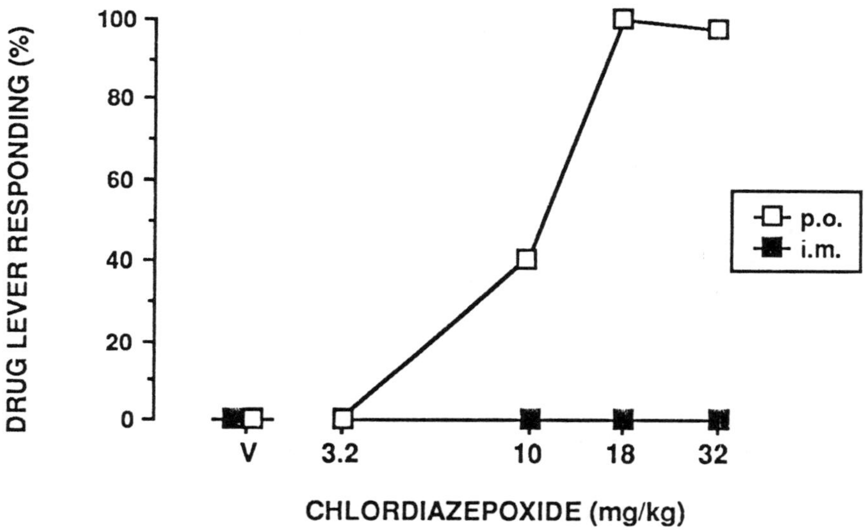

Figure 6. Results from a drug-discrimination study. The baboon was trained to discriminate pentobarbital (10 mg/kg) from the no-drug condition. In training sessions, a food pellet was delivered after 20 consecutive presses on one lever if no drug or vehicle had been given and after 20 consecutive presses on a different lever if pentobarbital had been given. In test sessions, 20 consecutive presses on either lever produced a food pellet. For the tests shown here, chlordiazepoxide or its vehicle (V) was administered either intramuscularly (i.m.) or orally (*per os*, p.o.) 60 min before the session. Points represent percentage of presses on the drug-paired lever.

---

in press). Dose ranges up to 56 and 100 mg/kg have been used with some of the less potent compounds, such as chloral hydrate, halazepam, meprobamate, methyprylon, and phenobarbital. Given that most of the baboons in these studies weighed 25 to 30 kg, a 100 mg/kg dose required that 2 to 3 g of compound be administered.

Figure 6 presents data from a representative subject in a recent drug-discrimination study in which chlordiazepoxide was shown to be much more potent when administered orally rather than intramuscularly (Ator &

Griffiths, unpublished data). In fact, conclusions regarding generalization to chlordiazepoxide in pentobarbital-trained baboons might have been erroneous had only the intramuscular route been used.

The method of oral dosing described above has been modified to train baboons to accept fixed oral ethanol doses. The taste of ethanol is apparently different enough from quinine that habituation to the ethanol taste itself is most effective in obtaining an oral ethanol dose-effect curve. In this case, baboons merely received increasing concentrations of ethanol (up to 20% w/v) daily in 100 ml distilled water. With baboons that had already been trained to drink quinine solutions, putting ethanol in the quinine vehicle facilitated acceptance of the higher ethanol concentrations (Ator & Griffiths, unpublished data). Baboons that had not been trained to drink the quinine vehicle, however, were successfully trained over a period of weeks routinely to accept doses up to 1.3 mg/kg ethanol or approximately 300 ml of 11% w/v, depending upon the weight of the baboon (Hienz, Brady, Bowers, & Ator, 1987).

CONCLUSION

To the extent that these new options reduce stress or discomfort, they can be said to improve the psychological wellbeing of nonhuman laboratory primates. Training animals to perform complex tasks so that they can be spared surgery and its sequelae may not at first glance appear to be an efficient use of research time (behavioral training in the blood-pressure measurement procedure requires an investment of at least 3-5 months). But the ultimate advantages not only include smoother progress of experiments at less expense (e.g., broken catheters do not unexpectedly delay a condition) but also assure the wellbeing of animals that are to serve as subjects over many months or years.

ACKNOWLEDGEMENTS

Development of the blood pressure measurement methods was supported by Grants HL34034 and HL40138 awarded by the National Heart Lung and Blood Institute. Development of the oral self-administration and oral dosing methods was supported by Grants DA01147 and DA04133 awarded by the National Institute on Drug Abuse. We would like to thank Mary Kay Story for assistance with the blood pressure training, Bettye Campbell and Susan James for assistance with the oral self-administration and oral dosing procedures and Kellie Leatherman for manuscript preparation.

Family: Hylobatidae (Gibbons)

— 21 —
An Inexpensive, Climate-Controlled Enclosure for Gibbons Utilizing Appropriate Technology

Jeremy F. Dahl

INTRODUCTION

The natural environments of apes vary with the season and with the time of day. This should be taken into account in designing, building, and maintaining environments suitable for captive apes (Dahl, 1982). Every attempt should be made to create large spaces that are maximally usable, that fluctuate with the weather outside but are buffered against extremes of temperature (Dahl & Smith 1985), and that can be built and maintained inexpensively. Appropriate technologies (Darrow & Pam 1976) can be of great value in achieving these goals.

An opportunity arose to utilize appropriate technologies in the design of new housing for Lesser Apes (Hylobatidae) when the Yerkes Regional Primate Research Center obtained a small breeding colony of white-handed gibbons (*Hylobates lar*) and initiated a pilot study of their reproductive characteristics. A suitable exterior enclosure was already available but new interior housing was needed. This paper describes the reasoning that went into the design of the new housing, how it was constructed, how it was operated, and the environmental biophysics that governed its main design features.

The theoretical basis for the design was drawn from the ethology and ecology of the gibbon rather than from psychology. Ethological explanations of behavior are based "solely on what was observed" (p.3), rather than on covert psychological processes and hypothetical mechanisms (learning, memory, etc.) (Mason, 1986). The design of primate enclosures should be based on observed behavior rather than on a covert state of "psychological wellbeing" that is difficult to define operationally or to measure. "Ethological wellbeing" is a more appropriate expression; it can be estimated from direct observation rather than by reference to mental processes that cannot be observed. The design of the gibbons' interior housing provided for their wellbeing by referring to the gibbons' morphology, positional behavior (*sensu* Prost 1965) and social organization, and the physical (particularly thermal) characteristics of their natural environment.

The goals of the pilot study, viz., to investigate reproductive physiology and behavior as well as other social behavior, also had to be considered. First, urine had to be collected daily from the female without disrupting her activity patterns (Dahl 1980).

Second, enclosures housing mixed-sex pairs have to be spaced as far apart as possible. Gibbons tend to live in territorial family groups (Carpenter, 1940; Ellefson, 1974), and when one pair is housed too close to another, the proximity of potential male or female rivals raises the possiblity of agitation or frequent agonistic displays. Hence enclosures for different pairs or family groups must be independent of each other.

DESIGN RATIONALE

Gibbon Morphology and Posture

Gibbons are the smallest and lightest extant hominoids. Adult males weigh about 4.3-7.9 kg and adult females about 4.1-6.8 kg. They are true brachiators (Napier & Napier, 1967), locomoting in a suspensory mode. They have one of the highest intermembral indices (129-137, Schultz, 1933) of all primates, and an extraordinarily large span. (Measures of span are not available in the literature, but my own measurements of adults yielded a range of 130-150 cm.) (Figure 1) This is critical in the structural design of a gibbon enclosure, for if the available space is to be usable for locomotion, it must provide the animals with suitable substrates appropriately placed (Prost 1965): plenty of small- to medium-gauge branch-like supports with clear space beneath (see Figure 1). Resting surfaces should be positioned one body length plus one arm length beneath the horizontal supports so the animals can sit with one arm suspended, a common resting posture (Ellefson, 1974, p. 71; personal observation).

The size and shape of gibbons, and the energy flow between their body surface and the ambient air (Campbell 1977), must also be considered in designing a suitable thermal environment for them. Gibbons have a high ratio of surface area to body mass, particularly when the limbs are extended (Patterson, 1986), so that heat loss can be rapid in cool weather unless the gibbons assume a nearly spherical posture. On the other hand, the potentially high ratio of surface area to body mass offers a large absorptive area to solar radiation, an advantage in sunny cool weather. The gibbons at the Yerkes Field Station assume a special sunning posture on cool, sunny, early mornings (personal observation): the arms are raised and partly flexed, the thighs are everted, and much of the ventral surface and face is turned toward the sun. It would be reasonable to assume, therefore, that if the gibbons were provided with cheap and energy-efficient *radiant* heat sources, they could maintain thermoneutrality at relatively low ambient temperatures by absorbing radiant energy at the body surface (Walsberg, Campbell & King 1978).

Usable Space vs Thermoneutrality

In the wild, primates do not occupy much space at any one time, but over time they range over a great deal more space by occupying a succession of smaller spaces one after the other (a "multiple space": Dahl, 1982). Gibbon groups typically occupy one or two trees at a time, but their home ranges comprise approximately 100 hectares containing hundreds of trees (Ellefson, 1974). The minimum space provided to a captive gibbon group, then, should equal the space within two trees; ideally they should be able to range over a space equal to that of a large number of trees.

These considerations prompted me to design a gibbon enclosure consisting of several separate, distinctive spaces, each equivalent to the space within two trees, that the gibbons could visit in turn. This design contrasts with the enclosure for a gibbon colony described by Breznock, Harrold and Kawakami (1977), which lacked almost all the key features of my design, and amounted to a poorly-differentiated single space having minimal variety within it. Their enclosure measured 10 x 12 x 7 ft high (available space 840 ft^3), and had a single resting ledge and ceilings made of 2x2-in. wire mesh. Temperature was allowed to fluctuate only within 6° F and the light-dark cycle was constant at 12:12 hr.

One way to increase usable space is to add a spacious outdoor enclosure whose microclimate fluctuates with the macroclimate (see Dahl & Smith, 1985; Yoshino, 1975). Increasing the volume and complexity of the indoor enclosure is also desirable, but it is costly to do so if traditional ways of heating, cooling and ventilating are used to limit temperature fluctuations and approximate thermoneutrality. An alternative is the technology of passive air conditioning: Maximize solar gain to the interior on cool days and minimize it on warm days. The interior can be allowed to fluctuate over a range of ambient temperatures provided that in cool weather radiant heat is available and humidity is kept high to reduce evaporative heat loss through the lungs (Campbell, 1977); in hot weather, on the other hand, radiant heat should be shut out and humidity should be kept low.

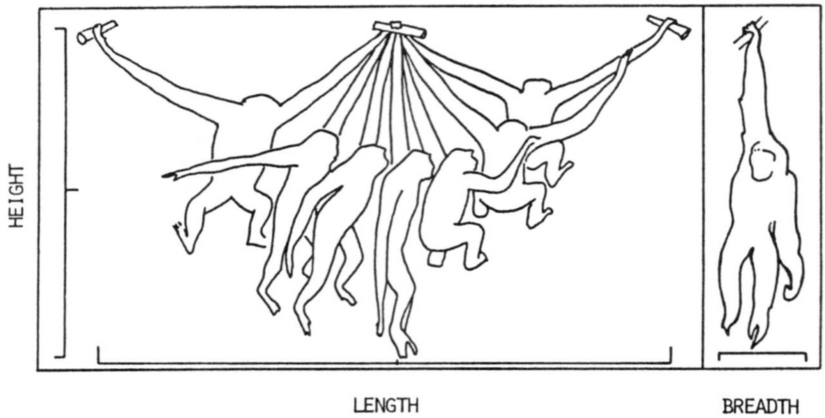

Figure 1. Relative dimensions of usable space for gibbon suspensory locomotion.

Figure 2. Exterior view of the gibbon enclosure.

The natural environments of *Hylobates* exhibit a wide range of ambient temperatures. Ellefson (1974) reported approximate maxima of 93° F in the habitat of *H. lar* in the Malay peninsula, while Carpenter (1940) reported daily minima of 45° F in China. Gibbons must survive such low ambient temperatures by huddling together (Dahl, Bernstein & Williams 1986). Comfort does not depend solely on ambient temperature; energy flow between body surface and air actually depends on a much wider range of physical factors (Campbell, 1977; Dahl & Smith, 1985; Patterson, 1986; Porter & Gates, 1969).

The same biophysical principles apply to the design of outdoor enclosures. These need sunny areas sheltered from the winter wind (Siple & Passel, 1945), and breezy areas shaded from the summer sun.

Existing Enclosures

An outdoor enclosure was already available for the gibbons, and its basic structure was not altered. It was selected because it had a number of desirable design features. *First,* it was a varied space consisting of a tower (15 x 6 x 17 ft high, 1,530 ft^3) continuous with another area with a lower ceiling (15 x 10 x 5 ft high, 750 ft^3). The tower had a central pole extending from floor to ceiling and was crossed by two horizontal poles (Figure 2). Walls and ceilings were made of 2x2-in. wire mesh, providing surfaces for climbing and brachiating. *Second,* the enclosure lay in a topographical dip, with a building (9.5 ft high) immediately to the north, so that the enclosure was exposed to the southern sun and, below 9 ft, was well protected from the wind.

Restraint System

Every primate facility needs some means of restraining animals for medical examination without undue physical harm or psychological trauma and stress. For our gibbon study I also needed to collect daily urine specimens from the female to assess fluctuations in reproductive hormones. I designed a tunnel, with a sliding door at each end, that connected indoor and outdoor areas; the gibbons could brachiate easily through the 4x2-ft opening. Below the raised wire-mesh floor of the tunnel we placed a urine-collection pan. In addition, a squeeze panel could be fitted to the tunnel for use when anaesthetizing recalcitrant or untrained animals. Critical to non-traumatic collection of urine and administration of anaesthetics was the use of positive reinforcement to train the female gibbon to enter the tunnel and tolerate confinement within it, as well as handling when it was called for.

FINAL DESIGN

Outdoor Enclosure

The Yerkes Center is located in northwest Atlanta, Georgia (approximately 34°N latitude, elevation 950 ft). Normal daily minimum temperatures range from 33° F in January to 69° F in July; maximum temperatures range from 51° F in January to 87° F in July. The climate at Yerkes is relatively mild to warm, with about 220 days of clear or partly-cloudy weather each year. Day length varies over the year from about 10 to about 14 hours.

The existing outdoor enclosure needed little alteration because it already had some structural complexity, enclosed a significant volume (2,250 ft^3), and had considerable usable space due to the lower 10-ft section and the southern exposure. We added more shade by putting corrugated iron sheeting (5x5 ft) over the center of the lower unit and over the highest cross-pole. The animals could sit in the shade where the cross-winds were strongest, and be protected from rain. In the mornings, there was some additional outdoor shade immediately to the west of the indoor area.

Indoor Enclosure

The walls of the indoor area enclosed an 8x5-ft floor area. They were constructed of an outer layer of fiberglass sheeting and an inner layer of 1/16-in. aluminum sheeting, with a 4-in. gap between them that was filled with cellulose insulation rated at about R14. To provide a large thermal mass for storing heat, the north wall was given an added layer of masonry bricks to a height of 6 ft. Access to the indoor enclosure for personnel was via a double-glazed Plexiglas door in the south wall. The east wall was fitted with a Lixit drinking device and a food hatch, 7 x 6 in.

The roof of the indoor enclosure consisted of a 40x40-ft section of double-glazed, highly-transmissive fiberglass sheeting with a 2-in. gap between the layers. The roof sloped 34° toward the south, an angle calculated to maximize solar gain during the winter at latitudes of about 34° N. The northern end of the roof was open, but was fitted with a 5x2-ft hinged metal hatch that could be raised or lowered over the opening. Foil-covered sheets of insulating material were hinged to the double-glazed section. They could be raised to allow sunlight to enter, or lowered to provide insulation in cool weather, and shade in hot weather (Figure 3).

The interior ceiling was made of 1x2-in. hardware cloth. It was installed a few inches below the roof over the southern two-thirds of the area, and farther below the roof at the northern end, to allow space to accommodate an exhaust fan and an array of infrared heating lamps mounted above the interior ceiling. The fan, controlled by a thermostat set to 75° F, provided added ventilation in summer. A space between the

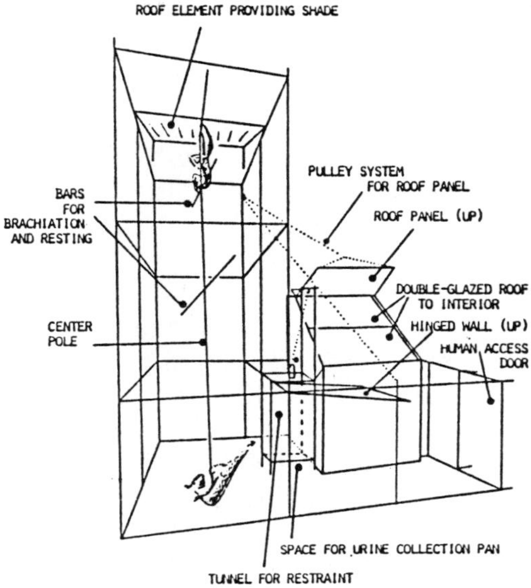

Figure 3. Functional parts of the outdoor gibbon enclosure.

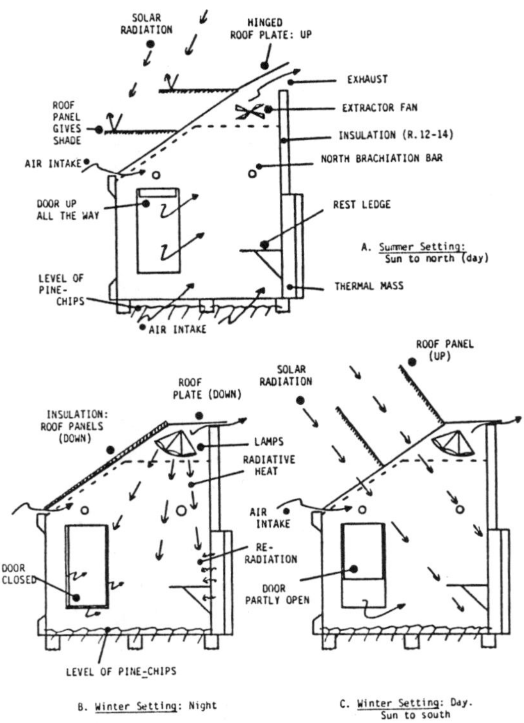

Figure 4. Interior enclosure. Structure and mechanisms of thermal regulation during (A) summer, (B) winter nights, (C) winter days.

bottom of the walls and the floor was left unsealed to provide additional summer ventilation, as well as passive cooling by adiabatic humidification (Kreith, 1958). In summer, the ground was kept as damp as possible: relatively dry air passing over damp ground will cool as it picks up moisture. Both the shape of the roof, and the gap between walls and floor contributed passive ventilation through the enclosure independently of the exhaust fan.

A black-bulb thermostat regulated the three 250-w infrared heat lamps. Black-bulb thermometers consist of a copper sphere, painted black, with a thermometer or thermistor centered inside the sphere. Black-bulb thermometers measure the resultant of ambient temperature and radiant energy falling on the sphere. By monitoring not only ambient temperature but also heat radiating from the sun or the infrared lamps, we were able to maintain the animals at thermoneutrality with greater energy efficiency.

Two brachiation bars (1/2-in. diameter) were placed across the enclosure, one to the north, under the heat lamps, and one to the south. The northern bar was placed 24 in. below the ceiling and 39 in. above a resting ledge. On cool nights the gibbons could rest comfortably on the bar, directly under the lamps, while grasping the mesh of the ceiling; or they could rest at a lower elevation, on the cooler ledge. The floor was covered with a deep layer of pine chips, providing the animals with yet another choice of resting place, still farther from the heat lamps. In winter, pine chips could be piled up to cover the gap between walls and floor to reduce ventilation and infiltration of cool outside air.

Restraint System

A tunnel (4 x 4 x 2 ft) extended from the west wall of the indoor enclosure a short distance through the low portion of the outdoor enclosure to the tower beyond. The roof of the tunnel was constructed of opaque Plexiglas to provide shelter from sun and wind, while the walls and raised floor were made of 1x2-in. hardware cloth. A pulley-operated guillotine door shut off the tunnel from the indoor enclosure; two aluminum panels, one at midpoint and one at the far (tower) end, could be slid into place to create a small confinement space. Two small brachiation bars were placed across the top of the tunnel, 2 ft apart. When all barriers were open, the animals could brachiate the length of the tunnel in one continuous motion.

When the female was initially being trained to enter and tolerate confinement within the tunnel, we first had to release the male to the outdoor enclosure as an inducement to her to go outside. So that staff could bait the tunnel, collect urine samples, and clean the outdoor enclosure without interference from the male, we installed a hinged wall that could be lowered to confine the male in the tower area (Figure 2); it was raised out of the way when not in use.

OPERATION OF THE ENCLOSURE

To function properly, the indoor climate-control system and the tunnel restraint system must be operated correctly. The indoor enclosure is a dynamic system that must be fine-tuned to suit the season.

Winter Operation

In winter, the gaps between walls and floor, and the roof's ventilation hatch are kept closed. Some ventilation is needed, however, to exchange the air and to prevent the accumulation of condensation. To remove moist air, curved piping was attached to the hatch, and the lid was curved slightly to prevent a tight seal when it was closed. A small slit between the roof and the top of the south wall provided another ventilation gap.

The gibbons were confined indoors at night, and the insulating roof panels were lowered to reduce heat loss. The infrared heat lamps were turned on automatically when ambient temperatures dropped *in the absence of radiant solar heat.* The red glow of the lamps signalled to staff that the system was working. On sunny winter mornings the insulating roof panels were raised to let in the heat of the sun, and the guillotine door to the outside was raised to a height of 18 in., high enough so the gibbons could go outside, but low enough to reduce drafts. Solar heat did not enter on overcast days, so the insulating roof panels were lowered to reduce heat loss. On sunny days, the pine-chip flooring was saturated with water during indoor cleaning in order to humidify the area. Contrary to popular belief, a humid interior is advantageous in winter because it reduces heat loss through respiratory evaporation. During very cold weather, when ambient temperatures fall below 40° F and there is a strong wind, the gibbons are not allowed outside and the guillotine door is tightly shut to prevent drafts (Dahl & Smith, 1985.)

Summer Operation

In summer the indoor climate-control system operates in reverse (Figure 4). The hinged roof is open and the pine chips are cleared away from the gaps between the walls and the floor. The guillotine door to the outside is raised to maximum height. The exhaust fan is on, and the insulating roof panels are lowered to provide shade and keep out solar heat. During cleaning, only the floor immediately below the walls is saturated with water, unavoidable when using a high-powered hose to clean the walls.

Urine Collection

The gibbons were confined indoors overnight and the tunnel was thoroughly cleaned. The following morning the guillotine door was raised briefly and the male immediately went outside to the tower. The sliding panel nearest the tower was then inserted to confine him in the tower, the tunnel was baited with pieces of orange, and a collection pan was placed under the raised floor of the tunnel. After 25 training sessions the female reliably entered the tunnel within 15 sec after the guillotine door was raised. She usually urinated within 30 min of entering the tunnel, whereupon the urine pan was taken away and all barriers to free movement through the enclosure were removed. While the female was confined in the tunnel, the male had potentially close contact with her through the wire-mesh north wall.

Medical Examination

As Carpenter (1940) and Matthews (1946-47) noted and Brody and Brody (1974) later demonstrated, gibbons are easily trained, with grooming as the reinforcer, to approach and maintain contact with humans. Once such regular contact has developed, it is very easy to anaesthetize an animal by injection while it is being groomed. A squeeze apparatus can be used if easy human-gibbon contact has not developed. In that case the gibbon is first confined to the whole tunnel and then to half of it by closing the midpoint partition. A false wall can then be pulled across the tunnel from north to south so that the animal can be injected while restrained against the south wall.

DISCUSSION

With the experience gained from several years' operation, the design of the enclosure can now be evaluated. The indoor enclosure was built in 1982 and a breeding pair of gibbons occupied it in December of that year. They have remained in good health through the subsequent years and at the time of this writing had produced two live offspring.

Even though the interior remained warm enough during Atlanta's coldest night of the century (a low of -9° F), a small convection heater equipped with a timer and a multiple-setting thermostat is an energy-efficient precaution against extremely cold weather.

The daily adjustments that had to be made in roof panels and floor litter to fine-tune the climate-control system proved to be problematic due to the number of different caretakers who rotated through the gibbon area. A climate-control system as carefully designed as this one requires the attention of a stable roster of well-trained and supervised personnel. The enhanced wellbeing of the gibbons and the system's fuel economy and efficiency are bought at the price of more labor-intensive attention to managing the system.

The design of the tunnel for urine collection was successful. We had no difficulty obtaining urine from the female almost every day through 11 menstrual cycles and the first 120 days of gestation. Once the female was trained and the male habituated to the collection routine, neither showed any signs of discomfort or stress. The male separated from the female without hesitation, and the female entered the tunnel within 5-15 sec. During confinement in the tunnel the female showed no sign of agitation; her only vocalizations were the typical low food calls that she uttered while eating the bait of orange slices.

The indoor enclosure was relatively inexpensive to construct (materials, including insulation, cost about $2,400). The climate-control system cost little in fuel.

The substrates (wire-mesh walls and ceilings; brachiation bars) permitted the gibbons to brachiate continuously over a fairly long line of travel. All animals were seen at times to brachiate in a continuous trip around the interior, through the tunnel, to the top of the tower and down again, and around the low outdoor area (a total distance of about 80 ft). Play between the adult male and the 3-year-old infant involved chase-and-grab through the tower and rough-and-tumble on the pine-chip flooring. Gauged by the frequency of its use, the pine-chip flooring provided a comfortable resting substrate, as well as something to forage through (see below).

Observations of gibbons at Zoo Atlanta and siamangs at the Audubon Zoo in New Orleans, both of which had ropes available as locomotory substrates, suggest that adults prefer stiff brachiation pipes and bars; they locomoted more hesitantly on swaying ropes. Immature animals, on the other hand, appeared to enjoy playing while suspended from the ropes. An automobile tire was suspended in our gibbons' tower but was seldom used. Adult gibbons appear to prefer well-anchored, stiff substrates over unstable substrates that sway during brachiation.

Since the prototype gibbon enclosure was first built, five additional enclosures have been built and are now in use. These Gibbon Appropriate Technology Structures (GIBATS) include some minor refinements of the prototypical design, but a more important difference is their placement within a densely-wooded area dominated by deciduous trees. This offers some additional improvements to the gibbons' wellbeing. First, there is a nearly continuous leafy canopy over the enclosures that provides shade in the summer; ambient temperatures at the GIBATS are as much as 10° F cooler at midday in summer than 100 m away, outside the woods. In the winter, of course, the trees lose their leaves and sunlight falls on the enclosures.

Second, the animals enjoy a view on all sides of the surrounding woods and their inhabitants (birds, squirrels, etc.), providing experiential enrichment throughout the year. In addition, leaves and various fruits fall into and around the exterior part of the enclosure, and the gibbons have been seen manipulating and often eating these. If arrangements were made to build future nonhuman primate enclosures in wooded areas similar to this one, considerable environmental enrichment, at no additional expense, would be assured.

SUMMARY

1. I set out to enhance the wellbeing of captive gibbons by keeping them in separate family groups in enclosures with much more usable and varied space than is traditional in gibbon enclosures (cf Breznock *et al.*, 1977).

2. I determined that an enhanced gibbon environment needed two structural components:

   a. A large (about 2,000 ft^3) outdoor area with vertical complexity, a rich supply of appropriate substrates, and regions that are sheltered from wind, rain and sun, as well as other regions that are exposed to the elements.

   b. A relatively large (about 275 ft^3) indoor area designed according to principles of alternative technology (Dahl, 1982) to maximize solar radiant-heat gain in winter, and minimize it in summer. In winter, solar heat is supplemented by infrared lamps, while ventilating air currents cool the enclosure in summer (a hybrid of passive and conventional systems).

3. A further aim was to minimize the trauma or stress of collecting urine for experimental purposes by building a simple restraining tunnel

between indoor and outdoor enclosures and training gibbons (by use of positive reinforcement) to enter the tunnel.

4. An outdoor enclosure was already available that had an appropriate mix of areas and volumes, substrates, and climatic variety. An indoor enclosure was built according to the principles of environmental biophysics to function as a simple and energy-efficient climate-control system. It provided shelter and a buffer against climatic extremes of heat, cold, and humidity while using a minimum of fuel (although the last has yet to be fully documented).

5. Personnel problems with operation of the indoor climate-control system showed that proper operation requires supervision of a stable staff of caretakers who must receive adequate training in the rudiments of environmental biophysics.

6. Captive gibbons responded well to an environment that was built to moderate rather than eliminate climatic fluctuations. Appropriate technology enables the construction and operation of inexpensive, energy-efficient, enriched environments for captive primates.

7. The wellbeing of nonhuman primates is further promoted by placing outdoor enclosures within deciduous woodland.

ACKNOWLEDLGEMENTS

I gratefully thank Dr. F. A. King, Dr. K. G. Gould, and Prof. H. Warner for their encouragement during this project. Mr. W. Stabrawa and Mr. D. McCumber provided indispensable technical input and skill during the practical design and fabrication of the indoor enclosure. I am particularly pleased to acknowledge input from the architectural firm of Hoss & Viscardi that was critical to many aspects of the final design. Thanks also to the veterinary staff of the Yerkes Center for advice on the design of the enclosure and for inexhaustible care of the gibbons. This work was supported in part by U.S. Public Health Service grant RR-00165 (Division of Research Resources, NIH) to the Yerkes Regional Primate Research Center, and by NICHD grant HD-19060 and NSF grant BNS 87-08406 to Dr. R. D. Nadler. The Yerkes Center is fully accredited by the American Association for Accreditation of Laboratory Animal Care (AAALAC).

Family: Pongidae (Great Apes)

— 22 —
Feeding Enrichment for Captive Great Apes

Mollie A. Bloomsmith

INTRODUCTION

Within the last decade, behavioral primatologists have become increasingly aware of the influence of feeding on the social behavior of captive primates (Plimpton & Rosenblum, 1983; Rosenblum & Smiley, 1984). Feeding strategies have great potential for rehabilitating animals showing certain behavioral pathologies (Chamove & Anderson, 1979; Chamove, Anderson, Morgan-Jones, & Jones, 1982; Gould & Bres, 1986) and for preventing behavioral problems that are unfortunately quite prevalent among primates in captivity. Reducing high levels of aggression and pathologic behavior such as appetite disorders (Capitanio, 1988) certainly qualifies as promoting psychological wellbeing. If we saw captive primates foraging and feeding roughly as wild primates do, this would be one piece of evidence that the animals were in an adequate captive habitat and that management strategies were appropriate. There are many practical constraints on the feeding of captive primates; we must find creative ways to overcome the constraints so that captive primates can feed and forage in a manner for which they are genetically prepared.

"Applied primatologists" (Maple & Finlay, 1986) who must solve management problems in zoos, laboratories and other primate facilities are developing feeding procedures as a management tool. The strategy is to identify potential factors that might contribute to a particular behavioral problem, and then experimentally alter them, one at a time, to document which are effective manipulations that can be recommended to colony managers. This approach requires prior knowledge of species-typical behavior in natural settings. In this chapter, I will detail information concerning the feeding of wild great apes as a model for improving their feeding in captivity, and then describe the realization of the model in the feeding enrichment program at The University of Texas Chimpanzee Breeding Colony.

THE FEEDING OF WILD GREAT APES AS A MODEL FOR CAPTIVE FEEDING STRATEGIES

The discrepancy between the feeding of captive and free-living primates may contribute to some of the other behavioral discrepancies between wild and captive primates. For example, in the London Zoo chimpanzees and orangutans spent 8% and 14%, respectively, of the day feeding. In contrast, wild apes spend 30% to 60% of the day feeding (Galdikas & Teleki, 1981; Goodall, 1986; Kano & Mulavwa, 1984; Rodman, 1979; Wrangham, 1977). This great reduction in time required to forage and feed in captivity may be responsible for some behavioral problems such as appetite disorders (e.g., coprophagy, repeated regurgitation), excessive self-directed behavior (e.g., overgrooming, rocking, self-mutilation) and abnormally high levels of aggression (Fritz & Fritz, 1979; Maple, 1980; Maple & Hoff, 1982). Effective management of feeding time may help to reduce these problems. As Dittrick (1976, p. 49)) wrote, "The animal requires something more than a thoroughly nutritionally adequate diet." Factors requiring attention include feeding schedules, presentation, diversity, quantity, color, size, manipulative characteristics, and novelty.

Feeding Time

Description: Feeding and foraging are the most time-consuming activities of wild apes; common chimpanzees give the largest portion of their day to them: six to eight hours daily (Goodall, 1968), or between 43% (Teleki, 1981 cited in Kano & Mulavwa, 1984) and 62% (Ghiglieri, 1984) of their waking hours. Wrangham (1977) compared the feeding of chimpanzees living in different habitats, and found that subjects living in areas as diverse as semi-evergreen forest, deciduous forest, and grassland spent very similar amounts of time feeding (54% to 57%). This consistency across environments underscores the importance of providing for this behavioral pattern in the captive environment, as well.

Other great apes also spend large portions of their waking time feeding. Pygmy chimpanzees devote approximately 30% of their day to feeding (Kano & Mulavwa, 1984), and orang-utans 46%

to 60% (Galdikas & Teleki, 1981; Rodman, 1979). An early estimate of mountain gorillas' feeding time, 25% (Fossey & Harcourt, 1977), has been revised upward to about 45% with improved experimental methodology (Harcourt & Stewart, 1984). Because even five-year-old gorillas reached that level of feeding (Harcourt & Stewart, 1984), it is probably an important consideration even for fairly young captive apes.

Application: Reviewing literature on 56 primate species, Clutton-Brock and Harvey (1977) concluded that proportion of day spent feeding was positively related to body weight and negatively related to proportion of foliage in the diet. Great apes are the heaviest nonhuman primates and the most prone to appetite disorders in captivity. Perhaps this is because in nature they spend more time feeding than do smaller primates, so their natural feeding pattern is more disrupted by captivity.

As a means of reducing appetite disorders and increasing environmental complexity, researchers have tried a variety of ways to increase feeding time, the element of feeding that is easiest to measure. Most feeding enrichment procedures have the effect of increasing feeding duration. Simply adding a straw floorcover for two spot-nosed guenons quadrupled foraging (Maple & Finlay, 1986). Feeding frozen produce increased feeding time, improved the distribution of food and reduced fighting in macaque groups (Chamove, Anderson, Morgan-Jones & Jones, 1982; Chamove & Anderson: this volume). Providing browse to captive gorillas caused feeding time to rise from 11% of the day to 27%, and reduced the incidence of repeated regurgitation (Gould & Bres, 1986). The authors first tried feeding blended food, but finding this ineffective in controlling regurgitation, they concluded that gorillas do not regurgitate because they need to further masticate their food. Prolonging feeding time appears to be more critical. Certainly more research is necessary to elucidate this point.

Diurnal Patterns in Food Choice

Description: Most primates, including the great apes, show similar temporal patterns of feeding bouts: a feeding peak early in the morning and another in late afternoon. Although the reason for this pattern is uncertain, speculations generally involve limits of the digestive process and the relative ease of restoring energy deficits quickly (Clutton-Brock, 1977). Similar patterns have been documented for common chimpanzees (Ghiglieri, 1984; Goodall, 1968; Sabater Pi, 1979), bonobos (Kano & Mulavwa, 1984), orangutans (MacKinnon, 1974) and gorillas (Sabater Pi, 1977). Wrangham (1977) found three feeding peaks for chimpanzees but cautioned that this finding might be invalid.

Some primates choose different foods at different times of day, even though items are available throughout the day (Clutton-Brock, 1977; Rumbaugh Washburn & Savage-Rumbaugh: this volume). Chimpanzees tend to eat fruit earlier in the day than leaves, for example (Kano & Mulavwa, 1984; Wrangham, 1977). Orang-utans have peaks in fruit-eating both early and late in the day; in the interim they consume larger amounts of leaves, bark, flowers, and insects (Rodman, 1977). Hladik (1977) found that chimpanzee feeding patterns varied seasonally. Fruits were consumed early and late in the day and leaves were consumed at mid-day during the minor dry season. But in the rainy season, fruits were eaten throughout the day, with a peak in leaf-eating in late afternoon. Again, this was independent of availability.

Most chimpanzees feed in uninterrupted bouts (Hladik, 1977; Wrangham, 1977). In one study, the subjects fed in a mean of 38 bouts daily, most lasting at least five minutes. Since animals moved between bouts to different food sources, "we may picture chimpanzee feeding behaviour as the composite of visits to about 14 major sources during a day" (Wrangham, 1977, p. 510). Orang-utans tend to eat fruit from one source in the early morning, then travel while eating leaves, bark and flowers, and then eat fruit from a second source late in the day (Rodman, 1977).

Application: The temporal patterns of ape feeding could be simulated in captivity by providing meals at different times of day that match in composition the meals just described. Two large meals of fruit would be available for orang-utans in early morning and late afternoon, with other foods being offered throughout the day. Chimpanzees might be given a major meal of fruit early in the morning, with more browse available later in the day.

Even though apes show feeding peaks, they often dine throughout their waking hours. Rodman (1977) reported that even during the midday low in feeding, orang-utans spent 29% of the time feeding. Thus, it is important to have foods available to captive great apes during most of their active hours. To avoid obesity, high-calorie foods should not be continuously offered, but small pieces of low-calorie foods, perhaps offered in ways to increase activity, could be constantly available. In contrast to this type of feeding regimen composed of many small meals, a survey indicates that 79% of captive gorillas are fed only once or twice a day (Akers & Schildkraut, 1985). Since foods are often used to entice apes to cooperate with management procedures, this may explain the few number of meals offered. Perhaps using small amounts of preferred foods for management routines would allow a balance between the need to use food as a reinforcer, and the need to have food available to apes during much of the day.

It would be helpful to identify the temporal patterning of behavioral abnormalities and agonism in captivity and then to carefully offer feeding enrichment at peak times to head off undesirable behavior. For example, most captive gorilla regurgitation/reingestion is exhibited in the afternoon (Akers & Schildkraut, 1985), so it would be reasonable to initiate more feedings during that time (*to head off* a regurgitation/reingestation bout, not to reinforce one already in progress!).

Diversity of Food Items

Description: Two of the great apes, chimpanzees and orang-utans, are primarily frugivorous, whereas gorillas are generally considered folivorous (Fossey & Harcourt, 1977; Rodman, 1984). However, Tutin and Fernandez (1985, cited in Tuttle, 1986) contested the classification of western gorillas as folivores because of the large amounts of fruit they consume. All apes' diets are quite diverse, and several researchers noted that apes may seek variety. For example, chimpanzees (Badrian & Malenky, 1984; Wrangham, 1977) and orang-utans (MacKinnon, 1974) were seen to leave plentiful food sources only to return later. There are striking variations in diet between different communities of great apes (Nishida, Wrangham, Goodall, & Uehara, 1983).

Chimpanzees have been observed to consume between 113 and 205 different foods at various field sites (Badrian & Malenky, 1984; Hladik, 1977; Kano & Mulavwa, 1984; Wrangham, 1977). The major component of their diet is fruit, generally followed in frequency of consumption by leaves and stems, seeds, other plant parts, animal matter, and inorganic matter (such as earth, ash, and rotten wood). Domestic plants make up a significant portion of some wild chimpanzees' diet (Kano & Mulavwa, 1984).

Orang-utans also have a highly varied diet, although Kortlandt (1984) claimed that proportional to the diversity of the available food supply, the chimpanzee diet is the most diverse. Orang-utans have been observed to consume between 56 (Rodman, 1979) and 227 (Galdikas, 1978 cited in Kortlandt, 1984) food types. The orang-utan hierarchy of importance of food categories is similar to that for chimpanzees: fruit, leaves, bark, flowers, and insects (Rodman, 1977). The major difference from the chimpanzee diet is the orang-utans' greater dependence on bark (Rodman, 1984). When fruit is not available, orang-utans eat more leaves, lianas, bark and epiphytes, whereas these are rarely eaten when fruit is plentiful, although they are always available (MacKinnon, 1974).

Gorilla diets show less variety than chimpanzees and orang-utans, but in certain regions their diets include over 100 items (Goodall, 1977). One group of mountain gorillas consumed portions of 33 vines, 25 herbs, 20 trees, 13 shrubs, 6 grasses, 4 ferns, and 3 semiparasites. At another site leaves, shoots, and stems composed 86% of the diet of a group of mountain gorillas (Fossey & Harcourt, 1977). Fruit made up less than 2% of this diet, in contrast to its importance in the chimpanzee and orang-utan diet. However, the western lowland gorilla (which is much more common in captivity) consumes a much larger proportion of fruit: 40% of its diet in one study (Sabater-Pi, 1977).

The variety of food eaten daily or monthly is, of course, much less than the total range of food types recorded. Chimpanzees eat 10 to 20 different foods daily (Hladik, 1977; Kano & Mulavwa, 1984; Wrangham, 1977), with a mean of 60 types eaten within a month in one study (Wrangham, 1977). At times, great apes seem to concentrate their feeding on a very few food types. An extreme example is bamboo, accounting for 90% of the diet of mountain gorillas at the peak of the bamboo shoot season (Fossey & Harcourt, 1977).

Application: There have been few published reports on the variety of foods in captive great ape diets (but see Aspinall, 1986). However, it is obvious that most captive apes have a much less diverse diet than that of their wild counterparts. A typical diet may consist of monkey biscuits, apples, oranges, bananas, carrots, potatoes, cabbage, bread, juices, and occasional peanuts or sunflower seeds. Restricting the diet to a few items also restricts the taste, texture, consistency, size, and color of food items. Increasing variety is one of the simplest ways to increase the complexity of captive primates' diets. Although it can be difficult to introduce new food items, we have successfully done so with socially-housed chimpanzees. Because young chimpanzees are more likely to consume a new food item (Nishida, Wrangham, Goodall & Uehara, 1983), introducing foods might be easier to accomplish in groups with young animals. The greater diversity of foods fed to chimpanzees during a feeding enrichment study I will describe later may have facilitated the increase in their consumption of monkey biscuits (personal communication: T.F. Beck, Colony Manager, University of Texas Chimpanzee Colony). This contrasted with the reduction in food intake that typically occurs during the hot Texas summer. Such appetite stimulation may be desirable in some primate populations, but not where obesity is a problem. In that case, the addition of low-calorie items such as browse and tree bark, which are important components of free-ranging apes' diets, might be especially appropriate.

Food Characteristics

Description: Diversity of foods entails great variety in the sensory characteristics of the food. All great apes feed from many plant taxa; these items vary in color, texture, smell, taste, shape, and size. These stimulus variables may be important in food choice (Goodall, 1978) and the "psychology of feeding" for great apes (Maple & Bloomstrand, 1983). Other food characteristics, such as the mode of preparation before it can be eaten (see the next section) and ability of the food to get away (animal sources), as well as individual preferences, may also affect feeding. Little information is available on the importance of various food characteristics, but Sabater-Pi (1979) has estimated that the diet of free-ranging chimpanzees' is composed of the following colors: 23% orange, 19% green, 18% yellow, 12% red, 12% light yellow, 10% green-yellow, 4% brown, and 2% purple. Goodall (1968; 1986) noted apparent chimpanzee preferences for certain food colors. Sabater-Pi (1979) also tasted chimpanzee foods and classified 58% of them as sweet, 45% as sweet-sour, and 3% as bitter. He noted that bitter foods made up 16% of the diet of gorillas of Rio Muni. Orang-utans also tolerate many bitter and sour foods, even compared with other sympatric primates (MacKinnon, 1974).

Application: The effect of food color has only been tested experimentally in one study (Barbiers, 1985). Artificially coloring foods or including foods of many colors may stimulate the appetites of some apes. Barbiers (1985) found this to be true for two juvenile orangutans and one of three adults. The possibility of habituation to novel colors has not yet been investigated.

Food Inspection and Preparation

Description: The literature on the feeding of wild great apes makes clear that rather than picking up abundant food from accessible sources, these animals *work* at obtaining their food. Much of their foraging time is spent choosing and preparing food items. Few items are eaten in their entirety (Fossey & Harcourt, 1977; Goodall, 1977). Feeding bouts, particularly when they involve large items, are often preceded by conspicuous inspection of the source, and feeding may not take place, after all, following inspection (Wrangham, 1977). Chimpanzees inspect individual food items by sight, touch, and smell, and many potential foods are rejected (Sabater-Pi, 1979; Wrangham, 1977). Kortlandt (1984) claimed that gorillas were less particular about their food choice than chimpanzees, but Goodall (1977) reported that gorillas meticulously pick out known food items from tangled masses of leaves and vines.

Many food items require time-consuming oral or manual preparation to obtain or consume the edible portions. For example, chimpanzees thoroughly masticate the fruits *Musa sapientium* and *Musa paradisica* for long periods before spitting them out in the form of "chewing wadges" (McGrew, 1974). Orang-utans use their teeth to tear off strips of lianas and bark that are then chewed and sucked, leaving a wadge of fibrous material (MacKinnon, 1974; Rodman, 1977). Some fruits have inedible shells (*Baccaurea* sp., *Diospyros* sp., *Durio* sp.); apes break these open with their teeth and hands and scrape out the edible contents (Kano & Mulavwa, 1984; MacKinnon, 1974; McGrew, 1974). Goodall (1986) reported seeing chimpanzees smashing a hard fruit against a tree or hitting it against a rock to break open the shell. Leathery pods from several food sources are bitten or bent open to expose edible portions (MacKinnon, 1974; McGrew, 1974). Tree trunks or woody vines are often processed by orang-utans or mountain gorillas by scraping them between clenched teeth or bending the plants to split them open (Goodall, 1977; MacKinnon, 1974). Mountain gorillas have also been observed pulling plants out of the ground or digging around their bases until roots were exposed before feeding on stem bases or roots (Casimir, 1975, cited in Tuttle, 1986; Fossey & Harcourt, 1977).

In some cases, common chimpanzees use tools to prepare food items. The "fishing" technique of stripping the leaves off a stem before poking it into an ant or termite mound (Goodall, 1968; McGrew, 1974) is a time-consuming type of food preparation. Leaves that are never consumed as food are sometimes chewed into "sponges" that are then used to remove the brain of prey or to clean out the inside of fruit (Hladik, 1977). Chimpanzees may pick green or dry leaves, combine them with meat, eggs, or bananas, and chew or suck the resulting pulpy mass (Goodall, 1986). In some areas, they have been seen to break nuts with tools (Boesch & Boesch, 1984; Hannah & McGrew, 1987. (Also see Chamove & Anderson: this volume, for photographs of monkeys breaking nuts with rocks. Ed.) Pygmy chimpanzees have not been observed using tools in food preparation, but indirect evidence of termiting, such as disturbed and broken mounds, has been reported (Badrian & Malenky, 1984).

Application: Providing foods to captive apes that require work to obtain, prepare or otherwise process is one of the most practical and promising ways to increase feeding duration and complexity. Browse is now routinely given to many captive primates, and they use and process it much as they would in the wild. When uprooted trees were given to chimpanzees at the University of Texas breeding colony, consuming the tree bark, leaves, roots, and dirt around the roots was the second-most frequent use of the tree (Maki & Bloomsmith, in press). The rate of regurgitation/reingestion by gorillas was reduced, but not eliminated, when browse was supplied (Gould & Bres, 1986). Foods such as coconuts, corn on the cob, sugarcane, artichokes, bamboo, pineapples and celery require processing to obtain the edible portions.

There would be a substantial market among zoos and laboratories for animal feed staples and food treats for captive primates packaged in new ways specifically intended to lengthen processing time. We who are responsible for the wellbeing of captive primates should encourage commercial feed producers to pursue the possibilities.

Requiring primates to search for small food items has also been popular as an enrichment technique. Scattering small food items reduces agonism and pathologies in some primates. Chamove and his colleagues (Chamove & Anderson, 1979; this volume; Chamove, et al., 1982) have reported that covering the floor of primate enclosures with woodchip litter seeded with scattered foods reduced aggression and abnormal behavior and increased play and foraging. Agonistic behavior declined among bonnet macaques when food was more widely distributed (Boccia, Laudenslager & Reite, 1984).

Attempts at feeding enrichment have also included the use of apparatus requiring specific manipulations to obtain food items. Markowitz (1979; 1982; Markowitz & Line, this volume) has pioneered the use of mechanical operant devices to increase the complexity of primate environments. He has designed several types of mechanical feeding units that promote locomotion and manipulation, and that offer the animal some control over its environment by allowing it to choose when it will work for food. One device that is particularly species-appropriate was designed for white-handed gibbons and gives these arboreal apes the opportunity to feed above ground.

It is quite a challenge to design and construct feeding enrichment devices for apes that are sturdy enough to withstand the apes' strength, are easy to clean and to restock with food, and that appeal to the animals. Two publications describe such devices for chimpanzees. An artificial termite mound provided to a group of chimpanzees at the Scottish National Zoological Park in Edinburgh "proved to be a viable simulation of naturally occurring termite mounds, with most chimpanzees showing tool use over the period of the study" (Nash, 1982, p. 221). The termite mound was constructed of concrete over a wire frame with three holes that gave access to plastic pipes which were filled with various foods. Seven of the nine subjects were seen using tools in the mound at some point. The effect of this device on other aspects of chimpanzee behavior was not measured. I will describe later a simple chimpanzee food puzzle that we evaluated at The University of Texas Chimpanzee Breeding Colony.

Animal Food Sources

Description: The prominence of animal foods in great apes' diets varies among species and location, although animal foods never make up more than a small portion of their diets. Chimpanzees eat more and a greater variety of animal matter than do orang-utans or gorillas. Chimpanzees have been observed consuming ants, termites (Hladik, 1977; Goodall, 1986; McGrew, 1983), scorpions (Hladik, 1977), grubs, beetles, bees, butterfly larvae, giant millipedes, snails, caterpillars (Badrian & Malenky, 1984), earthworms, eggs (Kano & Mulavwa, 1984; Wrangham, 1977), snails, snakes, (Badrian & Malenky, 1984), birds (Wrangham, 1977), a flying squirrel (Kano & Mulavwa, 1984), duikers (Badrian & Malenky, 1984), colobus monkeys, bushpigs, bushbuck, redtail monkeys, blue monkeys, a baboon, and other chimpanzees (Goodall, 1986; Wrangham, 1977). Some of the smallest items are apparently consumed incidentally along with plant foods (Wrangham, 1977), but some are sought out, hunted, and killed (Badrian & Malenky, 1984; Goodall, 1986; McGrew, 1983). Pygmy chimpanzees were seen apparently catching and eating small fish or shrimp in a shallow stream (Badrian & Malenky, 1984). In the Mt. Assirik region, chimpanzees eat arboreal, nocturnal prosimians that can be collected during daylight hours from their hiding places (McGrew, Tutin & Baldwin, 1979 cited in McGrew, 1983). There are regional differences in the importance of animal matter in chimpanzee diets (Wrangham, 1977); Hladik (1977) speculated that animals may supply essential amino acids or lipids not otherwise available.

Orang-utans have been observed eating several types of invertebrates and animal products including bees, termites, ants, wasps, honeycomb, (Galdikas & Teleki, 1981; MacKinnon, 1974; Rodman, 1977) and eggs (Kortlandt, 1984). They have been observed breaking off sections of termite nests, biting into them, and removing the insects with their lips and tongue (Rodman, 1977).

Among the great apes gorillas eat the smallest amount of animal matter. Goodall (1977) found no indication that mountain gorillas ate any type of animal matter even though they had the opportunity to do so, but Fossey and Harcourt (1977) reported seeing them consume slugs, larvae, worms, and grubs, both inadvertently and deliberately.

Application: There is no published account of systematic attempts to incorporate animal food sources to enrich the feeding experience for captive apes, although some apes have been given meat (Yerkes, 1943). Chimpanzees at The University of Texas Chimpanzee Breeding Colony are occasionally observed fishing for and eating the fire ants that manage to live in their outdoor compounds (despite attempts to exterminate them). Chamove et al. (1982) used mealworms as part of a successful feeding enrichment regimen for other nonhuman primates. (King & Norwood: this volume, describe fishing as feeding enrichment for captive squirrel monkeys. Ed.) Including some animal matter in the diet of captive great apes might be particularly appropriate in the case of chimpanzees since they display the greatest extent of this consumption in the wild. A simulated termite mound such as that described by Nash (1982) could be seeded with some appropriate invertebrates such as caterpillars or grubs. Another option would be to offer an animal food item that demanded a long search time. For example, earthworms might be raised in dirt-filled boxes that occasionally could be placed into chimpanzee enclosures.

Social Aspects of Feeding

Description: Great apes vary as to whether they feed alone or in company. The distribution of the food source is of course a very powerful influence on the likelihood of feeding socially. Chimpanzees usually forage alone or with their dependent offspring (Goodall, 1986), but larger foraging parties are observed at times (Badrian & Malenky, 1984). When feeding together, chimpanzees space themselves evenly, usually maintaining a distance of more than an arm's length from each other (Sabater-Pi, 1979; Wrangham, 1977), and even more when the food source allows it (Badrian & Malenky, 1984; Ghiglieri, 1984). Goodall (1986) reported that aggression is rare when chimpanzees feed some distance apart.

As for bonobos, arrival of a foraging group at a major fruit source generally leads to generalized "excitement" evidenced by increased frequency of chasing, begging for food, greeting, appeasing and copulating. In contrast, a "calm, lazy" manner is maintained when they feed on leaves and shoots (Kano & Mulavwa, 1984).

Orang-utans forage in small areas of the forest, either singly or in small, widely dispersed groups. In one study, over 80% of the foraging units consisted of a single animal (MacKinnon, 1974).

Gorillas maintain a loose group structure while foraging. Feeding gorillas stay farther away from partners than do nonfeeding gorillas, and adults spread out more than immatures (Fossey & Harcourt, 1977; Goodall, 1977). Agonistic interactions over food are reportedly rare, but those that do occur are associated with localized food sources (Fossey & Harcourt, 1977).

Application: Ideally, captive apes should be fed in circumstances that allow them to determine for themselves the distance they will keep from others, but this is often not possible because of limited space. Since we know that agonistic encounters are less frequent when social distances are greater, feeding techniques that maximize distances are desirable. Scattering small pieces of food throughout enclosures is the most obvious way to do this and has been part of successful enrichment protocols (Chamove & Anderson, 1979; Chamove, et al., 1982). Extensive use of visual barriers in enclosures of limited size might help reduce feeding-related agonism by increasing effective inter-individual distances.

Another consideration is the effect of supplying preferred foods. Since limiting availability of desirable items increases competition and aggression, it may be prudent to offer less- prized foods for many enrichment procedures. For example, rather than giving more fruit to increase feeding durations, it may be advantageous to feed large quantities of leaves or plant stems since these do not generally provoke aggression.

RESULTS OF ONE FEEDING ENRICHMENT PROGRAM

At The University of Texas Science Park Chimpanzee Breeding Facility, we have an ongoing program to evaluate our attempts at feeding enrichment. We believe that objective, quantitative evaluation is critical for improving the quality of life for the 125 chimpanzees in our colony. We have attempted to develop feeding enrichment strategies that are practical, economical, feasible on a large scale, and based on the feeding of free-ranging chimpanzees.

Food Puzzle Device

Evaluation of a simple food puzzle device was our first undertaking (Bloomstrand, Riddle, Alford & Maple, 1986). Riddle designed the food puzzle, which is made of sheet aluminum and in our colony is attached to the bars of ventilation windows in the outdoor enclosure (see Figure 1). The puzzle box consists of three shelves shielded by a front panel containing three rows of finger holes. In each shelf are holes through which food can drop to the next lower level. When the food item reaches the bottom of the box, it can be propelled toward a larger opening from which it can be removed. The hinged top of the box allows it to be filled from outside the enclosure. Small items like peanuts, sunflower seeds and popcorn work well in the food puzzle.

When tested with two social groups (N=17), the device, filled with peanuts, was used for a mean of 92 minutes per group. It induced time-consuming manipulations, even though some of these consisted of carefully working the peanuts out through the finger holes rather than down the "maze" of shelf holes to the bottom access hole. Our hopes were not met, however, for group levels of behavioral pathologies (e.g., coprophagy, regurgitation and hair-plucking) and aggressive interactions were not significantly altered.

An analysis of individuals indicated an asymmetrical frequency distribution of puzzle use moderated by dominance standing and perhaps by other factors as well. Some individuals showed significantly more and some showed significantly less aggression when the food puzzle was available. The importance of individual differences in responses to enrichment attempts has not been fully appreciated, but the possibility makes quantitative analyses even more vital (see Markowitz & Line: this volume).

Trees as Feeding Enrichment

To increase the complexity of our captive environments and to encourage species-specific behavior, uprooted bur oak trees (*Quercus macrocarpa*) were placed in outdoor enclosures. The reaction of three social groups of chimpanzees (N=28) was documented over a five month period (Maki & Bloomsmith, in press). As mentioned earlier, the second-most frequent behavior recorded was consuming the bark, leaves, roots and root ball of the tree, which accounted for 25% of tree-related activity. (The most common behavior was passive use of the tree such as sitting or locomoting on it.) The chimpanzees ate leaves, stripped bark and scraped the surfaces with their teeth, ate small twigs and portions of the roots. Subjects spent long periods of time manipulating twigs and roots before consuming them.

The chimpanzees continued to use the trees at a fairly consistent level even five months after their introduction. That fact makes uprooted trees a very practical enrichment procedure because the trees required little or no attention by the caretaking staff after their original placement, yet the animals continue to use them. We have recently placed the third set of trees into the enclosures; we simply pile a new tree onto the other trees about every six months. Sufficient space is needed to allow placement of such large items into ape enclosures, but on a smaller scale tree branches might be acceptable.

Simultaneous Use of Feeding Enrichment Strategies

Our most comprehensive feeding enrichment study to date analyzed effects of four feeding enrichment procedures on agonistic and abnormal behavior (Bloomsmith, Alford & Maple, in press). We studied a larger sample of a single species, over a longer time, than any previously-published feeding enrichment study. Six social groups of chimpanzees (N=41) were subjects; data were collected for 270 hours over six months.

Figure 1. Food puzzle being used by two young chimpanzees.

Figure 2. Adult male and juvenile female chimpanzee inspecting artichoke.

Figure 3. Chimpanzees foraging for sunflower seeds scattered in grass.

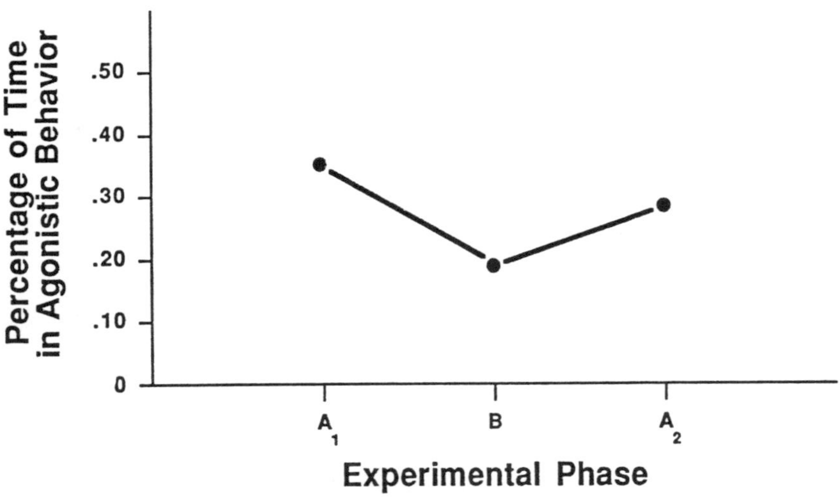

Figure 4. Mean reduction in agonistic behavior during the feeding enrichment phase (B).

The study consisted of three phases, a pretreatment phase (A1), an experimental treatment phase (B), and a posttreatment phase (A2). The baseline feeding strategy consisted of two meals of monkey biscuits, one feeding of vegetables (cabbage and carrots), and one meal of oranges, potatoes and bread. Small servings of juice and pieces of apple were also given. This routine is standard in many primate institutions. The experimental treatment incorporated the following feeding techniques implemented simultaneously. (1) An early morning meal was added that consisted of foods requiring processing time (e.g., corn on the cob, celery, and artichokes). (2) Monkey biscuits were replenished in outdoor dispensers twice daily (weather permitting) in an attempt to make the biscuits available at all times rather than only during discrete meals. (3) Each day small amounts (about six cups) of popcorn, sunflower seeds or peanuts were scattered in the grassy portions of each compound to encourage foraging. (4) Three food puzzles like the one already described were placed along the outer windows of each enclosure, and were filled with popcorn, sunflower seeds, or peanuts several times weekly.

The dependent measures were collapsed into four behavioral categories: agonistic, abnormal, feeding, and social grooming. The data were analyzed using a multivariate analysis of variance for repeated measures, with social group as a between-subjects factor and phase as a within-subjects factor. This overall test was significant ($p < .001$). Two statistical contrasts indicated that the subjects' behavior had returned to baseline in phase A2 and that each of the behavioral categories showed a significant change in the hypothesized direction in phase B. Agonistic and abnormal behavior significantly declined ($p < 0.003$) and feeding time significantly rose ($p < 0.001$). (See Figures 4, 5, and 6 for graphic representations.) Agonistic behavior declined by 40%, and abnormal behavior by 64% while feeding rose 500% above baseline.

Foods requiring extensive processing were the most successful in stimulating long feeding bouts. Subjects spent about twice the time consuming these fibrous foods as they spent with the other enrichment opportunities. Unhusked ears of corn were particularly effective as the chimpanzees formed a chewing wadge from the cob much as free-ranging chimpanzees might do.

Concurrently implementing several feeding enrichment procedures clearly succeeded in reducing two of the major behavioral problems in managing chimpanzees. The program was designed to approximate feeding in the wild in several respects: greater food variety, more frequent meals including one early in the morning, time-consuming manipulations to obtain foods, and greater total time devoted to feeding. Further work is required to clarify the separate effects of each enrichment procedure, and to identify the most important elements of this enriched feeding regimen.

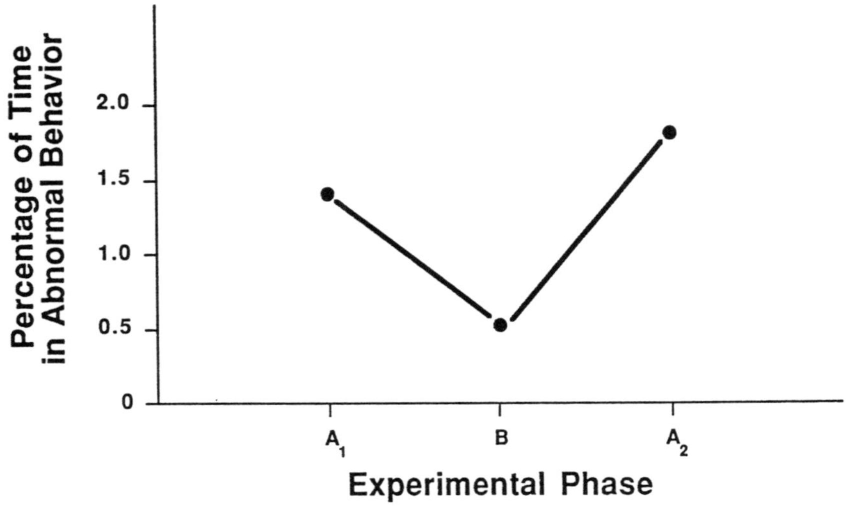

Figure 5. Mean reduction in abnormal behavior during the feeding enrichment phase (B).

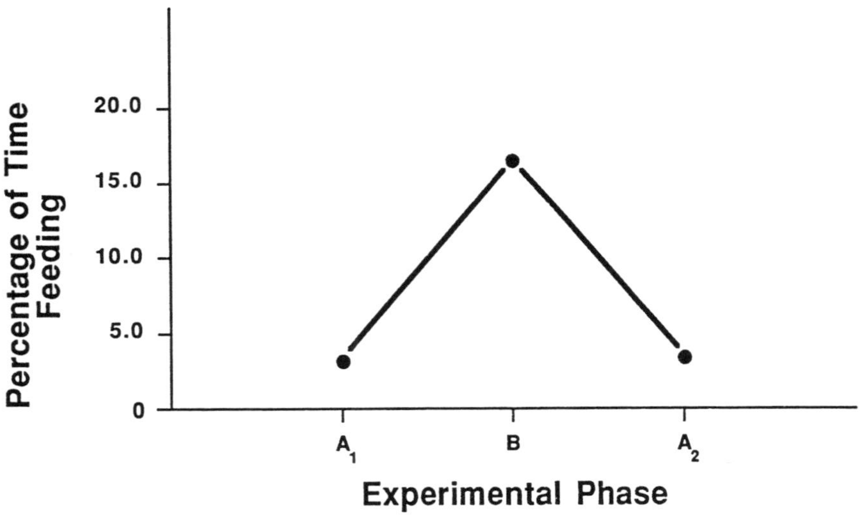

Figure 6. Mean increase in feeding during the feeding enrichment phase (B).

Pipe Feeder Device

The most recent addition to our feeding enrichment program is a simple device that simulates a termite mound. It is particularly appropriate for chimpanzees as they are known to dedicate long periods of time to termite or ant fishing. The pipe feeder does not look much like a termite mound (see Nash, 1982, for one that does), but it affords the animals a similar opportunity for activity (Figure 7). It can be attached to barred windows or to chain-link fencing so that the pipe openings are accessible to the animals. The plastic pipes may be filled with sticky foods such as spaghetti sauce, honey, mustard, and blended fruits, as well as dry, mealy foods that adhere to moistened sticks. The animals use sticks (readily available from the uprooted trees in their enclosures) to dip into these food sources. A pilot study of the chimpanzees' use of this device found that it is readily mastered and quite popular with adults (Maki, Alford, Bloomsmith & Franklin, 1988); 15 of the 18 subjects used it. Preliminary results indicate that chimpanzees living in more barren enclosures may use the device more than those in outdoor corrals (estimated means of 15 versus 5 min. per subject), and that abnormal behavior may be reduced during the short periods of time when the pipe feeder is stocked with food. As with our other food puzzle, it seems that multiple devices must be available to social groups of chimpanzees to avoid competitive aggression. A lack of age appropriateness may explain why a peer group of four infants (2 to 3 years old) did not respond well to the device, using it only once, briefly, although given multiple opportunities (unpublished data).

Practical Considerations in Feeding Enrichment

Practical considerations include cost of materials and labor for the additional time required to prepare and distribute foods, fill devices, and clean up the refuse. Items such as ears of corn, artichokes, and coconuts, which were not previously a part of our colony diet, have proven to be reasonably priced. During the summer of 1986 in Texas, a case of unhusked corn on the cob (approximately 50 ears) cost $14, and approximately 25 artichokes or coconuts cost $15. Our most economical enrichment food is popcorn. After the initial expense of a commercial popper, large quantities can be prepared quickly and cheaply. The food puzzle costs about $40 in materials and the pipe feeder requires about $125 in materials.

A

B

Figure 7. Front (A) and back (B) view of pipe feeder.

The food puzzle and chow dispenser are generally used outdoors. Because the devices are not waterproof, rain-dampened foods must be cleaned out. Foods in the pipe feeder are better protected from rain, but when moist perishable items are used, daily cleaning is required. The only food we have fed that significantly altered the cleaning routine is corn on the cob. The chimpanzees form a chewing wadge with the cob which they later discard and this sticky substance necessitates scraping before hosing down the area.

THE FUTURE OF FEEDING ENRICHMENT RESEARCH

Feeding enrichment has only relatively recently been examined from the scientific standpoint of applied primatology so it is not surprising that there are more questions than answers. Several topics deserve attention in the near future. Although most nonhuman primates share some basic feeding characteristics, there are species differences that must influence the appropriateness and utility of enrichment strategies. Documenting species differences will be important.

Differences between individuals and between social groups may also exist. In our major study of feeding enrichment there were distinct differences in the reactions of the six social groups. For example, the graphed data showed a great reduction in agonism or abnormal behavior for some groups, but no effect for others. Variation in group size, age/sex composition, and length of time since group formation do not explain these group differences. Individual differences are very likely an important factor: specific individuals may instigate changes in the behavior of an entire group, either by serving as models to be imitated, or as changed stimuli for agonistic or abnormal behavior. Dominance-modulated individual differences in response to a change in feeding have been reported (Bloomstrand, et al., 1986; Chamove & Anderson, 1979; Rosenblum & Smiley, 1984). Other variables to investigate include gender, age, rearing history, and individual food preferences. Identifying such influences would enable enrichment procedures to be individually tailored to particular animals.

The importance of novelty in the success of enrichment needs investigating. The enrichment techniques I have described have been incorporated into our management routine and, informally, we have not noticed a reduction in their usefulness. No trend toward declining effectiveness appeared during the six-week experimental phase of the formal study. If novelty does prove to be an important component of successful feeding enrichment, the imaginations of primatologists will be challenged to continually design variety into enrichment programs.

Applying the information from the available literature on feeding enrichment to primates that must be singly housed is also an important area for future research. Many of the techniques discussed in this chapter could be readily adapted for use in cages that house isolated animals. Certainly such animals deserve our special attention and concern.

Finally, it will be important to distinguish when a feeding system that develops a high or variable foraging demand actually improves quality of life and when it may do the opposite. Rosenblum and his colleagues (Rosenblum & Smiley, 1984; Plimpton & Rosenblum, 1983) found that requiring macaque mothers to work relatively hard for food caused them to display more agonism and less social grooming, and their infants grew to be insecurely attached. Such reactions are obviously undesirable.

The difference between a program that enriches and one that is too demanding may be whether the animals have options. When Rosenblum's subjects were in a condition of high-foraging demand, they apparently *had* to increase foraging time by 500% to maintain normal weight. This contrasts with the enrichment procedures at our chimpanzee facility, which are available in addition to the normal food supply, so animals can *choose* to work for food or not. Markowitz (1979; 1982; this volume) has discussed the issues of control and choice and their roles in behavioral enrichment. The influences of choice and scheduling of food delivery (i.e., fixed versus variable schedules) deserve more investigating.

Feeding enrichment is certainly fertile ground for further research and for improving captive primates' lives. It is already clear that incorporating feeding enrichment in management practices can improve our control over behavior considered detrimental and encourages species-appropriate behavior. Certainly this is an effective way to promote the psychological wellbeing of captive nonhuman primates.

ACKNOWLEDGEMENTS

I thank Dr. Terry L. Maple, Dr. Patricia L. Alford, and Dr. Michael E. Keeling for their assistance and encouragement in this project. Thanks also to Alice Fitch for the photographs and to Carolyn Crain for preparing the manuscript. This research was supported by The NIH/DRR grants 1 R24 RR03578 and 1 U40 RR03589 and The NIH/DRS contract NO1-RS-3-2116.

— 23 —
On the Care of Captive Chimpanzees: Methods of Enrichment

Duane M. Rumbaugh, David Washburn, and E. Sue Savage-Rumbaugh

INTRODUCTION

This is a paper by scientists who have worked extensively with captive chimpanzees in behavioral research on questions of language, learning, and cognition. We would like to share perspectives, ideas, new techniques, and data, when available, for the benefit of those who wish to improve management of captive chimpanzees in laboratories and zoos.

EARLY REARING

Early social and environmental stimulation is critical to the development of primates' efficient performance on complex tasks (e.g., Gluck 1979). This is particularly true for chimpanzees. Chimpanzees who had been born and reared in impoverished environs for the first two years of life were reared from age 2 on with others who had been group reared all along. When compared at age 14 on Transfer Index tests covering a variety of tasks, the chimpanzees who had been reared in impoverished environs for the first two years were clearly deficient in transfer skills. Indeed, they were so poor that there was essentially no overlap in the two groups' scores. Interestingly though, the disadvantaged chimpanzees did not differ from group-reared chimpanzees in simple discrimination learning (Davenport, Rogers, & Rumbaugh, 1973). These findings attest to the importance that experience during infancy has for the long-term development of chimpanzees' ability to learn complex tasks, and they carry obvious implications for the rearing of captive apes, and humans as well.

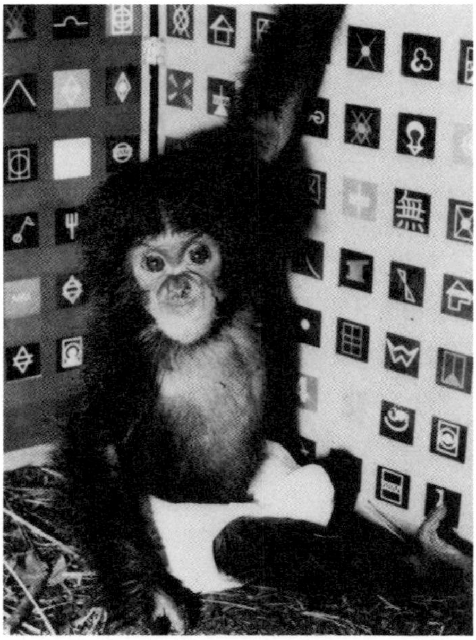

Figure 1. Mulika (*Pan paniscus*) at language keyboard. Early introduction to language facilitates long-term competence in language-related tasks.

Figure 2. Free transit between indoors and outdoors is made possible by design of doors that open easily and close automatically.

A

B

Figure 3. Examples of intellectual challenge that can be provided in the captive environment. Austin (*Pan troglodytes*) uses a socket wrench (A) and key (B) to open problem boxes containing hidden food.

GENERAL ASPECTS OF CAPTIVE HOUSING

Captivity can impose stress upon the chimpanzee by inadvertently denying it environmental, social, and intellectual challenge. The chimpanzee's natural forest environment provides both resources and risks, and has evolved in the chimpanzee the capacity to benefit from challenges in both resource procurement and risk management. Denied such challenges, the chimpanzee may become morose, short-tempered, superstitious, and dull-witted. Scientific conclusions based on such animals are likely to misrepresent the species' capacities.

In the wild, resources are obtained and risks are managed within a variety of social groupings, and the challenge to maneuver within a complex and varying social context contributes to developing the chimpanzee's functional intelligence. When dealing with animals removed from the richness of nature, it is important to simulate natural environmental pressures through tasks that engage the chimpanzee's attention and advance its learning and social competence.

All captive apes need to be given some choice in what happens to them each day, and they need challenges to alleviate the boredom and peevishness that otherwise typify captive chimpanzees, even when they are housed in "natural social groupings." Social tensions, even in a group that is well fed and given a variety of objects, can run high simply because there is little challenge other than that offered by bickering with companions.

Adolescent and adult chimpanzees benefit greatly from being able to *exert some choice over their companions*. They need to be able to move from one space to another and to choose when and how they do so. The ability to change location at will appears to do more to moderate aggressive encounters than does cage size per se. In the wild, chimpanzees vary their choice of companions as they move from place to place, yet most captive enclosures are designed so that chimpanzees are *always* together, either in a large arena, or in cages. Aggressive encounters diminish when chimpanzees have a series of areas they can go to, ideally furnished with doors they can shut behind them if they wish. Multiple enclosures with many doors are ideal for chimpanzees, who constantly move apart and rejoin one another in the wild. (Similar caging is not equally appreciated by another species of ape, the bonobo (*Pan paniscus*), who tends toward more stable social groups in the field. Bonobos become upset when a door separates any member from the rest of the group.)

It is easy to overlook how much stress may be placed on captive chimpanzees by forcing them to be within each other's sight all the time. The report by DeWaal (1982) of murder between captive adult males makes this point well. Similar episodes of males killing males *of their own group* have not been reported in the wild. It appears that DeWaal's males were responding in an extreme way to problems caused by captive housing, albeit housing that provided a lot of space, many objects, and dietary variety–but no external challenge nor opportunity to be selective about companions. Adult chimpanzees need companions, but they also need to get away from them at times. The ability to walk away from a disagreement and return later, when tempers have cooled, is a necessary part of their social fabric.

In managing the social relationships of chimpanzees in our laboratory (two adult males and an adult female with a male infant), we have found that nearly all problems can be handled by *frequently altering* caging arrangements. Sometimes the chimpanzees appear happiest when they are all together; other times, they like to be alone. As in the wild, they most often appear happy when mother and son pay short visits to the adult males and then leave, letting the adult males patrol the large part of the enclosure while mother and son have time together alone. This can change, however, depending upon other factors. When the female was in estrus, she wanted to be with both males, but her presence, in the swollen state, provoked aggression between them. In the wild, one of the males could have left the area, but since that was not possible in captivity, we arranged for the males to take turns with the female, and if they wished it, we gave the males time alone together. Had these males not been provided with flexible caging options, social tensions would have run unnaturally high, and excessive fighting might have resulted.

Management of chimpanzees should include the principle that they be allowed to control a substantial part of their day's routine and events. Management that takes this principle into account can avoid the "institutional syndrome" of depression and unresponsiveness. The chimpanzee also benefits from facing "danger" and the unknown from time to time, as it would in the wild. To encounter "danger" and to cope with it successfully in company seem to invigorate the chimpanzee, to mature social bonds, and to reinforce the chimpanzee's sense of control over its own activities.

CAGE DESIGN

Properly designed caging for apes need not be extraordinarily expensive nor extremely spacious. Indeed, large, expensive cages that do not address the nature of chimpanzee social behavior cause more problems than they solve. Our laboratory has three relatively small indoor cages (2.0 x 3.7 x 2.4 m) and two large outdoor cages (8.8 x 9.8 x 2.4 m). The cages are linked so that we can move any animal into another cage without its having to pass another another if he/she does not want to. Of course, adult animals who have not regularly been housed with other apes may not adapt at all to selecting partners or to alternating housing arrangements. These solutions are not advisable for apes who have experienced years of social deprivation. Nonetheless, as researchers strive for better ways of housing apes in future, they must be aware of the apes' need to vary companions *throughout the day.*

The fact that things happen around the chimpanzee's enclosure can be very helpful in preventing boredom. The adult males in our facility have developed a concept of "territory" that extends around their area of the laboratory. When strange people enter this "territory" (often enough to keep the males alert, but not so often as to become "fixtures"), the adult males expend considerable effort "patrolling." This functions to strengthen the bonds between the males. The adult female does not join such patrols, but she is very attentive to their patrolling activities, and seems (if we may anthropomorphize) quite appreciative of it, particularly since she is caring for an infant. The large outside area, being the place where one would first spy a stranger, is the one that is patrolled, while the small inside area is the core area that is protected. The female ventures outside only when things are calm whilst the adult males rush outside to patrol at the slightest disturbance. The female monitors their patrols from a high perch where she can look out a window. The patrolling routines of the males and their impressive displays toward strangers give them a clearly functional role. The female, too, has a role, to monitor the males' activities and take care of the infant.

In zoos, unfortunately, visitors are so frequent that males cease reacting appropriately to strangers, nor are patrols effective in a zoo as they cannot serve to alert others to the arrival of strangers. Research facilities should, however, allow apes to monitor and respond to the comings and goings of people in their immediate vicinity, as such monitoring is a very important function for them in the field, and the ability to monitor intruders in a captive environment helps to normalize social behavior.

Recognizing that it is not possible to redesign cages and environments everywhere, we offer some suggestions as to how laboratory housing can be adapted so that the chimpanzee still feels he/she has a functional role to play in a captive environment. We also suggest that functional social roles are more important to wellbeing than cage size or diverse objects. We believe that an enlightened view of the needs of captive apes must go beyond rote formulas that emphasize square feet and new play objects every day.

It is important for apes to be able to move freely between indoors and outdoors so long as weather permits. All three ape species in our laboratory are much calmer when permitted this freedom. When it is cold out, however (and temperature requirements differ for each species), animals may occasionally stay out longer than they should and it is necessary to restrict free access. In our experience, chimpanzees can safely have free access to the outdoors down to 50° F. Down to 30° F, they can have free access for limited periods. Bonobos and orangutans, however, cannot tolerate temperatures this low.

The outdoor area should be larger than the indoor area, and should permit running, brachiation, and leaping. Ideally, to encourage games of chase and keep-away, it should be divided into at least two areas, with doorways, to create a circular path for such games. A properly divided area will encourage much more play, and aid in keeping the animals physically fit. If possible, one or two small parts of the enclosure should rise higher than the main enclosure (up to 20 or 30 ft), so that the animals can climb up as if in a tree and view the surroundings. The ground should not be cement. Cement discourages vigorous physical play and activity and can cause joint injury if chimpanzees try to move vigorously around a large enclosure. At our laboratory, a deep pine-bark cover has worked well. A cover 10-12 in. deep decays naturally and can be replenished easily and inexpensively. The ground cover stays clean as long as all fecal material is removed daily. This does not take long. Outdoor enclosures of heavy-gauge chain link with bark ground cover are relatively inexpensive and, in our experience, very easy to maintain.

Doors between outdoors and indoors should be high off the ground so that insects, vermin, and material from the outdoor enclosure do not easily enter. Additionally, apes are most comfortable if, in passing from one area to another, they can go up high and survey the area they are about to enter. Doors between outdoors and indoors should close after the apes pass through. We have found that electronically-controlled hydraulic doors work best. A button can be installed on both sides of the hydraulic door. When the chimpanzee wishes to go in or out, he simply presses the button and the door lowers for him and stays down until he has passed through. Automatic closure can be overridden by flipping a switch when people must go into an area. Each cage door should have three locks for safety and a secured vestibule should be constructed around all cage entrances.

At our facility, high tunnels connect the indoor-outdoor area. These tunnels are the preferred resting areas. Outdoor tunnels are covered with lexan to allow the apes to rest in a high place, with a clear view of the surround, but protected from wind and wet on cool or rainy days. The tunnels extend the time the apes can be outdoors, since they are close to the building and remain warm. The closure of the hydraulic doors can be adjusted so that a small amount of warm air escapes from the building, giving the animals a high, small, warm place out of doors much of the year. It is the favored area of mothers with infants.

An important consideration in indoor cage design is that the area be easy to clean quickly with minimal disruption to the apes. In our experience this is best accomplished by placing the drains outside the cage rather than inside, so feces can be easily and quickly hosed *out* of the cage. If the floor is epoxy and properly sloped, all fecal material can be hosed out in a few minutes. If this is done at intervals throughout the day, the cage can be kept virtually *free* of fecal material. The apes do not then need to walk in it, get it on their hands, or smear it on the walls. Of course, one still needs to go into the cage once a day for proper scrubbing. It is also important that there be

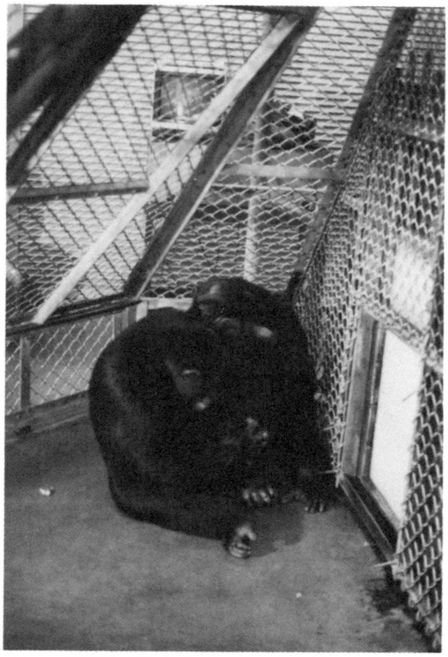

Figure 4. Sherman and Austin (*Pan troglodytes*) groom in a cage designed to reduce displays.

Figure 5. Kanzi (*Pan paniscus*), like other bonobos, but in contrast to the chimpanzee (*Pan troglodytes*), has an affinity for water.

high areas, well above the hose water, where the apes can go and stay while the hosing is done, rather than having to be constantly jumping out of the way. Hot water cleans better, but care must be taken not to scald the animals. In our experience, when indoor areas are kept free of feces and access is permitted to an outdoor area with bark cover, coprophagy does not occur, animals appear cleaner and they stay healthier.

Rumbaugh (1988) has reported a novel cage design for adult male chimpanzees that maintains them with fewer emotionally-taxing displays. The design provides for the front of the cage to slant inward toward the cage's center, so the chimpanzees cannot run against the front with great force or grasp and shake the structure. This, in turn, discourages displays in general and fosters a more tranquil social setting.

Clearly, existing facilities in zoos and research institutions cannot all be changed overnight. All facilities can, however, strive to alter conditions so as to decrease the frequency of serious aggression and stereotypies and to increase the range of normal behavior observed in the group. Sometimes, relatively minor changes, such as putting in a door to permit chasing around and around, and eliminating competition for food by scattering it liberally throughout the bark floor cover, can produce remarkable results. Each situation is different, however, making it unwise to try to prescribe universal, detailed guidelines for very different housing situations.

CARETAKERS

It is difficult to train personnel to be sensitive to the fact that the chimpanzee's companionship needs change over hours or days and that housing must be adjusted accordingly. A most stressful event for captive apes is a change in the caretakers who work with them. They often do not trust new caretakers, cannot communicate with them, and fear that their needs will not be properly met; and they may take out these frustrations on one another. New caretakers should be introduced gradually. It is our experience that this process can take up to several years.

Training staff about the normal range of ape behavior is critical. Staff also need training on how to monitor the apes' behavior and to assess whether changes in caretaking are having a positive impact in decreasing aggressive outbursts and increasing the range and variability of normal behavior. Teaching staff to give the apes as much choice as possible over significant events of the day can have a very positive impact. Even such a small thing as preparing the food in front of the chimp and letting the chimp determine by gesture which ingredients, in what quantity, are to be mixed together in a bowl, can have a very significant effect on the caretaker's relationship with the ape.

All too often, those who direct research with apes take no direct responsibility for their care and feeding. This is most unfortunate, because it relegates crucial aspects of animal care to those with little scientific training in interpreting their behavior, mood, and needs. It also tacitly suggests that the manner of care is a matter of no consequence, and has no relation to the data to be gathered.

It is important that both behavioral and biomedical researchers who employ apes as subjects have first-hand experience with their subjects. It is also important always to note and record behavioral data along with biomedical data, and to take seriously the import of the behavioral data, which provide important information on the animal's state of health and wellbeing. The obligation to take note of such data will insure that researchers are attentive to behavioral signs of wellbeing or distress. Many of the present housing inadequacies have arisen because it is all too easy for a researcher to design and direct a study while remaining remote from the day-to-day conduct of the research and the condition of the chimps involved.

DIET

Chimpanzees place a high premium on food, particularly tasty food. Variation in what they are fed is important to them. They enjoy the challenge of searching or working for food. They should have the opportunity to search or work for foods they particularly like, even if they are not essential to nutrition. Since chimpanzees do not frequently share plant foods with others, they should receive enough of these foods, dispersed throughout their living space, to preclude competition.

In our laboratory, we have found that when given a completely free choice in the matter over a long time, most apes eat little in the morning but snack steadily throughout the afternoon until near dusk. Feeding schedules have been built around this preference. Variety is built into the diet through a planned two-week menu, in which a choice of foods is offered five times a day. They may choose any one or all of the foods that are offered. With the exception of a few staple items, the foods offered at a given meal will not be offered again for two weeks. These non-staple foods are earned. Among other things, this precludes competition with other apes for the special foods. In addition, the tasks that earn the special foods serve as a challenge, much as having to traverse a considerable distance in the wild to obtain food serves as a challenge.

When the chimpanzees have no set tasks to do, one observes more grooming, more attempts to take the cage apart, and more hostility toward companions, even though the apes continue to enjoy optimal caging, companionship, play objects, and optimal diets. It seems that the chimpanzee, like the human, has a very busy mind, and if it is not occupied solving real problems, it will manufacture problems to solve. Deprived of challenges, the chimp becomes morose, depressed, and at times, overly aggressive.

a

b

Figure 6. Chimpanzees are highly social and smile through the course of positive social interactions, be they with (a) other chimpanzees, or (b) a friendly experimenter (E.S.S.-R.).

USING MODERN TECHNOLOGIES TO ALLEVIATE BOREDOM

It is rather straightforward to incorporate aspects of life in the wild into the laboratory. For instance, variety in feeding patterns or stability in social environments are relatively simple manipulations that may enrich the captive existence of chimpanzees. However, there is also potential gain in exposing captive chimpanzees to conditions unlike any in their natural habitat. Careful application of technological advances can supplement the stimulation and challenges commonly available to captive animals.

Television

Television can be a great source of environmental enrichment *if* the chimpanzee can perceive the relevance of what it sees on the screen to the world it knows. Not all chimpanzees will look at television and "see" scenarios, but most can learn to do so. Methods for accomplishing this need to be studied more carefully. In our experience, chimpanzees that are allowed to watch real-world events on television at the same time that they witness these events directly, "discover" how to read the television monitor. Savage-Rumbaugh (1986) has extensively discussed this and other factors relevant to developing an interest in television in two of our chimpanzees. Her studies make clear that chimpanzees can learn from television monitors the location of food and other incentives, and what is happening in areas beyond their view. On one occasion when they were frightened by the sound of chain saws coming from an outdoor area that they could not see directly, their fear abated when they could watch on television the work crew that was cutting up a tree that had fallen in a storm the night before. Their fear gave way to intense fascination as they watched activities that were coupled with sounds that a little while before had frightened them.

Video Tasks

Savage-Rumbaugh (1986) reported that two sophisticated chimpanzees spontaneously learned to use joysticks to play a video game simply by watching the experimenter play the game. The speed and skill that characterized their performance spawned a host of subsequent experiments to determine the parameters of performance on video tasks in nonhuman primates. A wide range of subjects, from rhesus monkeys to mentally handicapped and normal human children, have been trained to use the joystick in various video tasks.

The common requirement of most video tasks has been to manipulate a joystick to bring a cursor on the monitor screen into contact with a target on the screen. The target may be stationary, as in matching to sample and discrimination learning, or it may be moving, requiring the subject to pursue and contact the target by skillful manipulation of the joystick. Each of these types of task has been

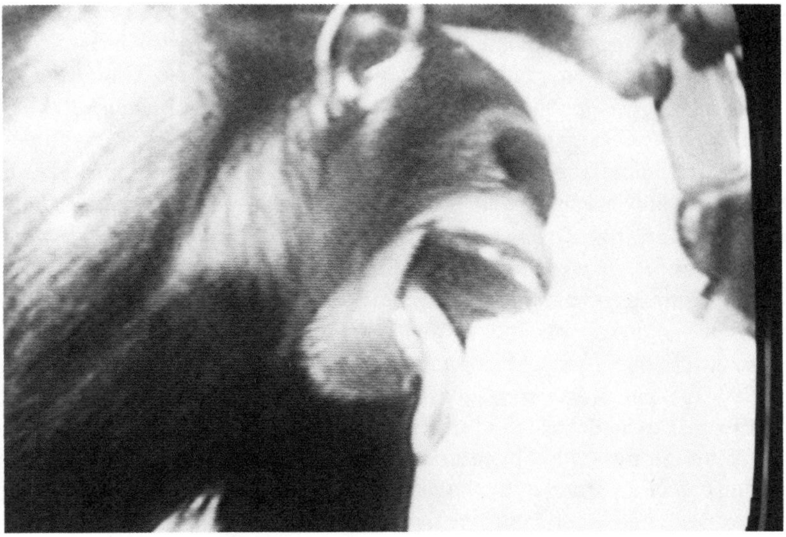

Figure 7. Chimpanzees can make use of televised images in solving problems. Austin (*Pan troglodytes*) is using a TV camera to try to see down his throat. On his own initiative he got a flashlight (held in his left hand) and is trying to shine it down his throat to improve the image on the TV monitor.

Figure 8. Austin uses a joystick to control the movement of a cursor on a monitor to play "Chase" and to work on a variety of computer-controlled tasks.

employed to good effect with nonhuman primates in our lab. Animals have demonstrated the ability to match to sample various computer-generated stimuli, including blocks of color, Arabic numerals, and letters of the alphabet. They have proven quite adept at pursuing and touching moving stimuli with a cursor and will even track the movements of targets over a sustained period of time (Rumbaugh Savage-Rumbaugh Hopkins Washburn, & Runfeldt 1987; Rumbaugh Richardson Washburn Savage-Rumbaugh & Hopkins 1987). In video tasks, psychological questions are translated into computer-mediated, interactive video problems, similar to popular video games but differing from them in being designed to address behavioral questions.

Computers are valuable tools for enriching both the environment of laboratory animals and also the research capabilities of the behavioral laboratory. Although at one time it was difficult to justify the expense of computer systems for many scientific purposes, the contemporary computer industry has erased this roadblock. Inexpensive personal computers, much more powerful than their costly predecessors, can be dedicated to simple tasks with great benefit.

Although numerous benefits may be realized by application of computer technology to comparative psychology, we will underscore just two. First, the computer-mediated video task greatly enhances the potential and facilitates the execution of research with nonhuman primates. It opens a whole new range of tasks, variables, stimuli, and reinforcers that can be exploited and studied. Moreover, experiments can be executed with greater efficiency and less strain upon both human and nonhuman participants.

Neuringer (1969) and Menzel (1987) have noted that animals prefer to work for food rather than to receive identical food free. Research at our laboratory with two five-year-old rhesus monkeys (*Macaca mulatta*) supports these observations. These animals were given a "video menu" from which they could select any one of six video tasks, each of which they had previously mastered. Once selected, the task remained available to them for five minutes, whereupon the menu appeared once again for the subject to choose either the same or some other task for the next five minutes, and so on.

Of significance both to husbandry and to theory is the fact that we did ***not*** rely on food privation to motivate either training or testing. The monkeys were given full rations of lab chow and oranges each afternoon, and they sustained healthy growth throughout months of training. Inevitably, there was leftover food in their cages the next morning, which was removed about 9:00 o'clock when video tasks became available. Reinforcement for training generally consisted of banana-flavored pellets, either of sucrose or chow.

From the menu of video tasks the monkeys could choose tasks requiring them to make contact with stationary targets, chase moving targets, sustain contact with erratically-moving targets for varying durations up to 7 or 8 sec, "shoot at" moving targets, or match to sample. Alternatively, they could opt for free food. Whatever their selection, five trials were administered, separated by five-sec intertrial intervals, and food pellets were dispensed for correct performance at the completion of each trial. In the free-food option, the monitor screen was blank and a pellet was dispensed automatically *every* five seconds.

Although the project is still in progress, it is clear that even after several months the animals have favor the free-food option only two-to-one over all others combined. When a task is selected, it is a relatively easy one, such as contacting a stationary target.

We have found that overall patterns of behavior vary greatly depending on whether video tasks are available to the monkeys or not. A series of 30-min formal observations was made of two rhesus monkeys at times when they did and times when they did not have access to a video task. When they did not, the animals foraged in their bedding for food, directed their attention toward features of the cage (e.g., manipulated the bolts that hold the cages together, etc), engaged in stereotypies like rocking, or otherwise occupied themselves by grooming, staring out the window, and so on. In contrast, both animals worked a video task over 80% of the time when one was available, and invested little time in the other listed activities. In particular, stereotypies declined to less than 25% when video tasks were available.

These observations, combined with other preliminary data, support the suggestion that application of computer technology can enrich the environment of captive primates and serve to reduce aberrant behavior. Although further study is necessary, it seems probable that, given access to appropriate food-rewarded video tasks, monkeys will spend a great deal of time on them and will continue to improve in skill even though not deprived of food.

SCIENTIFIC JUSTIFICATION FOR ENRICHMENT TASKS

Beyond the interest of such accomplishments per se, the attractiveness of video tasks lies in their application to the investigation of cognitive processes in nonhuman primates. Video tasks have been adapted to studies of attention, memory, and lateralized brain function in nonhuman primates. With each application, one gains ready access to an aspect of primate cognition that is only accessible through arduous efforts by other methods. Transfer Index data (unpublished) recently obtained from rhesus monkeys tested in video format were surprisingly comparable to those obtained with conventional, mechanical apparatus (Rumbaugh & Gill 1975).

Another example of the investigation of a complex cognitive operation in nonhuman primates is a study (Rumbaugh Savage-Rumbaugh Hopkins Washburn & Runfeldt 1987) of counting by a chimpanzee. The subject of the experiment was Lana, a chimpanzee (***Pan troglodytes***) with extensive language training and experimental experience (Rumbaugh 1977). Lana confronted a video task in which a numeral (1, 2, or 3) appeared on the screen, along with a number of computer-generated boxes. She had to use the joystick to touch the number of boxes corresponding to the numeral, and then indicate by another response that she was finished. Lana did quite well in this task, but her learning was by no means instantaneous. Many modifications had to be made in the basic video task before a numerically naive chimpanzee became numerically competent, at least within this limited range. Structured and systematic changes in feedback, salient cues, and task demands were made as we learned how to teach number skills to Lana. These changes would be difficult or impossible to duplicate outside the video format, but with our computer technology, they could typically be made in a matter of minutes.

The end results of this effort include increased scientific knowledge about the basic operations of counting and about the cognitive capabilities of our subject. The benefits to Lana are especially salient in the present context. She has been challenged daily by new demands, new opportunities to learn, and by the need to work for much of the food she receives. At the same time, we believe that these data have been obtained without undue stress upon her and by methods that gave her more control over her activities.

Scientists readily acknowledge the utility of computer technology to research, but the argument that computer technology can enrich the lives of research animals is less familiar. (But see Markowitz & Line: this volume.) Although controlled investigations with groups of animals that either do or do not have access to video tasks have yet to be conducted, our preliminary data, such as those discussed above, support our view that computer tasks may afford captive primates important opportunities for cognitive stimulation and challenge.

HUMAN REARING

In some cases, either because their mothers reject them, or because experimental protocol demands it, chimpanzee infants are reared by humans instead of their mothers. This is a very difficult task, and can, if done improperly, cause great harm to the infant. Infants reared without their mothers have grown up to display numerous stereotypies, abnormal sexual and social behavior, and paranoia (Davenport & Rogers 1970).

The most important experience of an infant chimpanzee's life is its nearly constant bodily contact with its mother for most of the first year. Infants raised without this contact invariably exhibit a wide variety of stereotypes (rocking, eye-poking, bobbing, etc.) that interfere with normal emotional and cognitive functioning. If chimpanzees are to be removed from their mothers, it is essential that they receive round-the-clock care from a small, stable group of caretakers who are sensitive to their needs for contact and motion. The infants should not be put down quickly or passed quickly from person to person, nor should they be handled by strangers. They need to be fed on demand.

It is very important for such infants to have regular contact with other chimpanzees and with their mother if possible. In our laboratory, infants who are separated from their mothers are raised by a small group of humans who provide round-the-clock care, contact, and social interaction, and who give the chimpanzee daily exposure to its mother and other apes. Every caretaker must have a good relationship with the mother or other apes so that the infant can interact with the other apes without being traumatized by separation from the caretaker. These rearing conditions have produced infants who show no stereotypies (even when stressed), who are more confident and out-going than mother-reared animals, whose social, sexual, and maternal behavior is normal, and whose relationship with human beings is positive. These infants weigh more than mother-reared infants, achieve larger adult size, gesture earlier and in greater variety, and manipulate objects more often and more creatively.

Chimpanzees who are exposed to both other apes and humans during infancy, and who are not caged, rejected, left alone, or physically abused or traumatized, become remarkable adults. As adults, they remain friendly toward humans they know, and show unusual patience and tolerance toward their own youngsters.

It is very difficult to provide this quality of care for a chimpanzee infant. All too often, mother-separated infants are simply placed in a cage and taken out when it is convenient for the experimenter to feed and work with them. Care-giving responsibilities are often shared by a large number of people, only a few of whom express attachment to the infant. Consequently, the infant is not able to develop normal emotional responses and, as he/she grows older, becomes unpredictable and dangerous both to humans and to other apes. Human rearing is likely to be harmful unless 1) the laboratory can count on a stable group of caretakers who are well acquainted with chimpanzee behavior and 2) the caretakers have good relationships with other apes so they can comfortably provide the infant with the essential experience of daily contact with other apes.

In our laboratory, such interspecies rearing enables us to study the development of language skills in young apes. Infants who are reared in daily contact with both humans and other apes develop strong attachments to both species. They also experience a very rich social and physical environment. Their exposure to humans allows their communicative skills to develop far beyond what they would otherwise be. They can describe what they are doing, and what they plan to do. They can say where they would like to go and why. They can express what they would like to see others do, and what they do not want to happen. These communications they share with their human companions, or with other chimpanzees similarly reared. They become accustomed to traveling through dense forest, and they do not know confinement and its attendant boredom. They give us a glimmer of what the chimpanzee can do when it has few limits set upon it, and much opportunity given to it. They permit us, as human beings, to relate to chimpanzees who are neither "captive" nor "caged," nor yet truly "wild."

CONCLUDING REMARKS

Captive primates need psychological stimulation to substitute for the challenges that in nature would support their wellbeing. They benefit by being given control over important aspects of their environment, and by being challenged to learn and to work for food as they would in the wild, yet in ways unlike any in their natural habitat. Additionally, attention to their need for variety of social contexts and modes of feeding can, along with the incorporation of computer-controlled video tasks, provide important and needed stimulation.

We have successfully developed video tasks that enable us to study a range of cognitive skills. The same sort of tasks could be exploited for the study of cooperation and competition with other apes, as well. These tasks provide an array of activities and stimulation otherwise unavailable to the animals. We have preliminary evidence to support the claim that the animals work on these tasks for reasons intrinsic to the activity itself.

There are, however, important questions still to be answered. What is it about video tasks that so challenges and captivates the animal and enriches its life? Are there social as well as individual benefits of such stimulation? How long-lived are they? Are there negative consequences that have not yet been discerned? The answers to these and related questions require systematic empirical research. Openness to the possibilities is likely to be repaid by the uncovering of ever new ways that computer technology can assist scientists in the acquisition of knowledge and in improving the care of their animal subjects.

It will be very difficult for institutions housing large numbers of apes to make rapid, radical changes in the apes' psychological and physical environments. The reasons are both financial and pragmatic. In place of uniform, detailed regulations designed to make every ape facility alike, we suggest that changes be instituted stepwise. One of the first things to strive for at every research institution or zoo is the establishment of a group of happy, healthy, reproductively-competent, well-adjusted apes who fight very little but play a great deal, and whose intelligence, as manifested in performance on psychological tasks, is evident to the entire staff. Even if *all* apes in a primate facility can not at once be shifted to this kind and quality of maintenance, the existence of at least one such group in every facility will have a positive effect on all personnel, and on people's perception of apes. Seeing apes perform difficult psychological tasks will produce a new respect for the animals and a willingness to attempt communication with them.

Once the ape's intrinsic value becomes accepted, the search for means to improve the housing and husbandry of other apes is more likely to succeed. Ways to minimize problems associated with biomedical testing can also be sought. A chimp could perhaps keep some contact with its group by means of live video, for example. In any event, the greater the respect for primates that can be generated by firsthand experience of advantaged specimens, the greater the likelihood that the psychological wellbeing of these animals will become a high priority for everyone concerned with them. We should remember, also, that while there are many ways to improve upon current care and housing of captive apes, ***there is no one best way.***

We acknowledge certain unavoidable constraints in implementing some of the recommendations and ideas set forth in this chapter. A variety of priority, peer-reviewed biomedical research projects with implications for public health entail procedures that may not be compatible with the housing and management procedures we use in studying primate cognition and language. We cannot hold that what is requisite for research on primate cognition and language can always be the standard in every research context. Notwithstanding, we support the careful, considered, rational review of all protocols for maintaining and using primates in research. Vigilant scrutiny and analysis of primate care before, during, and after research, regardless of its priority and authorization, are necessary both for good research and for primate wellbeing.

ACKNOWLEDGEMENTS

Preparation of this chapter and the research described were supported in part by National Institutes of Health Grant RR-00165 from the Division of Research Resources and by Grant HD-06016 from the National Institute of Child Health and Human Development to the Yerkes Regional Primate Research Center, which is fully accredited by the American Association for Accreditation of Laboratory Animal Care. Additionally, grant NAG2-438 from the National Aeronautics and Space Administration to Georgia State University supported software developments herein reported and the preparation of this chapter. The authors gratefully acknowledge the assistance of William Hopkins and Susan Hall in the preparation of this chapter.

— 24 —
Signs of Enrichment Toward the Psychological Well-Being of Chimpanzees

Roger S. Fouts, Mary Lee Abshire, Mark Bodamer and Deborah H. Fouts

INTRODUCTION

In the Animal Kingdom chimpanzees are our closest living relatives at a behavioral as well as a biochemical level. The biochemical similarity is especially profound: chimpanzees are biochemically and genetically closer to humans than they are to gorillas. Twenty years ago the chimpanzee, gorilla, and orang-utan were thought to be very similar to each other but distant from humans by as much as 20 to 30 million years of separate evolution. But in the early 70's new evidence from the University of California at Berkeley shattered this long held view. The work of King and Wilson (1975) on the biochemical similarities of the blood (immunologies and amino acids) of the great apes and humans gave us a new perspective on our relationship to the great apes. Blood characteristics of gorillas, chimpanzees and humans are within 1% of each other, whereas the orangutan's are within 3% of the others. The chimpanzee shows only a .2% difference from human blood, while the gorilla's blood is .8% different. In fact, successful blood transfusions from humans to chimpanzees have been performed. King and Wilson were so struck by the biochemical similarity between chimpanzees and humans that they referred to chimpanzees as our "sibling species".

More recent research by Sibley and Ahlquist (Lewin 1984) has supported the findings of King and Wilson with studies on DNA comparing nucleotides between the species. Whereas King and Wilson looked at 467 different characteristics of blood, Sibley and Ahlquist have pushed the comparisons even further by comparing entire genomes. Their results suggest that the chimpanzee/human line split off from the gorilla line about 10 million years ago, and then humans and chimpanzees parted ways 6 to 8 million years ago.

The behavior of the chimpanzee is just as similar to human behavior as are the biochemical and genetic characteristics. The work of Jane Goodall (Goodall 1968 1975 1986) has shown us that wild chimpanzees are not so different from non-technological humans. Wild chimpanzees live and hunt in communities. They are attached to their mothers, even to the extent of mourning to death over a mother's loss as the young Flint did when his mother Flo died. They make tools. Perhaps most important of all, they suffer emotional as well as physical pain.

In addition to the social similarities, chimpanzees bear cognitive similarities to humans. Research over the past 20 years by B. T. and R. A. Gardner (1969 1971 1978 1984) has shown that chimpanzees have the capacity to acquire the signs of American Sign Language. Other researchers (Fouts Hirsch & Fouts 1982; Fouts Fouts & Schoenfeld 1984) have demonstrated that signing chimpanzees can pass such skills on to the next generation and that they sign spontaneously to converse with each other. The difference between apes and humans is one of degree, as the Darwinian notion of continuity leads one to expect. Donald Griffin (1976) has gone so far as to argue that even so human a characteristic as awareness is shared with other animals. These similarities have ethical implications for animal research. Because animals are aware, because they feel, and because they are capable of thought, there are ethical aspects to consider before doing any sort of animal research.

CAIN AND ABEL

Their similarity to humans is not particularly beneficial to chimpanzees. Their biochemical similarity makes them attractive to the biomedical research community as human surrogates in research. But this has put our species in a very uncomfortable "Cain and Abel" relationship with our sibling species and has raised grave ethical questions. In the United States the response to this dilemma was to enact federal legislation to at least ensure the psychological wellbeing of laboratory primates. An amendment to the *Animal Welfare Act* (1985) requires that all captive primates have "...a physical environment adequate to promote the psychological well-being of primates." In addition, the current *Public Health Service Policy on Humane Care and Use of Laboratory Animals (1986)* states: "...the living conditions of animals should be appropriate for their species and contribute to their health and comfort."

Sensitivity to this need is widespread. For example, Frederick King (1986a), director of the Yerkes Regional Primate Research Center, has stated that: "Human awareness of the sentience of other species and our ability to empathize with suffering leads us to a responsibility, recognized by our society, to treat animals with a compassion and concern for their sentience" (pp. 406-407).

Concern for the psychological wellbeing of primates goes beyond humane treatment. There is also the question of the empirical validity of research on primates who have been housed under conditions of social or environmental deprivation. Davenport (1979) noted that laboratory primates are typically housed in stressful, unnatural environments conducive to abnormal behavior patterns. Standard primate housing result in both social and environmental deprivation. Such conditions confound the research and result in inaccurate findings. Frederick King (1986b) states:

> There are many ways in which non-experimental stress can occur in the laboratory or in the animal housing situation and many ways in which stress can affect experimental results....environmental and social stress can influence neurological, endocrinological and immunological processes .... Among the stresses which can produce abnormal biological and behavioral changes are: ... immobilization, isolation, aggression, trauma and crowding. (p. 10)

Finger and Stein (1982) have noted that the restrictive conditions of standard laboratory housing have the potential for producing neurophysiological damage in animals. Studies with rats on the effects of enriched and deprived environments "...provide strong support for the contention that the central nervous system can be modified by the environment, and for the idea that subjects raised under different conditions differ both behaviorially and biologically" (p. 179). Standard (impoverished) laboratory caging for rats results in the following neurophysiological effects: 1) fewer dendritic branches in selected cortical and subcortical sites, 2) smaller synaptic boutons, 3) less cortical depth and weight, 4) differences in RNA-DNA ratios and 5) a wide variety of brain enzymatic differences (acetylcholinesterase levels, etc.). Finger and Stein argue that these differences between enriched and restricted animals are not simply because the enriched animals had greater opportunity for activity, but because they actively interacted with their enriched environments.

Another stressful aspect of laboratory housing concerns the social nature of most primates. Chimpanzees are highly social, yet as Pereira and Altman (1983) state:

> ...laboratory-born primates usually begin life with their mother in a small isolation cage, get separated from her early, spend their juvenile years either alone or with a small group of peers in other barren cages, and then spend most of their adolescent and adult lives in a cage about one meter tall that offers about three quarters of a square meter of floor space....In such environments, we assert, nonhuman primates are not permitted to develop and behave normally. Indeed, we predict that under these conditions primates will exhibit aberrations not only in their behavior but also in all of their physiological systems. (p. 3)

Jane Goodall (personal communication, January 21, 1987) furthers this point with her statement:

> ...It seems unlikely that these sick individuals will provide responses to any drug assays or baseline physiological or biochemical testing that are in any way representative of those normal healthy individuals. Experimental results, therefore may sometimes be misleading...

It can be concluded that experimenting with laboratory primates who have been housed in the impoverished conditions that are standard in many laboratories is not a scientifically sound procedure. Subjects of experiments should be representative of the population to which the research is intended to apply. Deprived subjects are representative of a deprived population not a normal one.

Although we need positive information on how to promote primate wellbeing, what past research has given us is mostly negative information – about the lack of wellbeing. Most of this negative information was gathered in deprivation experiments (Davenport & Menzel 1963; Davenport Menzel & Rogers 1966; Davenport & Rogers 1968 1970; Davenport Rogers & Rumbaugh 1973; Menzel 1963 1964; Menzel Davenport & Rogers 1963a 1963b 1970; Harlow 1962 1964; Harlow & Harlow 1962 1965 1969; Evans 1967; Elias & Samonds 1973). These studies tell us what to avoid in housing captive primates, yet only implicitly suggest what might be beneficial. For example, we know that isolation is deleterious, but not which types of social housing are optimal. Too often, the selection of positive procedures and conditions conducive to psychological wellbeing are left to the educated guesses of individual researchers. Line (1987) makes this point in regard to cage size: "Standards concerning the amount of living space required for laboratory animals are controversial and generally are based on experience and professional judgment rather than objective considerations" (p. 857).

Science and humane treatment need not oppose each other, in fact they should complement one another. The needs of the animal subject and the needs of science can be filled simultaneously. This paper will focus on space requirements, the benefits of play and exercise, and social and environmental enrichment for the chimpanzee. But first we will very briefly review some of the behavioral damage that isolation can do to primates.

DELETERIOUS EFFECTS OF ISOLATION AND DEPRIVATION

Davenport (1979) found that chimpanzees reared in socially and environmentally deprived conditions exhibited many behavioral abnormalities and problems not exhibited by chimpanzees reared by their mothers or by humans in a nursery. The deprived chimpanzees typically developed stereotypies such as rocking, head banging, and so on. Rearing in a restrictive environment for 2 to 2-1/2 years produced enduring deleterious effects. Several years after the deprivation conditions had ended some chimpanzees were still showing intellectual deficits. He states that "The persistence of cognitive deficits in the restricted-reared chimpanzees, even after 12 years of environmental enrichment, prolonged testing, and group maintenance, is interpreted to mean that deficits so acquired are not readily corrected" (p.351). Davenport also noted other long-term behaviorial dysfunctions produced by deprivation. Female chimpanzees reared in restricted environments did not show appropriate maternal behavior as adults, and reproductive behavior in adult male chimpanzees was also disrupted. There were deficits in other kinds of social behavior as well. Restricted chimpanzees were either unresponsive or responded inappropriately to social stimuli for the first few years after their isolation. In summary, restrictive environments deprive chimpanzees of the necessary sustenance for healthy intellectual, maternal, sexual, and social development.

As we said earlier, in addition to psychological damage restrictive conditions have the potential to produce neurophysiological damage (Finger and Stein 1982), and this, indeed, may underlie the behaviorial, intellectual, maternal, and social dysfunctions evidenced by the chimpanzees studied by Davenport (1979).

Comparable deleterious effects of isolation have been demonstrated in other primates, including rhesus macaques (Harlow 1962 1964; Harlow & Harlow 1965 1969), pigtailed macaques (Evans 1967), bonnet macaques (Rosenblum & Smiley 1984), and cebus monkeys (Elias & Samonds 1973). These effects are like those reported by Spitz (1946) in his famous study of human children raised with minimal social stimulation in an orphanage. There are, of course, species as well as individual differences in the effects of deprivation (Barnett 1981, p.411-412). For example, Sackett Holm Ruppenthal and Fahrenbruch (1976) found that complete social isolation affected rhesus (*Macaca mulatta*) and pigtail (*Macaca nemestrina*) monkeys very differently. The rhesus monkeys rocked, hit and bit themselves, clutched themselves, and failed to explore their environment, while the pigtail monkeys showed no such abnormalities. An important lesson for primate wellbeing is that we need specific information on optimal conditions for each primate species held in captivity or studied in the research laboratory.

CAGE SIZE AND SPACE REQUIREMENTS FOR CHIMPANZEES

Cage size is an important variable for the psychological wellbeing of chimpanzees. Clarke Juno & Maple (1982) examined the behavior of three chimpanzees who were moved from a laboratory (8 x 20 ft cages) to an island (125 x 40 ft). Two weeks prior to the move to the island the male had a mean hourly stereotypy rate of 26.5 (week 1) and 45.75 (week 2). On the island this hourly rate of stereotypies dropped to 8.5 the first week, and thereafter ranged between 0 and 4.0 for the following 21 weeks. Pfeiffer and Koebner (1978), in a similar study, moved eight chimpanzees from laboratories to an island and found that stereotypy rates in two of the chimpanzees who originally had the highest rates were down 70% six months after their move to the island.

In a related study, Berkson Mason & Saxon (1963) experimented with varied enclosures and measured the rate of stereotypies using Davenport and Menzel's (1963) deprived chimpanzees. The Berkson study began after the chimpanzees had been living in the standard laboratory environment at Yerkes for approximately three years. Stereotypy rates were highest (stereotypies took up 85% of the time) when the chimpanzees were placed in a wooden isolation cubicle 81" x 59" x 78" where, in addition to being isolated, they could not see out. When they were placed in an outdoor barred cage 69" x 72" x 85", still isolated but now able to see out, stereotypies occupied a smaller 52% of the time. The lowest stereotypy rate, 15%, occurred when the chimpanzees were placed with other chimpanzees in an outdoor enclosure measuring 39 ft x 57 ft.

A second experiment investigated the effects of giving isolated chimpanzees objects to play with, as well as the effects of a third cage, a small cage measuring 60" x 33" x 32" but having only three rather than all four sides occluded. While in this cage the chimpanzees engaged in stereotypies 34% of the time when they had play objects and 36% of the time when they did not. These stereotypy results comparewith 59% stereotypy in the completely-occluded wooden cubicle when objects were present and 67% when there were no objects. The researchers concluded that cage size by itself does not necessarily affect the incidence of stereotypies. Nevertheless, comparison of these results with the results of Experiment I reveals a very noticeable difference in the incidence of stereotypic behavior: 15% when the chimpanzees were housed in a group in the large outdoor enclosure of Experiment I, but 34-36% when they were housed in the small isolation cage (only three sides occluded) of Experiment II.

What is important about the Berkson Mason & Saxon (1963) study is that the larger (33.2 sq. ft) wooden cubicle of Experiment I, which was opaque and isolated the chimpanzee completely, was worse than the smaller (13.75 sq. ft.) cage of Experiment II from which, however, the chimpanzee could see out. Any positive effect of increasing cage size was cancelled by total (visual as well as social) isolation. Too often when this study is cited the offsetting negative effect of total and profound isolation is overlooked and instead what is stressed is that the larger cage did not make a difference to the incidence of stereotypic behavior: a smaller (but not visually isolating) cage induced less stereotypic behavior than a larger (but totally isolating) cage. For example, Line (1987) states: "...stereotyped behaviors were not related to cage size, but they were reduced by the opportunity to manipulate objects" (p.857). This was true only when an opaque cubicle was used to totally isolate the chimpanzee. Even in the barred solitary outdoor cage of Experiment I (34.5 sq. ft.) the chimpanzee displayed repetitive stereotypic behavior approximately 52% of the time. The really dramatic improvement occurred when these six chimpanzees, who had been raised in total isolation for the first 2-1/2 years of their lives, were put in a 39 ft.x 57 ft. outdoor group enclosure; their stereotypic behavior was reduced dramatically. Clearly, housing chimpanzees in social groups in large outdoor areas very positively affects their psychological wellbeing whereas housing them in isolation or in small cages affects them very negatively.

In addition to cage size there are other critical variables that affect psychological wellbeing. Wilson (1982) examined activity levels of gorillas and orangutans in 41 zoos in relation to seven environmental factors: size of the enclosure, usable space, frequency of feeding, number of animals, and the number of objects (stationary, temporary, and movable). Contrary to what was expected, size of enclosure, usable surface area, and frequency of feeding were not related to activity level. Instead, high activity levels were related to increased numbers of animals in the enclosures and many movable and/or temporary objects. A study by Traylor-Holzer and Fritz (1985) also confirms the need for a complex environment. It was found that adult chimpanzees used upper cage levels, cage perimeters, and small areas more often than open cage areas. Juveniles, on the other hand, tended to use all available space evenly. The authors concluded: "The incorporation of vertical structures, visual barriers, and multiple small areas to create complexity in captive chimpanzee housing is often practical and may aid in creating an environment more conducive to normal social interaction" (p.126).

Van Hooff (1967) describes a chimpanzee colony in which the chimpanzees spent most of the day close to the building although there were 25 acres to roam in. In all this space, the only shade was next to the building, and activities such as feeding occurred in the same area. The chimpanzees remained clustered in this area, and the level of aggression was high. If shade and feeding stations had been spread out, the usable space would have increased, the chimpanzees would have spread out and aggression surely would have decreased. This is further evidence of the need for a complex environment, not simply a large space.

DESIGN OF HABITAT FOR CAPTIVE CHIMPANZEES

Wild chimpanzees range over large areas, both on the ground and in trees (Goodall 1986). In our facility, which has four enclosures in four separate rooms connected by an overhead tunnel system and a tunnel on the floor, we have tried to recreate some of this diversity. Wild chimpanzees are social yet at times travel alone, so our design gives the chimpanzees their choice of being social or alone. It provides opportunities to avoid contact (visual as well as physical) when desired by going to another room. Benches in each room add more diversity and increase usable space. The chimpanzees nest in the overhead tunnels at night. Other researchers hold that such escape outlets, privacy refuges and visual barriers to allow the animals to avoid contact help to control aggression and social stress (van Hoof 1973; Maple 1978; Maple & Stine 1982; Fritz & Nash 1983; and Traylor-Holzer & Fritz 1985). Chimpanzee housing must be large enough and varied enough to encourage social interactions but also also the opportunity for privacy. (Also see Rumbaugh, Washburn & Savage-Rumbaugh: this volume; and Miller-Schroeder & Paterson: this volume.)

In research in our laboratory involving 899 one-minute observation periods, we have noted preferences for areas as a function of the activity. Fifty-seven per cent of the play-and-chase games took place in rooms 3 and 4 (a total of 70 sq. ft.) and another 14% in the floor tunnel connecting these two rooms. The overhead tunnel to room 2, and room 2 itself, accounted for 29% of play. Rooms 1 and 3 were the preferred rooms for affinitive social behavior (67%) and mutual grooming (60%). We also found individual differences among the five chimpanzees. Moja took her food to room 4 to eat 92% of the time, Dar always ate in the floor tunnel between rooms 3 & 4, whereas Loulis ate in rooms 1 (23%), 2 (15%), 3 (39%) and the tunnel between rooms 2 and 3 (23%). As for grooming, Moja groomed only in in room 1, while 91% of Loulis's mutual grooming occurred in room 3. Moja played only in room 3. All the chimpanzees preferred the tunnels for sleeping at night, resting during the day, and observing the humans in the laboratory. The diversity of design allows these chimpanzees to fully use their environments in a wide variety of different activities.

EXERCISE AND SOCIAL INTERACTION

Critical to the psychological wellbeing of primates is the opportunity to exercise and to interact socially. One form of social interaction that has been extensively studied is play. Although adults as well as young primates play, playing has been studied mostly in the young, where it has an important role in psychological and neurophysiological development. Mason (1965) found that young chimpanzees spend 70% of their time playing and have as many as 16 times more social contacts than adults, the majority of these play interactions. Harlow and Harlow (1962) found that peer play was important in the development of young rhesus; it promoted social development and helped compensate for maternal deprivation. In human children play has been found to be critical in cognitive development, especially of creativity and divergent thinking (Lieberman 1965), and it facilitates problem solving in children (Rosen 1974). To develop properly, in short, primates need to be housed in species-typical groups and to have the opportunity to play with conspecifics during exercise periods.

SIGNS OF ENRICHMENT

We have already discussed the kinds of environments that produce psychological distress in primates. The question remains what type of environment promotes psychological wellbeing. When we turn for answers to ethological research in the wild to complement research on captive chimpanzees, a few clear points emerge. One of the first things that stands out to the ethologist who studies wild chimpanzees is the lack of diversity in the lives of captive chimpanzees (Goodall 1986; Kortlandt 1966). As Kortlandt (1966, p.6) states, this lacuna exists in the design of the quarters, variety of foods, and environmental conditions and events.

Nothing is more disturbing to a sensitive ethologist than to see a chimpanzee isolated in a small cage. From Kohler (1927) to Yerkes (1943) the paramount importance to a chimpanzee of belonging to a compatible social group has been recognized; it is essential to their very nature. More recently Maple (1978) has reiterated this point. Wild chimpanzees live in large communities ranging from as few as 20 to as many as 120 individuals (Goodall 1987). Nishida (1979) found that the average size of a group within a chimpanzee community was 12.5 if the alpha male was in it and eight if he was not.

Different management techniques and housing designs based on the chimpanzee's natural environment have been implemented to promote the development of normal behavior and successful reproduction (Reynolds and Reynolds 1965; van Hoof 1967; Brambell 1975; Clarke Juno and Maple 1982). Variables such as temporal variability in abiotic conditions and food availability, social opportunities, and structural diversity have been identified (Kortlandt 1966; Maple 1978; Dahl 1982). But the social environment is perhaps the most critical factor in chimpanzee wellbeing, more important even than design of the living space. Unfortunately it is seldom given the prominence it deserves in considerations of chimpanzee wellbeing. The social environment includes not only the chimpanzees who live together, but also the careproviders and other humans who visit their area. In captivity the human is part of the chimpanzee social group. This is especially critical where a chimpanzee is isolated from conspecifics. Providing an aesthetically pleasing environment that looks more or less natural is not enough. If you lived in a lavishly furnished condominium with a beautiful view, but were surrounded by people who reviled and shunned you, you would be very miserable indeed.

PERSONNEL TRAINING

Each new member of our laboratory staff is put through an extensive training program and given a course in "chimpanzee etiquette" prior to any direct contact with the chimpanzees. Beyond teaching our method of collecting data, we give extensive training on the behavior of chimpanzees in general and our chimpanzees in particular, using the data of Goodall (1965 1968 1971 1975a 1975b 1986) on the behavior of wild chimpanzees as a guide. We require extensive reading, including our behavioral taxonomy which lists over two hundred different kinds of chimpanzee behavior. We also use training videotapes and films to aid in analyzing and explaining the different types of chimpanzee behavior.

Staff members have to be accepted by the chimpanzees before they are allowed to stay around them. We know they have been accepted when the chimpanzees no longer direct charging or banging displays toward them. The duration of training depends on the chimpanzees' responses to the new person. Throughout the training process these responses are observed and discussed with experienced staff members. If the new person continues to receive a negative reception after extensive training, we do not allow that person in the chimpanzee areas. We reason that since the chimpanzees cannot leave the laboratory, the least we can do is ensure that they are surrounded by people they like. No one is rushed into a relationship with the chimpanzees; it takes time for good relationships to develop. Moreover, each of the chimpanzees is an individual with a distinct personality, so it may take differing amounts of time before a new person is accepted by all the chimpanzees. Impatient people are reminded that Jane Goodall was at Gombe for over eight months before she could get within 100 meters of the chimpanzees.

Trained staff spend the major part of their time on daily chores and caring for the individual needs of the chimpanzees. We act as maid, cook, and trusted family butler. We clean their rooms, cook three meals, prepare their snacks, make sure they have good magazines available for browsing, and that both the children and adults have interesting toys to play with. We also play the role of understanding and reassuring friend when a family member becomes upset with another. As members of the chimpanzees' extended social group we try to respond to them as another chimpanzee would. For example, in the wild if chimpanzees are unexpectedly frightened or startled by an unaccustomed sound they typically seek physical contact with companions, touching, embracing or kissing (Goodall 1986). Mason (1965) showed that physical contact had the same calming effect on captive laboratory chimpanzees.

The roles of researcher and careprovider are not separate in our laboratory. Our method of data collection is not much different from the way trusted family servants find things out about the family they serve. They unobtrusively and sometimes covertly observe the family members interacting. We also explicitly train new personnel in the servant attitude, to insure that human arrogance is avoided.

Our goal of providing variety in a socially-enriched environment can be illustrated by describing three aspects of the chimpanzees' daily schedule: meals, cleaning and maintenance, and free time.

MEALS

Variety in meals is one of the easiest forms of enrichment. We prepare three meals a day, serving primarily fruit for breakfast, protein for lunch, and carbohydrates for dinner. This seems to suit the circadian rhythm of chimpanzee digestion, as evidenced by a reduction in diarrhea when they are on this regimen. Within these rough guidelines we achieve variety by using different combinations of ingredients. The day's menu is as follows: for breakfast several different fruits are blended with herbal tea for liquid, along with vitamins C and E and nutritional yeast for the B-vitamin complex. For lunch we serve a soup of pinto beans, garbonzo beans, lentils, or split peas cooked with a variety of vegetables; or tuna, eggs, nuts or cheese, all in addition to raw vegetables. For dinner we serve rice, oatmeal, barley, cream of wheat, potatoes, corn, popcorn, etc.

The manner of serving the meal also adds variety. We usually serve in plastic bowls with spoons, but sometimes in plastic bags. We try to make each meal an interesting and enjoyable social event. All the chimpanzees come into one room to be served. Lab members often serve meals with a soft food grunt, sharing in the excitement of a good meal. As in the wild, the most dominant member of the group is usually served first.

Snacks are sometimes offered between meals. This not only provides more dietary variety, but also serves as social enrichment. Some examples of chimp snacks: fruit, fruit leather, decaffeinated coffee, herbal tea, sugarless gum, cattails (entire plant), corn stalks (large tough leaves removed), iced treats (iced tea, Kool Aid, decaffeinated coffee, and such items as gum, raisins, bananas, and grapes frozen in a bowl of ice), raisin boards (2"x4"x2" board with drilled holes stuffed with raisins, always accompanied by willow-branch tools), flowers, snow, water-balloons (occasionally filled with fruit drinks and perhaps frozen), willow branches, and so on.

CLEANING AND MAINTENANCE

The chimpanzees' enclosures are cleaned at least twice a day. The chimpanzees have unrestricted access to the entire enclosure while cleaning is going on and are free to do whatever they choose, which is often to interact with the people who are doing the cleaning. This interaction can range from passive watching to elaborate attempts to engage in play-and-chase games. The chimpanzees often request drinks or even help to clean. Cleaning gives the humans a chance to observe closely how the chimpanzees interact with each other. This illustrates how a required maintenance activity can double as social enrichment for the chimpanzees and triple as scientific research.

TOYS, GAMES, ACTIVITIES, AND EVENTS

A variety of play objects is provided for the chimpanzees throughout the day: magazines, brushes, toys, stickers, unbreakable plastic mirrors, perfume samples, whistles, rubber-bands, cardboard, cloth sacks, and so on. Many of these items can be hidden inside another; for example, balloons can be hidden in shirt pockets. The chimpanzees often request certain items or activities. In an effort to provide them with a stimulating and varied environment we try to do different things each day. A valuable lesson we have learned is that often the simplest objects are used the most, for example, rubber-bands, cardboard, empty paper, or cloth sacks (used in games of chase: The chimpanzees will cover their heads with a sack or other object and then play a game similar to our human game of Blind Man's Bluff).

Unstructured time is the occasion for a variety of social activities. Sometimes we offer the chimpanzees drawing or coloring materials (we give them non-toxic paints and crayons). Looking at a magazine or a photo album is an activity conducive to interaction between chimpanzee and human. We also play lively games of tickle and chase with the chimpanzees. Since they are in cages and we have only minimal physical contact with them, we use toys like plastic dinosaurs and long branches to tickle or touch them. To be an effective play partner the human uses chimpanzee gestures and vocalizations such as the breathy chimpanzee laughter, playfaces, playwalking, or foot stomping. Television is a special treat and is always shared with a human. A Sesame Street videocassette was a recent favorite.

In addition to everyday variety there is also the variety of special occasions: the chimpanzees' birthdays as well as most of the calendar holidays. This provides for at least one and usually more extra-special days each month. Birthdays include cake with candles, streamers, party hats, cards, balloons, and a variety of tasty treats. The Christmas holidays start shortly after Thanksgiving when we get the Christmas tree and begin to make the edible ornaments. The tree is covered with different edible treats such as nuts, dried fruit, popcorn, and gum. The tree is a popular chimpanzee topic of conversation throughout this period; they make daily, sometimes hourly, inquiries about the availability of the tree's ornaments. These special celebrations are just another example of our constant battle to break the deadening routine so typical of captivity.

CONCLUSION

Many of the activities that occur in our laboratory may be peculiar to us and inappropriate in other settings, but the critical issue has general application: to avoid boredom by giving the chimpanzees things to do. Since chimpanzees are very social beings we make a special effort to devise situations conducive to positive social interactions among the chimpanzees and between us and them. Human/chimpanzee relationships require that the humans understand chimpanzees both as a species and as individuals. We have an extensive training program to accomplish this. A unique social relationship can develop between the chimpanzee and his or her human companions. We believe such a relationship is not only possible but is essential for the chimpanzees' psychological wellbeing given their extremely social nature.

By providing the chimpanzees with a wide variety of things to do we have also created an environment that is conducive to interesting social interactions. This is exactly the type of environment one would expect to help a chimpanzee achieve some measure of psychological wellbeing. While the chimpanzees benefit by having an enriched and stimulating environment in which they can play and manipulate objects, science also benefits by having healthy chimpanzees to study.

If careproviders and researchers take the time to understand the chimpanzee's nature and needs and if they actively strive to accommodate their procedures to those needs, then they begin to learn something very important: that "human" and "chimpanzee" are just names for our shared beingness.

— 25 —
Environmental Influences on Reproduction and Maternal Behavior in Captive Gorillas: Results of a Survey

P. Miller-Schroeder and J.D. Paterson

INTRODUCTION

The difficulty of providing for the wellbeing of captive gorillas has been recognized for a long time (Yerkes & Yerkes 1929). Often described as morose and bored in zoos and laboratories, these intelligent social primates seem easily depressed and adversely affected by the common conditions of captivity. As a species, the gorilla has proven difficult to breed in captivity; when it does breed, maternal care is often lacking or inadequate.

The zoo community has become acutely aware of the problems involved in setting up self-sustaining captive populations of gorillas. In 1980, a symposium on great ape fertility emphasized the need for more research into this problem before the species disappears (Johnson 1981). In 1982 an international symposium on great ape infertilty and insufficient breeding reflected growing concern. Also in 1982, a Species Propagation Group for the lowland gorilla (*Gorilla gorilla gorilla*) was established by the American Association of Zoological Parks and Aquariums to study reproduction problems and to set up a Species Survival Plan. This group heard reports in September 1983 that a major crash of the North American gorilla population was likely. According to these reports, many captive animals are now over 20 years old, an age when reproductive senescence starts to become apparent. It is not known whether this is a physiological or a behavioral problem, although a combination of factors is likely. Meetings of the Gorilla Propagation Group in 1985 and 1986 documented minor improvements in captive breeding, but insufficient reproduction and inadequate maternal care remain major problems to be overcome.

The *International Gorilla Studbook* (Kirchshofer 1979) showed that in the period from 1962 to 1979, 66% of all captive-born gorillas (40 animals) died within their first year of life. This figure does not include miscarriages died within their first year of life. This figure does not include miscarriages or stillbirths. By 1983 the infant survival rate in North American zoos was better but still low (8% of females and 13% of males died before age one), and the rate of population decline was about 3% per year (Seal 1983). In the wild, Schaller (1963) had found infant mortality within the first year of life to be 23% and Fossey (1979) found a similar figure of 27%. A portion of the wild mortality is due to intergroup infanticide and human poaching (Fossey 1979), factors that do not influence captive mortality. Captive infant mortality may be related to the fact that few gorilla mothers raise their own offspring. Nadler (1975) reported that 80% of captive-born gorillas are removed for hand-rearing (also see Nadler: this volume). Maternal abuse and neglect, or the *expectation* of such behavior, is the most common reason given by curators for hand-rearing. The situation has improved little since 1975, and this disturbance of the mother/infant relationship is likely a manifestation of captive conditions.

The behavior of primates cannot be properly viewed outside their environmental context. Animals evolve to survive in a particular environment and have intrinsic social and biological needs because of this. If circumstances force them to live in a different physical and social environment (as often occurs in captivity), many will not adapt. The most adaptable may survive and their descendents may even flourish; this is one way evolution works. Their descendents will be different animals, however. For other captive primates, a cycle of reproductive failure, including infertility and disturbances in parental care, may reflect the poor fit between the species' adaptations and the alien captive environment. This is what appears to be happening with the captive gorilla population.

THIS STUDY

Study of the environmental influences on reproduction and, especially, maternal care, has great potential for improving the wellbeing of gorillas in captivity. In the attempt to establish healthy, self-sustaining populations of captive gorillas, all conditions that impinge upon a female must be evaluated, both singly and together, to estimate their influence on reproductive and maternal behavior. To this end, in 1982 we began a study of the reproductive and mothering status of a majority of the female gorillas in captivity. Individual characteristics such as maternal parity, temperament, rank, history, and sex of infant, as well as variables in the physical and social environment, were examined for their effect on mothering.

The research addressed six interrelated questions:

1. What factors in the physical and social environments of captive females influence their ability to reproduce successfully?
2. What factors influence the tendency to abuse or neglect infants?

3. What characteristics of the mother affect her maternal competence?
4. Are there differences between abusive and neglectful mothers, and if so, are they attributable to environmental or individual differences or both?
5. Do gorilla mothers treat male and female infants differently?
6. Based on environmental and individual differences, can we develop a statistical tool to predict a female's potential aptitude as a breeder and mother?

Methodology

The main research instrument was a questionnaire inquiring about the reproductive and mothering status of adult female gorillas in captivity, as well as factors in their social and physical environment that might influence their status. A total of 79 zoos in 20 countries provided data on 192 females, 139 of whom were of breeding age.

Descriptive statistics and discriminant analyses were used to identify differences among individuals and their environments that might distinguish parous versus nonparous adult females and good versus poor mothers. Descriptive statistics were compiled using the Statistical Package for the Social Sciences (SPSS) (Nie Hull Jenkins Steinbrenner & Bent 1975). As our analytic tool we chose discriminant analysis, a method of multivariate analysis for predicting group membership by identifying critical differences between groups. Unfortunately, the use of discriminant analysis precluded consideration of variables for which values on some gorillas were missing, even though this meant leaving out some potentially important discriminators. A total of 17 variables was included in the final analyses; Table 1 lists them. Table 2 lists the group dichotomies for which discriminators were sought. The kappa statistic (K) (Cohen 1960; Titus & Williams 1984) was employed to test statistical significance of the differences uncovered by discriminant analysis. A full description of these procedures appears in Miller-Schroeder's (1985) MA thesis.

As an important adjunct to the statistical analyses of the questionnaire data, we also undertook a descriptive case study of three primiparous gorillas at the Metro Toronto Zoo, using field notes, keeper records, and film and video recordings.

INFLUENCES ON PARITY

The Social Environment

It is now recognized that the gorilla is not the slow, lethargic, simple creature it was once thought to be. The gorilla is revealing itself as an intelligent, sensitive animal with a previously unsuspected depth and richness to its social system (Viet 1982; Harcourt 1979; Fossey 1979 1982). Considering findings from studies of wild gorillas along with data garnered from this study should allow us to reach a better understanding of problems in the captive situation.

Virtually all wild studies have concentrated on the highly endangered mountain subspecies of gorilla (*G. g. berengei*), whereas captive studies, including this one, deal primarily or totally with the western lowland subspecies (*G. g. gorilla*). The eastern lowland subspecies (*G. g. graueri*) is virtually unstudied both in the wild and in captivity. We cannot rule out subspecific differences influencing maternal behavior.

Relation of Group Composition to Breeding

In wild gorilla societies, a single male and female do not constitute a stable social group (Viet 1982). Viable breeding units include at least two adult females (Viet 1982; Yamagiwa 1983). A lone female will not normally stay with a lone male unless he can attract other females; rather, she will emigrate (Viet 1982; Harcourt 1979; Fossey 1982).

Of the 192 females in this study, 139 were eight years of age or older and classed as breeding age. Of these, slightly more than half (72) had given birth; 67 were nulliparous. More nonbreeding females (28%, 19 animals) than breeding females (22%, 15 animals) were kept in male/female pairs rather than in larger groups. Although the difference was not statistically significant, it agrees with trends in the wild.

Transfers between groups for female gorillas in Harcourt's (1979) study varied from one to seven and averaged four. In 13 years at Karisoke, Fossey (1982) saw 26 females transfer 43 times. Intergroup transfer may delay conception; transferring females tend to be older at the birth of their first infant (Fossey, 1982). In our study, nonparous females were on average younger than parous females (Figure 1). This could be both because the youngest females of breeding age had not yet bred, and because transferring females tend to be older at first breeding. Our study showed that parous females as a whole have been in their current social groups longer than nonparous females, while a significantly larger number of nonbreeders ($p<.005$) had been in their social groups less than three years.

Table 1
Environmental Variables Used in Discriminant Analysis

Subset: Social Variables	age of leaving own mother
	social group during pregnancy
	social group during/after birth
	agonistic relationships
	alliances or friendly relationships
	sex of infant
Subset: Physical Variables	inside cage area
	outside cage area
	access to live vegetation
	access to privacy
	provisions for vertical cage use
	nesting material supplied
	provisions for activity/play
	changes in diet during pregnancy
	type of caretaking arrangements
	positive response to people

Table 2
Degree of Discrimination Between Maternal Groups

Group Dichotomies	% Discrimination
infant deaths vs none	74
rear infants vs do not	89
infant removals vs none	85
abuse or neglect infants vs neither	78
abusive vs neglectful	96

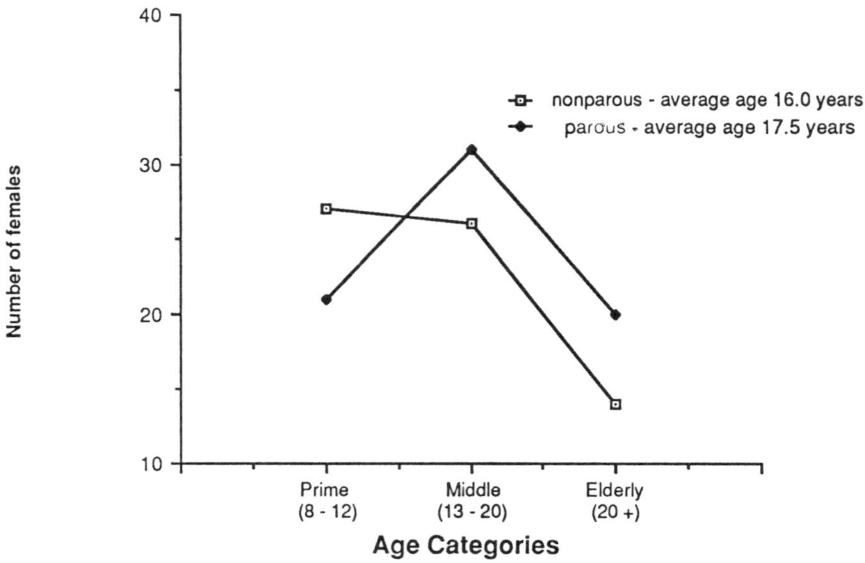

Fig. 1. Age distribution of captive adult female gorillas by parity. Average age of nonparous females: 16.0 yr; parous females: 17.5 yr. Differences likely due to young nulliparous adults who may eventually breed.

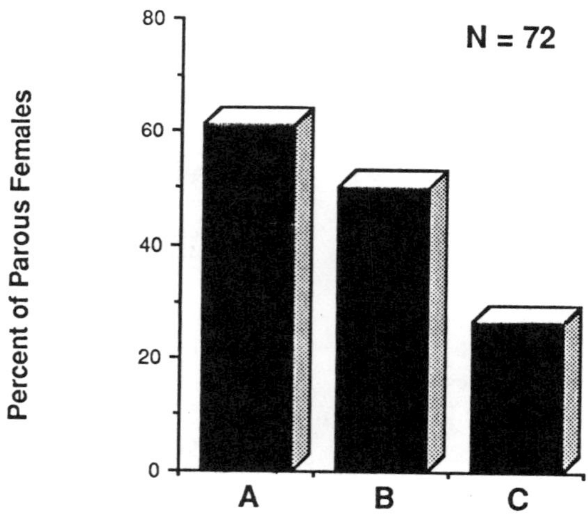

Fig 2. Variation in mothering skills of parous females. A: Never reared infants (due to illness, zoo policy, or poor mothering). B: Abused or neglected at least one infant. C: Good first time mothers. (Many mothers never reared their infants; fully half abused or neglected at least one. Few adequately mothered their first infants.)

Natural Bonds

In the natural social system of gorillas, adult females within a group are usually unrelated and, moreover, may not have known each other for long due to emigration (Harcourt & Stewart 1983). Females prefer to join newly-forming groups over established ones, probably because their status upon entry to an established group is likely to be low (Foster 1982). When related individuals such as mother and daughter do remain together in the wild, they evidence friendly bonds that are stronger than the male/female bond (Harcourt 1979; Harcourt & Stewart 1983). When unrelated females live for a long time in one group, as often happens in captivity, similar bonds may develop. Female coalitions may occur, at least in captive groups (McGrath 1986). It would, in fact, be surprising if both alliances and agonisms did not occur in such a socially complex animal as the gorilla. In the wild, long-term or severe agonism would likely result in the departure of one of the antagonists for more congenial surroundings, but this is precluded in captivity. In our study, less than 25% of the total female population had agonistic relationships and these were usually with an adult male; almost 50% had alliances and these were usually with other females. Nonparous females had slightly more agonistic relationships and fewer alliances than parous females. Perhaps these females are less socially adept.

The strongest social bonds holding gorilla groups together are said to be those between females and the leader male (Harcourt & Stewart 1983), mothers and their young infants (Fischer 1983), and dependent youngsters and the leader male (Yamagiwa 1983). Mothers with infants are a strong social focus for the group (Harcourt 1979; Stewart 1984). Youngsters retain close ties to their mothers even after the birth of new siblings (Fossey 1979) and siblings play together more than might be expected from size and age differences (Harcourt 1979).

Age of Separation From Mother

It is likely that the age of leaving her own mother, and the amount of exposure she has had to other gorillas, will have a lasting influence on a captive female's behavior. Statistics on these experiences were difficult to obtain since the majority (57%) of adult females in the study were wild-born and captured as infants. Because in the past infants were normally collected by killing their mothers, it is likely that the majority of wild-born females were separated from their mothers before age two. For adult females whose history is known, 77% of parous and 86% of nonparous females were removed from their mothers before they were two. It is likely that the vast majority of adult females in zoos today are deficient in this crucial childhood bond with their mothers.

Relationships With Peers and Other Adults

Peer interactions are also important for proper social learning (Kondo Minami & Itoigawa 1981), and other individuals who contact and care for infants play an important part in infant socialization as well (Lee 1983). In our study, the data on early exposure to other gorillas are sketchy because of the same ignorance of early history. Among adult females whose history is known, significantly more parous than nonparous females had early exposure to gorillas other than their mothers (83% parous, 70% nonparous, $p<.05$). More parous than nonparous females had exposure to both peers and adults when young; nonparous females were more often raised with no conspecifics or with adults only. It appears that a mix of ages and sexes in a social group is important for proper socialization of young gorillas, and for appropriate reproductive behavior later.

Bonds With Humans

Human caretakers make up another part of the social system of captive animals (Hediger 1950). Most female gorillas in this study, regardless of parity, had generally stable caretaking arrangements. Slightly more nonparous females had one main keeper, and they tended to respond more positively to keepers and other familiar people than did parous females. The differences were not statistically significant but may indicate a trend among nonparous females to consider humans as part of their social group, making them less likely to breed with other gorillas. For example, if a human keeper has the leader role normally held by the silverback male, females may be unwilling to let gorilla males less dominant than the keeper mate with them.

THE PHYSICAL ENVIRONMENT

Home Range

We found interesting trends in the physical environments of captive gorillas. Estimates of home range in the wild average from 5.6 km^2 to 30 km^2 (Jones & Sabater Pi 1971; Goodall & Groves 1977). Differences are likely due to habitat types and population density. Gorillas tend to wander a great deal, both in search of food (especially preferred foods) and to avoid other gorilla groups (Fossey 1974). In view of the large natural home ranges and the gorillas' tendency to wander, even the largest zoo enclosures are restrictive. Within the limits afforded by zoo habitats, figures that were obtained show that parous females had on average more space, both outdoors and indoors. Aside from a few very large enclosures, parous females had 20% more indoor area, 35% more indoor volume and 6% more outdoor area on average than nonparous females. If we include the very large enclosures, parous females had 33% more indoor and outdoor area and almost 70% more indoor cage volume.

Variety of Stimuli and Opportunities for Activity

By the standards of the natural environment all captive habitats are deficient to some extent in stimulation. In the wild there is a multitude of things to stimulate an intelligent and inquisitive animal. But too often in captivity there is very little. Moreover, gorillas are not strictly terrestrial; young gorillas are at home in the trees and adults like to climb on occasion. In captivity even adult males have been observed high in trees where these were provided (James Foster, Woodland Park Zoo, personal communication; William Mager, Alpenheul, The Netherlands, personal communication), but too often their captive surroundings provide only limited opportunity for climbing.

Provision of usable vertical space and play objects to stimulate activity have been suggested as important to the wellbeing of captive apes (Maple 1979; Maple & Stine 1982). We found only small differences between parous and nonparous females in usable vertical space and cage furniture, but what differences there were favoured parous females. On the other hand, small differences in play articles favored nonparous females. Behavioral engineering devices are as yet not used much to enhance the captive habitats of female gorillas. Although still controversial (Quick 1984), their use in future may help to enrich captive ape habitats (see Markowitz & Line: this volume).

Refuges for Privacy

In the wild, gorillas are not normally in close proximity to humans. In fact, humans are to be avoided as one of the gorilla's few predators. In captivity, the extent and type of space between animal and human can either add to, or alleviate stress (Hediger 1950; Maple 1979; Maple & Stine 1982; Maple & Hoff 1982; Stoskopf 1983). Noise also can be a potent stressor in the captive situation (Stoskopf 1983). In our study, almost half the nonparous females had completely or partially open indoor cage barriers, such as moats or wire mesh, that allowed human scent and noise to reach them. More than 60% of parous females had completely closed barriers, such as glass, that partially filtered out human stimuli. These differences were statistically significant ($p<.05$). Differences outdoors were less substantial, with 75% of parous and 90% of nonparous females having open-air enclosures. Enclosure size is a complicating factor since in a large open enclosure an animal can distance itself from unwelcome human stimuli while in a small enclosure it cannot.

Privacy from conspecifics can also add to wellbeing (Maple 1979; Maple & Hoff 1982; Fossey 1983). In the wild, the lush rain forest provides a natural screen allowing animals privacy from others of their group. In our study of captive habitats, provision for privacy, such as sight barriers and dens, varied. Fully 20% of nonparous females had no private refuge at all while only 1.5% of parous females were so deprived. This was a highly significant finding ($p<.0005$) (Table 3). Almost 85% of parous females had more than one means of privacy while only 65% of nonparous females were so provided.

Vegetation and Nesting Material

Preferred gorilla habitat in the wild is primary or secondary forest and woodland that is relatively moist and lush with a carpet of herbaceous forage (Wolfheim 1983). In view of the importance of live vegetation in the life of the wild gorilla, it is surprising how many female gorillas did not have access to any vegetation (27% of nonparous and 19% of parous females). Gorillas are also superb nest builders, constructing both day and night nests of varying complexity, yet 15% of nonparous and 17% of parous females were given no nesting materials. Of those who did receive nesting material, more parous than nonparous females received a variety of materials. Fossey (1983) and Maple and Hoff (1982) have suggested that access to vegetation and nesting materials can be important sources of environmental enrichment for captive apes.

Temperature and Humidity

In the gorillas' natural habitat, temperatures vary from 13°C to 37°C and humidity from as low as 45% in some areas during the dry season to 98% in most areas during the rainy season (Goodall & Groves 1977). Dahl (1982; also see Dahl: this volume) argues that captive ape habitats should mimic the features of the natural environment, including fluctuations in temperature and humidity. Data on climatic factors proved difficult to come by, however; these questions were not well answered on our questionnaire. On the whole it appears that nonparous females have a slightly wider range of indoor temperatures but a narrower range of outdoor temperatures than parous females. Inside and outside heat gradients, both natural (skylights, shade) and mechanical (heated floors), were provided almost equally for parous and nonparous females. Data on humidity were too few to analyze.

Diet

The diet of wild gorillas is extremely varied and they devote a large proportion of their time to feeding. Over 200 types of food plants have been identified as gorilla food by various researchers, and proto-cultural food preferences have been reported (Maple & Hoff 1982). Disappointingly, the data we received on diet were inadequate for analysis. Some institutions sent detailed descriptions whle others sent fragmentary accounts and others did not answer at all. This question is worthy of further pursuit. The majority of adult females receive some form of vitamin preparation but the number is greater for parous than nonparous females.

Discriminant Analysis of Parity

The initial discriminant analysis attempted to distinguish the characteristics of parous from nonparous females but correctly classified only 70% of cases, and the kappa statistic (K) indicated that a relatively large part of the variance could be attributed to chance. Many variables could not be included in the analysis either because they did not apply to nonparous females (social group during pregnancy, for instance), or because values were missing for too many animals (such as experience with other gorillas). Thus, the usable variable set was reduced, particularly for social variables, and this contributed to the low level of discrimination.

The final variable set consisted of the social factors of age of leaving own mother, agonistic relations, and alliances; and the physical factors of inside cage volume, inside cage area, outside cage area, access to vegetation, refuges for privacy, usable vertical space, nesting material, provisions for play and activity, caretaking arrangements, and positive response to people. Among these, the most important variables proved to be access to vegetation, outside area, inside area, privacy provision, and usable vertical space. These variables together correctly classified 68% of the animals. It must be remembered that any one of these variables may by itself be of minor importance, but as a group they all contribute to the female's probability of breeding successfully.

Table 3
Privacy Refuges Available to Adult Female Gorillas

Type of Refuge	% Parous	% Nonparous
None	2	20
Open dens/rooms	21	23
Sight barriers	24	23
Dens/barriers	39	19
Ladders/shelves	8	12
Other unspecified	6	3
TOTAL	100 (N=66)	100 (N=65)

Table 4

Classification Results for Mothers Who Rear Some vs No Infants

	Group	N	Predicted Group Membership None Reared	Predicted Group Membership Some Reared
TOTAL	None Reared	32	30 (94%)	2 (6%)
VARIABLES	Some Reared	22	4 (18%)	18 (82%)
		54		

Percentage of grouped cases correctly classified: 89% (30+18)

	Group	N	Predicted Group Membership None Reared	Predicted Group Membership Some Reared
SOCIAL	None Reared	43	36 (84%)	7 (16%)
VARIABLES	Some Reared	29	11 (38%)	18 (62%)
		72		

Percentage of grouped cases correctly classified: 75% (36+18)

	Group	N	Predicted Group Membership None Reared	Predicted Group Membership Some Reared
PHYSICAL	None Reared	32	23 (72%)	9 (28%)
VARIABLES	Some Reared	22	4 (18%)	18 (82)
		54		

Percentage of grouped cases correctly classified: 76% (23+18)

This part of the study indicates that, as a group, parous females resided in richer captive environments than did nonparous females. While for many individual variables the differences were not large, parous females in most instances had *more* (space, privacy, friendships, etc.) than nonparous females. To fully account for some females' failure to breed, more variables, especially social variables, will have to be extracted and tested, and individual personalities, mate preferences, and social rank of females and their mates will have to be considered.

MATERNAL ABUSE AND NEGLECT IN CAPTIVE GORILLAS

In the natural condition, mother gorillas have been known to slap, push and threaten their infants during socially stressful situations (Fossey, 1979), but the amount of abuse is minimal, and the mothers generally reassure their infants with an embrace. Infanticide occurs among wild gorillas; Fossey (1979) saw three out of six infants killed during their first year. The infants were not killed by their mothers, however. Deaths usually occurred during intergroup confrontations or during attempts by lone males to induce females to join them (Fossey 1979 1983).

Repeated Episodes of Abuse and Neglect

Although some abusive mothers do improve with subsequent offspring, a larger proportion show no improvement with experience or show some improvement only after abusing several infants. Half (36) of the 72 parous females in our study have abused or neglected an infant (Figure 2), and 17 of these mothers repeated their abuse with subsequent offspring. Over 40% of the abusive mothers (15 animals) showed no improvement in maternal behavior (Figure 3), while less than a quarter (8 animals) did improve, but not necessarily with their second or even third baby. One female did not improve until her fifth infant. Suomi and Ripp (1983), in their review of macaque mothering, mention that the duration of a poor mother's exposure to her infant influences her treatment of the next infant. Mothers who kept their babies for at least 48 hours, no matter how poorly they treated them, tended to be at least adequate with their second-born. Poor mothers who kept their infants for less than 24 hours were at very high risk of abusing their next infant. The data that would have provided insight into this situation for gorillas were, unfortunately, not available to this study.

Sex of Infant

This was an important variable distinguishing mothers with and without infant deaths (Figure 4). Of 175 sexed births in this study, 83 were male and 92 were female. More of the males were abused or neglected by their mothers (48% males vs. 42% females) and more males were stillborn or died as neonates (29% males vs 21% females). More females were raised by their mothers to at least one year of age (19% males vs 28% females). Although the differences were not statistically significant, they were suggestive of trends.

Our study suggests that maternal inexperience *per se* is unlikely to be the decisive factor in infant abuse and neglect. Since 42% of abusive mothers do not improve with subsequent offspring, factors in both the social and physical environment, as well as the sex of the infant, seem to influence maternal performance.

Maternal Experience of Abuse

The majority of captive female gorillas were wild-born and captured as infants. Although their early background is therefore unknown, it is not unreasonable to presume that they were traumatized by capture and transportation to dealers and zoos, and that they experienced some abuse and deprivation, intentional or not. Gucwinski (1981) has commented on the poor physical and psychological condition of young gorillas coming from the wild to the Wroclaw Zoo in Poland in the 1970's. Although these youngsters may not have been abused or neglected by their natural mothers, they did indeed suffer during infancy. Kotelchuk (1982), studying human mothers at high risk for child abuse, pinpointed an unhappy childhood as a key distinguisher between abusive and nonabusive mothers.

Social Isolation

Kotelchuk's study indicated that social isolation was the other predictor of abusive human mothers. This has interesting implications for captive female gorillas. Almost 40% of the parous females in our study were isolated with their infants following birth. In cases where females gave birth within their social group, the birth tended to excite group members who evidenced a desire to see or touch the baby. Fossey (1979) reported that in the wild curiosity is the most common response of group members toward a neonate. Thus, gorilla mothers and infants form a social focus for their groups. Stewart (1984), who observed wild births, emphasized that females about to give birth did not separate themselves from their group and may even have sought out conspecifics. She witnessed an excited, aggressive response by one female toward another who was having a difficult labour, but no harm resulted. Stewart suggests that if the captive quarters are small, it might be best to separate a labouring female, but to reintroduce her to the group soon after delivery. Based on their studies of macaques, Shapiro and Mitchell (1983) suggested that the presence of other adults and playmates diffuses infant-related stress and may be the wisest strategy to prevent infant maltreatment in captivity. (Also see Nadler: this volume.)

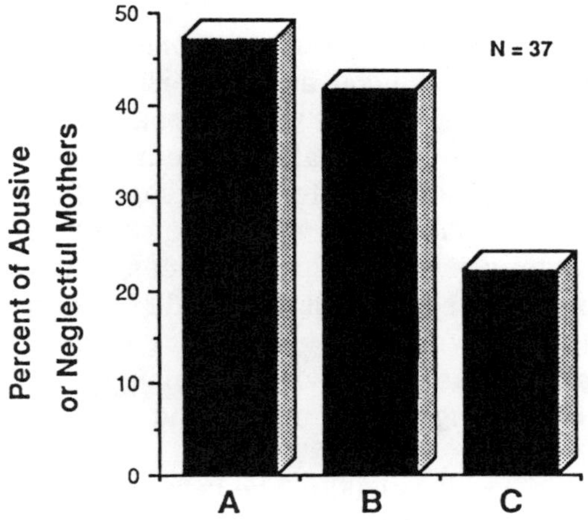

Fig 3. Improvement in infant care by abusive or neglectful gorilla mothers. A: Persistently poor mothers who abused or neglected two or more infants. Some mothers eventually improved. B: Persistently abusive or neglectful mothers who did not improve with later infants. C: Abusive or neglectful mothers who improved with second or later infants. Some abusive mothers improve with later offspring, but more do not, or improve only after abusing several infants.

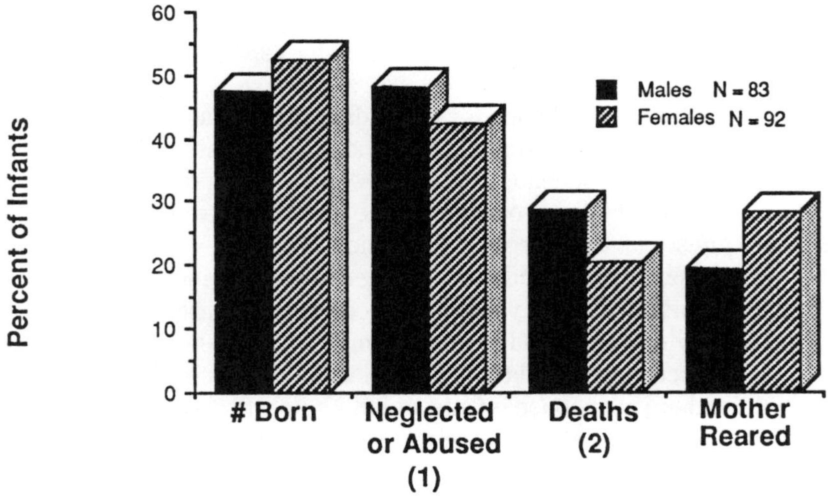

Fig 4. Fate of infant gorillas by sex. (1) Includes infants killed by mothers. (2) Includes deaths from illness, birth defects, & stillbirths. More females were born and more were reared by their mothers to age 1 yr. More males were abused or neglected, or died as neonates.

Discriminant Analysis of Maternal Abuse

The discriminant analysis was designed to weigh the importance of such conditions as mother's childhood, and social isolation of mother and infant, for infant abuse and neglect. As with the analysis of parity, many variables could not be used because values were missing for too many females. The final variable set consisted of the social factors of age of leaving own mother, social group when pregnant, social group following birth, agonistic relationships, alliances, and sex of infant; and physical factors of inside cage volume, inside cage area, access to vegetation, access to privacy, provisions for vertical cage use, nesting material, provisions for activity and play, diet supplements during pregnancy, caretaking arrangements, and positive response to people

Some of the variables that could not be included because of lack of information, especially childhood experience with other gorillas, and diet, are undoubtedly important and deserve further study.

The discriminating variables were entered into the analysis in sets of 1) social environmental variables; 2) physical environmental variables; and 3) total environmental variables including both social and physical subsets. The complexity of the interaction among environmental factors that influence quality of maternal behavior is evident in the fact that the best discrimination usually resulted from using all variables, both social and physical. There was an indication in all of the analyses that social factors best classified one group of mothers while physical factors best classified the contrasting group. This again shows how essential it is to consider both social and physical spheres when trying to understand the causes of infant maltreatment.

The calculation of kappa for these discriminant analyses indicated that the chosen predictor variables were accurate beyond chance for all of them. The canonical correlations also showed that the predictor variables accounted for a substantial proportion of the variance between groups. The analyses suggest that there are real differences between:

1. mothers who do and don't succeed in rearing some young
2. mothers who have young removed (for any reason) and mothers who are allowed to keep them;
3. mothers who are abusive or neglectful and mothers who aren't;
4. mothers who abuse infants and mothers who neglect them.

They also suggest that the chosen variables are good predictors of the group differences.

Infant Deaths

The least successful discriminant analysis concerned mothers some of whose infants died and mothers none of whose infants died. Although 75% correct classification was obtained using only social variables (the only case where a partial variable set was better than the total), kappa indicated that much of the variance between groups could be due to chance.

Successful Infant Rearing

The analysis of mothers who succeeded in rearing some infants to at least one year of age and mothers who did not gave very good discrimination of 89% (Table 4). Social variables alone correctly classified 75%; they were more accurate at classifying mothers who reared no infants. Physical variables were accurate to 76% and were better at classifying mothers who reared some infants. The most important variables here were agonistic relationships, pregnancy social group, vertical cage use, access to vegetation, supplements to pregnancy diet, outside cage area, inside cage volume, provisions for activity/play, age of leaving own mother, and inside area. Here, as in all the dichotomies tested, some females could only be used in the social subset because of missing information for variables in the physical subset (or vice versa). This caused differences in numbers of cases between subsets but did not affect the validity of the analyses.

Table 5
Results for Groups of Mothers with Some vs No Infant Removals

Group		N	Predicted Group Membership: None Removed	Predicted Group Membership: Some Removed
TOTAL	None Removed	23	20 (87%)	3 (13%)
VARIABLES	Some Removed	31	5 (16%)	26 (84%)
		54		
Percentage of grouped cases correctly classified: 85% (20+26)				
SOCIAL	None Removed	28	19 (68%)	9 (32%)
VARIABLES	Some Removed	44	9 (20%)	35 (80%)
		72		
Percentage of grouped cases correctly classified: 75% (19+35)				
PHYSICAL	None Removed	23	18 (78%)	5 (22%)
VARIABLES	Some Removed	31	11 (36%)	20 (64%)
		54		
Percentage of grouped cases correctly classified: 70% (18+20)				

Table 6
Classification Results for Groups of Abusive-or-Neglectful Mothers versus Mothers Who Were Neither

Group		N	Predicted Group Membership No Abuse	Some Abuse
TOTAL	No Abuse/Neg	34	29 (85%)	5 (15%)
VARIABLES	Some Ab/Neg	38	11 (29%)	27 (71%)
		72		
Percentage of grouped cases correctly classified: 78% (29+27)				
SOCIAL	No Abuse/Neg	34	24 (71%)	10 (29%)
VARIABLES	Some Ab/Neg	38	9 (24%)	29 (76%)
		72		
Percentage of grouped cases correctly classified: 74% (24+29)				
PHYSICAL	No Abuse/Neg	34	30 (88%)	4 (12%)
VARIABLES	Some Ab/Neg	38	15 (39%)	23 (61%)
		72		
Percentage of grouped cases correctly classified: 74% (30+23)				

Infant Removals

The analysis concerning mothers some of whose infants were removed versus mothers none of whose infants were removed correctly classified 85% of cases using all variables (Table 5). These were not the same groupings as the ones in the previous analysis. Although mothers who reared no infants to age one were likely to be the mothers some of whose infants were removed, mothers who reared *some* infants to age one were not necessarily the mothers who had no infants removed. Physical variables alone gave correct classification of 70% and were more accurate for mothers who had no infants removed. The important variables here were **pregnancy social group, age of leaving own mother, caretaking arrangements, vertical cage use, sex of infant** (in this case male), **nesting material** and **provisions for play/activity.**

Abusive Versus Good Mothers

The discriminant analysis for abusive or neglectful mothers versus mothers who were neither provided a discrimination of 78% using all variables (Table 6). Social variables alone correctly classified 74%; they were more accurate for classifying the abusive/neglectful mothers. Physical variables alone also correctly classified 74% but were more accurate at classifying mothers who were neither abusive nor neglectful. Important variables here were **supplements to pregnancy diet, provision for activity/play, inside cage volume, pregnancy social group, vertical cage use, inside cage area, caretaking arrangements,** and **nesting material.** The two variables that best predicted whether a mother would mistreat her infant were pregnancy diet supplementation (vitamins, milk, etc.) and provisions for the female to be active. This emphasizes the importance of maternal physical and mental health. A study by Shapiro and Mitchell (1983) showed that over 50% of abusive macaque mothers were in poor health.

Abuse Versus Neglect

The discriminant analysis for abusive mothers versus neglectful mothers was extremely interesting (Table 7). An excellent discrimination level of 96% was achieved using the complete variable set. The number of animals in the analysis was small (25) and the majority of these mothers were neglectful rather than abusive, rousing suspicion that the accuracy of classification here was mainly due to chance. However, the calculation of kappa indicated otherwise. There were very real differences between these groups, far beyond what could be attributed to chance. At Yerkes, Maple and Warren-Leubecker (1983) found that gorillas neglected infants far more commonly than they abused them, and our analysis confirms this. Reidy (1980) states that while both abused and neglected children are unusually aggressive, the abused child tends to be more aggressive than the neglected one and this pattern often endures into adulthood. Williams (1980) believes that abuse by the mother can be the result of social learning in her childhood, while maternal neglect may be due to her fear of novelty or to her inexperience. At any rate, there appear to have been genuine differences between the abusive and the neglectful mothers in our study. The physical variables alone were able to correctly classify 80% of these mothers, and correctly classified all the abusive mothers. The social variables alone correctly accounted for 69% of cases and were more accurate at classifying neglectful mothers. The most important variables were live vegetation, inside cage volume, inside cage area, privacy, number of males born, vertical cage use, nesting material, outside cage area, pregnancy social group, and age of leaving own mother.

Main Discriminators

Patterns emerged through these discriminant analyses. Certain variables served again and again as important discriminators. The variables appearing most often in the four analyses that discriminated the best were usable vertical space, pregnancy social group, inside cage volume, age of leaving own mother, nesting material, and provisions for play and activity. Each of these appeared in at least three of the four analyses; none was consistently most important; and the relative weight of each varied across analyses. However, their recurrence is significant. The repeated appearance of usable vertical space should alert us to the fact that providing different levels within a captive habitat can add immensely to its complexity, utility, and even privacy.

Other Important Discriminators

Among the other variables, some were nearly as heavily weighted as the ones just mentioned or were as heavily weighted but only in some analyses. Chief among these were agonistic relationships, access to live vegetation, pregnancy diet changes, inside and outside areas, caretaking arrangement, sex of infant, and privacy. Their influence cannot be discounted. Social group following birth showed up as moderately significant only once, in separating abusive from neglectful mothers. We expected it would be more significant. It must be remembered that collapsing variables into compressed categories for purposes of discriminant analysis may have obscured some potential discriminators. Future analyses should make use of categories divided as finely as possible and should try to incorporate more variables than only the ones chosen here. Information on early exposure to other gorillas and on diet would be especially desirable, if it could be obtained. Other ways of testing social group of mother and infant following birth would also be useful.

Table 7
Classification Results for Mothers Who Abuse vs Neglect Infants

Group		N	Predicted Group Membership Abuse	Predicted Group Membership Neglect
TOTAL	Abuse	5	5 (100%)	0 (0%)
VARIABLES	Neglect	20	1 (5%)	19 (95%)
		25		

Percentage of grouped cases correctly classified: 96% (5+19)

Group		N	Predicted Group Membership Abuse	Predicted Group Membership Neglect
SOCIAL	Abuse	10	7 (70%)	3 (30%)
VARIABLES	Neglect	26	8 (31%)	18 (69%)
		36		

Percentage of grouped cases correctly classified: 69% (7+18)

Group		N	Predicted Group Membership Abuse	Predicted Group Membership Neglect
PHYSICAL	Abuse	5	5 (100%)	0 (0%)
VARIABLES	Neglect	20	5 (25%)	15 (75%)
		25		

Percentage of grouped cases correctly classified: 80% (5+15)

The discriminant functions do not reveal the value of each variable on its own, but each contributes to the outcome, and the more implicated variables that are present, or the more powerful the variables that are present, the greater the risk for abuse, for the necessity of removing the infant from its mother, and for death. The variables identified in our study as important make sense for they indicate that the size and complexity of a female's environment, along with the richness and species "normalcy" of her social background, influence the success of her maternal behavior.

THE CASE STUDY

The case study of three gorilla mothers, all living in one group, supported the surmise that variables within a female's environment may differentially affect the maternal caregiving of individual females. These females, Samantha, Josephine and Caroline, were all primiparous and all gave birth to daughters. Their early rearing, social experiences, and ages were similar. All were wild-born and came to the zoo as infants about two years old. Since that time they have experienced the same physical and social environment in a relatively stable, long-term social group. Zoo husbandry techniques, as well, were the same for all animals. (For a detailed description of the social and physical environment of these animals see Miller-Schroeder 1985.) Despite strongly similar background, each female differed markedly in how she treated her first-born. Samantha initially appeared to be a good mother but became increasingly neglectful over time. Josephine was initially hesitant but progressively improved in mothering skills. Caroline was an ambivalent mother, giving her infant such basic care as nursing and cleaning but often ignoring her for long periods. At last report Caroline's infant was over two years old and still with her mother. She appeared to be developing normally and was receiving excellent aunting care from other females in the group, including Josephine and Samantha.

This case reveals a continuum of maternal care, from neglectful to borderline to competent, among females of similar ages who were reared in the same social and physical environment from infancy. This may indicate the importance to maternal behavior of temperament and rank of individual females.

Josephine and Samantha gave birth to daughters within 48 hours of each other, in November, 1980. Immediately following the birth of Samantha's infant they were separated from the main group but kept with each other. Josephine, who reared her infant, was the lowest-ranking female in the group. She is, however, a calm, placid individual who avoids conflict and seems to accept her position. Early problems arose when Josephine prevented her infant from nursing. This was probably related to Josephine's health. She had a long, difficult, and tiring labor that apparently disoriented her. Later she had problems with engorged breasts. When this was relieved by her keeper's manually expressing milk, she no longer sought to prevent the infant from nursing. Josephine's maternal behavior continued to improve over time even though she developed a severe cough and cold. She showed concern for Samantha's infant before it was removed, and became an excellent aunt to Caroline's infant. She cleaned, cuddled and tried to nurse her second, but unfortunately stillborn, infant until its removal. Josephine has been a good example of a young, primiparous, wild-born female who was able to mother her infants correctly almost from the start.

Samantha, who eventually neglected her infant, is an excitable, active gorilla. She is low-ranking but dominant to Josephine. Unlike Josephine, Samantha is not content with her situation and will periodically try to assert herself, attacking the more dominant Caroline. Fights usually result, and Samantha is generally frustrated in her attempts to change her situation. In the wild she would likely emigrate. Her first infant was removed after three weeks, and she has not given birth again. This may suggest that she is unlikely to breed in her present social situation.

Samantha, like Josephine, suffered from engorged breasts and a bad cough or cold postpartum. She seemed more bothered than Josephine by the separation from the group and also reacted more to noisy disturbances from the other gorillas and human visitors. Her personal set of internal and external stresses, along with infant-related stresses, seemed to overwhelm her. Samantha may have the potential to become a good mother; she properly mothered her infant for the first few days and has since become a good aunt to Caroline's baby. Samantha's behavior might have improved had she been re-introduced to her social group, and motherhood might have elevated her social position as it seems to have done for Caroline (see below). In the wild, giving birth stabilizes a female's position in the group and her relationships with the leader male and with other females (Yamagiwa, 1983). This, of course, cannot happen if a female is separated from her group during birth and infant rearing, as was done with Samantha.

Caroline gave birth to a daughter in May, 1983. She was separated from the group prior to birth but had some limited contact with the male with whom she has a special relationship. Several weeks following the birth, the keepers began reintroducing her to the other group females, one at a time. After several months the adult male was reintroduced to the all-female group.

Caroline was the third-ranking female at the time her infant was born, and she has since become the dominant female in her group. She was, and is, an ambivalent mother, alternating good with mediocre-to-bad maternal care. She is generally reserved with the other gorillas but can be tough and assertive, and seems secure in her position. She has the advantage of being the male's long-time favorite among the females.

Studies with captive chimpanzee indicate that how much a mother cradles her infant in the days following birth may predict how good a mother she will be (Allen, 1981). Mothers that eventually abuse their infants cradle them very little after birth. Caroline cradled her baby less and less as she became more agitated at her separation from the group. The introduction of social buffers in the form of doting aunts likely prevented Caroline's maternal behavior from deteriorating into outright abuse. The intense interest of the other females in Caroline's infant, and the resultant rise in her social position, seemed to positively reinforce her mothering. In addition, the child care provided by the other females likely diffused Caroline's infant-related stress, allowing her to relax and accept her infant.

Measures of Maternal Competence

Two behavioral measures often used to judge maternal competence are brought into question by these case studies. Nursing, and maternal transportation of infants, are often monitored to judge if an infant should be removed for hand-rearing. It is now time to ask if more flexibility is not needed in evaluating maternal competence.

Nursing. This has long been thought to be a measure of good mothering; indeed, the infant must be nurtured to survive. However, frequency and duration of nursing are variable. All mothers in the case study nursed their baby on the first day for several minutes at a time. Fossey (1979) notes that in wild gorillas suckling bouts during the first two weeks are very short, usually not more than a minute. All the females, by this standard, should be judged to have been good mothers. Fossey noted that nursing sessions are initiated by infants and terminated by mothers. If a mother does not actively place her infant's face on her breast but holds it in a generally correct position it will eventually find the breast by rooting.

It may be that the short suckling bouts in the wild help prevent sore nipples, especially in primiparous females. Engorged, sensitive breasts were definitely a problem for at least two of the females (Josephine and Samantha) and this has been reported for other captive females (Mager 1981). The pain of milk engorgement may be a potent stressor adding to other stress the female may be suffering such as that caused by social isolation. If engorgement can be relieved (manually if necessary), there is a greater likelihood that the mother will accept her baby.

Maternal Transport of Infants. This is also necessary for infant survival, at least in the wild. Fossey (1979) notes that while most infants are carried in a supported ventral/ventral position for the first three months, they are sometimes carried unsupported. She notes that primiparous mothers are much less successful than multiparous mothers at carrying newborns, their babies often slipping into awkward positions. If constant ventral/ventral support is a measure of good mothering, as often thought, all three mothers in this study should be judged to have been good mothers. Josephine early and persistently carried her infant in the ventral/ventral position without support. Samantha initially carried her baby in several positions more appropriate for older infants, which prevented the infant from nursing. Caroline also carried her baby in inappropriate positions that were generally more abusive than Samantha's. For example, she often dragged her baby, yet she would still position it to nurse. Both Samantha and Caroline often placed their babies on the ground and Caroline frequently left hers unattended. The fact that Caroline received social support from other females and Samantha did not was likely crucial to the different outcomes for their infants.

Variability in ways of transporting older infants is common in the wild, with some infants dangling from their mother's necks or riding backwards (Fossey 1979). A study of captive female gorillas at Yerkes (Hoff Nadler & Maple 1983) reported idiosyncratic modes of maternal transport in three primiparous females. Ventral/ventral unsupported and dorsal/dorsal transport occurred fairly early. Fossey (1979) believed that differences in the way mother gorillas handle their babies are due more to temperament than experience. Mode of transport should be judged along with other facets of maternal care to decide if a gorilla is a poor mother. For example, a group that provides buffers in the form of aunts who model good mothering and give extra care to the baby may enable an indifferent mother to rear her infant successfully.

Normal Differences In Maternal Behavior. There are differences among successful gorilla mothers as among humans. Questions of what constitutes child abuse have proven problematic for researchers (Williams 1980). Kotelchuk (1982) urged caution to prevent misclassifying adequate human parents as abusive, for there is as yet no foolproof way to differentiate them. If infant survival, adult social adjustment, and reproduction are reasonable measures of good mothering, then what counts as good mothering may vary a great deal, even within a single human or nonhuman society (Altmann 1980). To identify abuse and neglect requires that we know what is adequate mothering. Since we lack norms for gorilla mothering, abuse and neglect have often been presumed (Maple & Warren-Leubecker 1983) where it might be more accurate to speak of acceptable idiosyncratic variation within a normal range. This may be especially true of neglectful mothers who, with extra training or social support, may turn out to be adequate if not optimal mothers.

Beck (1983) has argued that to improve captive propagation in zoos certain risks must be taken, including, at times, actions that may be potentially harmful to individuals. An example is the decision to let a borderline mother like Caroline rear her infant. The rewards for the future captive gorilla population will be great if mothers learn to care for their infants and are provided with a physical and social environment where this can occur. Even a gorilla who was poorly reared by other gorillas may have a better chance to live in a captive gorilla group than a human-reared gorilla who cannot adjust to gorilla society.

CONCLUSIONS

1. There are factors in the physical environment of captive female gorillas that can influence their ability to reproduce successfully. **Privacy refuges** from conspecifics in the form of **sight barriers and dens** inside the enclosures, and refuges from the public, in the form of partially or completely closed barriers to filter out unwelcome human noises and odors, are important, and especially so in small enclosures. **Cage area and volume** are also important. Parous females have more room, especially in the height of indoor enclosures. What is provided to enable the animals to use **vertical space** is also important. **Shelves, ladders, ropes, trees, catwalks,** etc. can all add significantly to the extent and complexity of usable space.

2. Socially it appears important that as female gorillas grow up they not only have access to their mothers but also to other adult gorillas and to peers. A **mix of ages and sexes** in a group appears to be optimal for socialization and later breeding success.

3. There are factors in the physical environment that appear to influence tendency of mothers to abuse or neglect their infants. Again, the important factors include cage area and volume and usable vertical space. **Opportunities for activity and play** reduce stress and boredom, as do refuges for privacy. Access to **live vegetation and adequate nesting material** can add to the interest and comfort of the enclosure. *Generally, it seems that the larger and more complex a female's environment, the better her chances of rearing her infants normally.*

4. There are factors in the social environment that influence mothers' tendency to abuse or neglect their infants. The type of social group during pregnancy is important. **Residing in a social group is important** since gorillas are highly social animals. Solitude and separation from familiars are stressful for any primate. If a pregnant gorilla sees or has seen other females giving birth or caring for infants, it increases the likelihood that she will care adequately for her own. Whether or not the female was separated from her own mother before age two and whether or not she currently has agonistic relationships will influence how she treats her young. Generally, the more species-typical the social grouping and the background of a female, the better are her chances of normally rearing her offspring.

5. There are individual characteristics that may affect mothers' tendency to care for or abuse their infants. Female gorillas of similar age and background, and with identical parity, social group, physical surroundings and caretaking arrangements, can differ markedly in how they treat their offspring. **Temperament and personality, including excitability and frustration tolerance, as well as social rank,** are influential factors. **Maternal health** is also strongly implicated. Differences in maternal handling of infants may be idiosyncratic and still remain within normal range, however odd by Western human perceptions. Mothering that we may judge as abusive or neglectful may yet fall within the range of normal maternal

behavior for gorillas.

6. There appear to be real differences between mothers who physically abuse their infants and those who merely neglect them in some way. Abusive mothers make up a relatively small group and seem strongly influenced by factors in the physical environment, chiefly cage size and complexity. The environment may act differently on these females due to some innate or acquired qualities of temperament. Neglectful females seem to be a diverse group and may include many inexperienced and socially naive animals. These females may be rehabilitated if given the **opportunity to learn mothering** and if provided with **social buffers to reduce infant-related stresses.**

7. There appear to be differences in both maternal treatment, and neonatal vulnerability, of male and female infants. More males than females are abused and neglected and more are stillborn or die before age one. More females than males are raised by their mothers to the age of one. Although the differences we found were not statistically significant, they carried the hint that males may be more vulnerable or fragile than females. Whether there are differences in male and female infant temperament or irritability, or whether males are less adaptable and suffer more in stressful environments, will require further study.

8. Discriminant analysis has the potential to be useful as a predictive tool in identifying potentially abusive or neglectful mother gorillas. However, it must be used with caution, at least until we have identified variables that can predict poor vs. adequate mothers with a greater degree of accuracy. We must take care not to label a female as a poor mother before the fact, and so bias the decision as to whether to allow her the opportunity to rear her infant.

RECOMMENDATIONS

1. Nonbreeding adult females should be allowed to "emigrate" to other groups. A newly forming group is ideal, as the female has a better chance of establishing a high rank than if she joins an established group. Although emigration may improve her overall breeding success, the start of breeding may be delayed, as transferring females tend to breed later, after they have had time to adjust to their new group.

2. Gorillas, especially breeding, pregnant and nursing females, should be kept in enclosures as large as possible. Efforts should be made to increase both the size and complexity of habitat space, especially usable vertical space, by providing shelves, climbing apparatus, ropes, trees, cat-walks, raised pathways or even a second story to the cage.

3. Provisions for privacy are important at all times but especially during breeding, pregnancy and following parturition. Increasing usable vertical space and floor area, and adding a variety of sight barriers, can serve this purpose.

4. Attention to diet during pregnancy, especially by vitamin and dairy

supplementation, may be important to the production of healthy infants and healthy, less stressed mothers to care for them. Health problems of nursing mothers, especially engorged breasts and sore nipples, should be promptly attended to before nursing is seriously disrupted. In some cases manual expression of milk by caretakers may be necessary to ease breast soreness.

5. The environment of new mothers with infants should be both physically and psychologically stimulating. If a female must be separated with her infant (*this is not recommended*), there must be ample opportunity for play and activity to reduce stress and boredom.

6. Social groups are important to gorillas. Animals should be in species-correct groupings (more than one female per adult male). Agonistic relationships can affect maternal caretaking. Where this is a problem, transfer of mother and infant to a more compatible group may be advisable. Alternatively, remove the antagonist. Female emigration is what would happen in the wild as females search for the best group in which to rear offspring.

7. Females should be with their compatible group during pregnancy, following birth and during child rearing. If they must be separated for medical or other reasons the separation should be as brief as possible and should be relieved by the companionship of a compatible animal.

8. *The presence of other group members as a buffer against child-related stress cannot be over-emphasized.*

ACKNOWLEDGEMENT

This study was supported by a research scholarship from the Alberta Mental Health Advisory Council.

— 26 —
The Psychological Well-Being of Captive Gorillas

Ronald D. Nadler

INTRODUCTION

A major difficulty in addressing the issue of "psychological" wellbeing in captive nonhuman primates is the absence of a generally agreed upon definition. *Webster's Seventh New Collegiate Dictionary* (1967) defines general wellbeing as health, wealth (or success), and happiness, the last most nearly related to the concept of psychological wellbeing. In this chapter, I assume that whatever psychological wellbeing represents in a nonhuman primate, it is 1) a component of general wellbeing and 2) dependent, in part, on the other components. Psychological wellbeing requires good health as well as, for animals living in a mixed-sex group, success in the sense of reproductive success. In this chapter I will emphasize two points: 1) Wellbeing requires consideration of species differences; and 2) the natural habitat is the appropriate environment for defining species-typical behavior relevant to wellbeing. Harcourt (1987) has similarly stressed the relevance of field data for the management of gorillas in captivity.

IN CONSIDERATION OF SPECIES DIFFERENCES

Species evolve different adaptations based on the particular environmental pressures that confront them. As a result, animals that are otherwise closely related may require substantially different maintenance. Although an uncommon species in laboratory research, the gorilla illustrates some useful points about the relevance of species differences and the value of naturalistic observations to the maintenance of nonhuman primates in captivity.

A number of years ago, a survey of more than 90 gorilla births in captivity (Nadler 1975) confirmed an earlier report (Kirchshofer 1970) of a major problem in infant-rearing of gorillas in captivity. Kirchshofer reported that of the first 29 infants born in captivity world-wide, most were rejected by their mothers, some were even injured and only one was reared beyond one year of age. The second survey (Nadler 1975), however, found that several primiparous female gorillas did rear their infants in a reasonably competent manner. This refuted Kirchshofer's hypothesis that inadequate maternal behavior of gorillas was due to the mothers' inexperience with infants and infant-rearing during their own early development (as a consequence of their capture at an early age and rearing in captivity without adult conspecifics from whom to learn maternal behavior). Instead, the data demonstrated that adequate maternal behavior occurred provided that at the time of the birth and thereafter the mother remained with the male (sire) or with her familiar group (Nadler 1980). Female gorillas living under rather modest physical conditions, such as a renovated lion cage (Nadler & Green 1975), exhibited adequate maternal behavior as long as their preexisting social relationships were preserved.

At the time of the first gorilla births in captivity, it was common practice at many zoos to separate the pregnant female from the male or from her group because of fears that the others might injure or kill the infant, either by design or inexperience. This well-intentioned approach by zoo officials, which had been successful with the closely related chimpanzee and orang-utan, nevertheless was probably responsible for the problems in gorilla maternal behavior. The most important consideration, however, is the character of the mother's prior relationships with other animals. Social isolation of mother and infant may be entirely appropriate if it appears that the infant would otherwise be in jeopardy.

A relevant case study was carried out at the Yerkes Regional Primate Research Center of Emory University with a female gorilla who had been separated from her first infant after one week because she was seen acting abusively toward the infant. Based on zoo practice with gorillas at that time, and on successful infant-rearing by chimpanzees and orang-utans at the Yerkes Center based on similar practices, the gorilla female had been separated from conspecific companions during the latter stages of pregnancy. The same female, again separated from companions prior to her second delivery, showed adequate maternal behavior toward her second infant for about three months, after which she began neglecting the infant. After the mother's former female cage-mate was reintroduced into the cage where mother and infant were living (it was known that the mother was dominant over the other female), the mother retrieved the infant and reared it adequately over the next several years that they lived together (Nadler 1980).

This case and a few others in captivity (cited in Nadler 1975), demonstrate that 1) social factors play a significant role in infant-rearing and thus, in the wellbeing of this species and 2) if social factors are ignored, even the most enriched of physical conditions may not compensate for the social deficit. This conclusion was subsequently supported in a prospective study conducted at the Yerkes Field Station. Three female gorillas were maintained in a social group with a male prior to and following the birth of their infants (Nadler 1976). All three were primiparous, i.e., inexperienced maternally, and yet, living within their customary group they all reared their infants adequately for a number of years (Hoff Nadler & Maple 1981).

Further clarification of this issue was obtained in a study of gorillas in their natural habitat (Harcourt 1979; unpublished observations [RDN]). Under natural conditions the distance between the gorilla mothers and the silverbacked leader was related to the ages of the infants. Mothers with the youngest infants stayed nearest the leader male, whereas mothers with older offspring and females without infants stayed farther away. In both field studies, moreover, the birth of an infant was followed immediately by a dramatic change in the spatial relationship between the leader male and the new mother. Whereas previously these females had been on the periphery of the group, following the births of their infants they became the animals in closest proximity to the leader. As a result of their attraction to the leader, the mothers with new infants also came into closer association with other mothers and infants in the group. These observations suggest that the typical pattern of behavior following the birth of an infant gorilla is for the mother to seek the proximity of conspecifics, especially the male that sired the infant. Although the basis for such proximity-seeking is not known, that it occurs under natural conditions supports the proposal that it is the preclusion of such behavior in captivity that compromises adequate infant-rearing

Separating captive chimpanzee and orang-utan mothers and infants from companions does not introduce the problem of inadequate mothering that is found in gorillas. At the Yerkes Center the same type of caging that was used for the gorilla mother-infant dyad mentioned above had no deleterious effect on maternal behavior and infant-rearing by chimpanzees and orang-utans (Miller & Nadler 1981). This difference in social requirements in captivity is reflected in the behavior of these species in their natural habitat (Nadler 1984). Whereas new gorilla mothers seek social proximity under natural conditions, new chimpanzee and orang-utan mothers do not. The wellbeing of gorillas under these circumstances is enhanced by companionship, whereas wellbeing of the other great apes is not, and in fact, may be adversely affected. Social grouping, therefore, is not a universal requirement (or panacea) for all primates under all conditions.

BEHAVIOR IN THE NATURAL HABITAT AS YARDSTICK

The preceding discussion illustrates the usefulness of naturalistic observations for improving the wellbeing of captive nonhuman primates. A further illustration is found in the mating behavior of gorillas. When observed in traditional laboratory pair-tests (subsequently referred to as free-access mating tests – both animals freely accessible to each other in a single cage), some gorilla pairs mated more frequently than the two days per cycle reported for gorillas in the wild (Harcourt Stewart & Fossey 1981). In one case a captive gorilla pair mated for 11 consecutive days (Nadler Collins Miller & Graham 1983). Significantly, the increased mating produced no pregnancies. This increased frequency of mating in the free-access tests is common also in the chimpanzee and orang-utan (Nadler Herndon & Wallis 1986). It was originally ascribed in the gorilla to male aggression because the males charged around the cage a great deal during the tests (Nadler & Miller 1982). During the field study described above (unpublished observations), however, a different perspective was obtained on the function of charging.

The field study confirmed an earlier report by Harcourt et al. (1981) of a dramatic change in distance between the leader male and females at the time of estrus. Except during the few days of estrus each cycle, cycling female gorillas remain mainly peripheral to the core of the group consisting of the leader male and mothers with dependent offspring. Because of the dense vegetation, the leader may not even see the cycling females except for those few days every month or so. These data suggest that daily free-access mating tests in the laboratory bring males and cycling females into proximity much more frequently than would occur in the wild. (Although the western lowland gorilla (*Gorilla gorilla gorilla*) was the subspecies studied in captivity and the eastern mountain gorilla (*G. g. beringei*) the subspecies in the wild, there is no evidence of behavioral differences in this respect.)

The most important finding of the field study, not previously reported, relates to the initiation of mating in the wild. Initially, the most conspicuous component of behavior preceding copulation was the female's assertive backing into the male in a typical primate present posture. The male merely stood up and assumed a tight-lipped, stiff-legged stance, and the female approached and presented. On a few occasions, however, the male charged a short distance parallel to the female before assuming the stiff-legged stance, and on still fewer occasions, the male performed a chest-beat display prior to the charge. It appears, therefore, that some variant of the chest-beat display, generally abbreviated to the stiff-legged stance that normally occurs at its termination, serves as a cue for presenting by estrous females in the wild. This has obvious implications for interpreting the laboratory results, and indirectly, for the wellbeing of captive gorillas.

If charging (followed by the stiff-legged stance) by the adult male gorilla is a normal response to the approach of a cycling adult female, then the frequent charging by captive males in free-access mating tests can be seen as a natural response to an unnaturally frequent (daily) condition of proximity to the female, instead of proximity on only the few days of estrus. If charging normally stimulates the female to present, then the increased presenting and copulation in the free-access tests can also be seen as natural sequelae to an unnatural condition of proximity between the male and female.

This interpretation received support in a second series of laboratory tests, restricted-access mating tests with female choice. In these tests, the male and female are placed in adjoining cages with the doorway between them set at a height that allows only the smaller female to pass through. This simple arrangement gives the female the choice of whether and when copulation may occur. Under these conditions, the number of days of mating during the cycle was reduced and closely resembled the pattern in nature. Use of the two different tests, free-access and restricted-access, permitted the experimental assessment of different aspects of mating in this species, and in conjunction with data from the field provided some insight into their maintenance in captivity.

The two examples of maternal and mating behavior of gorillas are interesting for showing that opposite spatial relationships exist among wild gorillas according to the reproductive status of the female. A female seeks close proximity with the leader male when in possession of a dependent infant and for a few days of estrus when cycling, but otherwise avoids proximity. Making these options for social spacing available to female gorillas in captivity will likely enhance their wellbeing.

ACKNOWLEDGEMENTS

Preparation of this chapter was supported in part by NIH grant RR-00165 from the Division of Research Resources to the Yerkes Regional Primate Research Center and by NIH Grant HD-19060 and NSF Grant BNS 87-08406 to the author. The Yerkes Center is fully accredited by the AAALAC.

REFERENCES

Abee, C. R. (1985). Medical care and management of the squirrel monkey. In L. A. Rosenblum & C. L. Coe (Eds.), *Handbook of squirrel monkey research*. New York: Plenum.

Adams, S. & Babladelis, G. (1977) An ecological approach to animal groups in zoos. *International Zoo News 24,* 14-22.

Akers, J.S. & Schildkraut, D.S. (1985) Regurgitation/reingestion and coprophagy in captive gorillas. *Zoo Biology 4,* 99-109.

Albright, J.L. (1982) Production changes improve cow veal welfare. *Feedstuffs 12,* 23-33.

Alexander, B. K. & Roth, E. M. (1971). The effects of acute crowding on aggressive behaviour of Japanese monkeys. *Behaviour 39,* 73-89.

Alford, P.L. (1985) The prevention and control of infectious disease in the nursery. In C.E. Graham & J.A. Bowen (Eds.), *Clinical management of infant great apes* (pp. 35-42). New York: Alan R. Liss.

Allen, M.L. (1981). *Differential engagement in and overdetermination of active maternal/neonatal ventral-ventral contact in captive chimpanzees (*Pan troglodytes). Paper presented to 4th Annual Meeting of the American Society of Primatologists, San Antonio.

Altmann, J. (1974). Observational study of behavior: Sampling methods. *Behaviour, 49,* 227-265.

Altmann, J. (1980). *Baboon mothers and infants*. Cambridge, MA: Harvard University Press..

Altmann, S.A. (1962) A field study on the sociobiology of rhesus monkeys (*Macaca mulatta*). *Annals of the New York Academy of Sciences, 102,* 338-435.

Altmann, S.A. & Altmann, J. (1970) *Baboon ecology: African field research.* Chicago: University of Chicago Press.

Alvarez, F. (1973). *Comportamiento social y hormonas sexuales en* Saimiri sciureus. Monografias de la Estacion Biologica de Donana, Nm. 2, 74 pp. Madrid: Consejo Superior de Investigaciones Cientificas.

Alvin, S. (1965). *Music for the handicapped child.* London: Oxford University Press.

Alvin, S. (1978). *Music therapy for the autistic child.* London: Oxford University .

American Association of Zoological Parks and Aquariums (1986) *Annual Report.*

Anderson, D. F. (1970). *Backward masking of complex patterns in monkeys.* Unpublished Master's thesis, University of South Dakota, Vermillion.

Anderson, D., Hartley, H., Hartley, C., Reid, J., Schneiderman, N., Skinner, J., Spelman, J. & Turkkan, J. (1984). Report of task group on restrained animals considerations. In J.A. Herd, A.M. Gotto, P.G. Kaufmann, & S.M. Weiss (Eds.), *Cardiovascular instrumentation: Proceedings of the working conference on applicability of new technology to biobehavioral research* (pp. 291-296). DHHS/NIH Publication No. 84-1654. Washington, D.C: U.S. Government Printing Office.

Anderson, J.R. (1983) Responses to mirror image stimulation and assessment of self-recognition in mirror- and peer-raised stumptail macaques. *Quarterly Journal of Experimental Psychology 35B,* 201-212.

Anderson, J.R. (1986) Mirror-mediated finding of hidden food by monkeys (*Macaca tonkeana* and *Macaca fascicularis*). *Journal of Comparative Psychology 100,* 237-242.

Anderson, J.R. & Bayart, F. (1985) Les effects du miroir sur le comportement des macaques: habituation, evitement et facilitation sociale. *Biologie du Comportement 10,* 159-167.

Anderson, J.R. & Chamove, A.S. (1981) Self-aggressive behaviour in monkeys. *Current Psychological Reviews 1,* 139-158.

Anderson, J.R. & Chamove, A.S. (1984a) Allowing captive primates to forage. In *Standards in laboratory animal management.* Symposium Proceedings, Vol. 2. Potters Bar, England: Universities Federation for Animal Welfare.

Anderson, J.R. & Chamove, A.S. (1984b) Early social experience and responses to visual social stimuli in young monkeys. *Current Psychological Research & Reviews 3,* 32-45.

Anderson, J.R. & Chamove, A.S. (1985) Early social experience and development of self-aggression in monkeys. *Biology of Behavior, 10,* 147-157.

Anderson, J.R. & Chamove, A.S. (1986) Infant stumptailed macaques reared with mirrors or peers: Social responsiveness, attachment, and adjustment. *Primates 27,* 63-82.

Animal Plant Health Inspection Service, U.S. Department of Agriculture (1985). Subchapter A: Animal Welfare. Part 3: Standards. Subpart D: Specifications for humane handling, care, treatment, and transportation of nonhuman primates.

Animal Welfare Act. (1985, December 23). Subtitle F (*Animal Welfare*) of Food Security Act of 1985. Public Law 99-198, Title 17, Subtitle F, Section 1752. (See *Federal Register 52(61)*, 10303, Proposed Rules.)

Aquino, R. & Encarnacion, F. (1984). Characteristics and use of sleeping sites in *Aotus* (Cebidae: Primates) in the Amazon Lowlands of Peru. *American Journal of Primatology, 11(4)*, 319-331.

Archer, J. (1979) *Animals under stress.* London: Edward Arnold.

Aspinall, J. (1986). The Howletts gorilla bands. In K. Benirschke (Ed.) *Primates: The road to self-sustaining populations* (pp. 465-470). New York: Springer.

Ator, N.A. & Griffiths, R.R. (1983a). Lorazepam and pentobarbital drug discrimination in baboons: Cross-drug generalization tests and interaction with Ro 15-1788. *Journal of Pharmacology and Experimental Therapeutics, 226,* 776-782.

Ator, N.A. & Griffiths, R.R. (1983b). Oral self-administration of methohexital in baboons. *Psychopharmacology, 79,* 120-125.

Ator, N.A. & Griffiths, R.R. (1985). Lorazepam and pentobarbital discrimination: Interactions with CGS 8216 and caffeine. *European Journal of Pharmacology, 107,* 169-181.

Ator, N.A. & Griffiths, R.R. (1986). Discriminative stimulus effects of atypical anxiolytics in baboons and rats. *Journal of Pharmacology and Experimental Therapeutics, 237,* 393-403.

Ator, N.A. & Griffiths, R.R. (1987). Self-administration of barbiturates and benzodiazepines: A review. *Pharmacology, Biochemistry & Behavior, 27,* 391-398.

Ator, N.A. & Griffiths, R.R. (in press). Asymmetrical cross-generalization with lorazepam and pentobarbital training conditions. *Drug Development Research.*

Ausman, L. M., Gallina, D. L., & Nicolosi, R. J. (1985). Nutrition and metabolism of the squirrel monkey. In L. A. Rosenblum & C. L. Coe (Eds), *Handbook of squirrel monkey research.* New York: Plenum.

Bada, J. L., & Brown, S. (1985). *In vivo* racemization in teeth and ocular lens nucleus. In R.T. Davis & C.W. Leathers (Eds.), *Behavior and pathology of aging in rhesus monkeys* (pp. 91-100). New York: Alan R. Liss.

Badrian, N., & Malenky, R. K. (1984). Feeding ecology of *Pan paniscus* in the Lomako forest, Zaire. In R.L. Susman (Ed.) *The pygmy chimpanzee: Evolutionary biology and behavior* (pp. 275-299). New York: Plenum.

Baldwin, J.D. (1968). The social behavior of adult male squirrel monkeys (*Saimiri sciureus*) in a seminatural environment. *Folia primatologica, 9,* 281-314.

Baldwin, J. D. (1985). The behavior of squirrel monkeys (*Saimiri*) in natural environments. In L. A. Rosenblum & C. L. Coe (Eds.), *Handbook of squirrel monkey research.* New York: Plenum.

Baldwin, J. D. & Baldwin, J. I. (1971). Squirrel monkeys in natural habitats in Panama, Colombia, Brazil, and Peru. *Primates 12:*45-61.

Baldwin, J. D. & Baldwin, J. I. (1972). The ecology and behavior of squirrel monkeys (*Saimiri oerstedii*) in a natural forest in western Panama. *Folia Primatologica, 18,* 161-184.

Baldwin, J. D. & Baldwin, J. I. (1976). Effects of food ecology on social play: A laboratory simulation. *Zeitschrift fur Tierpsychologie 40,* 1-14.

Banerjee, B. N. & Woodward, G. (1970) A comparison of outdoor and indoor housing of rhesus monkeys. *Laboratory Animal Care, 10,* 80-84.

Barbiers, R. B. (1985). Orangutans' color preference for food items. *Zoo Biology, 4,* 287-290.

Bareham, J.R. (1972) Effects of cages and semi-intensive deep litter pens on the behaviour, adrenal response and production in two strains of laying hens. *British Veterinary Journal 128,* 153-163.

Barnett, S. A. (1981). *Modern ethology: The science of animal behavior.* New York: Oxford University Press.

Barrett, A. M. & Stockhon, M. A. (1963). The effect of housing conditions and simple experimental procedures upon corticosterone level in plasma of rats. *Journal of Endocrinology, 26,* 97-105.

Bauchop, T. (1971). Stomach microbiology of primates. *Annual Review of Microbiology, 25,* 429-436.

Bauchop, T. (1978). The significance of microorganism in the stomach of non-human primates. *World Review of Nutrition and Diet 32,* 198-212.

Bauchop, T. and Martucci, R.E. (1968). Ruminant-like digestion of the langur monkey. *Science, 161,* 698-700.

Bearder, S. K. (1974). *Aspects of the ecology and behavior of the thick- tailed bushbaby,* Galago crassicaudatus. Ph.D. Dissertation, University of the Witwatersrand, Johannesburg.

Beck, B.B. (1983). Managerial strategies in the captive propagation of endangered species. *Zoo Biology 2 (3)* :253-257.

Bell, R. W., Miller, C. E., Ordy, J. M. & Rlsten, C. (1971). Effects of population density and living space upon neuroanatomy, neurochemistry, and behavior in the C57B1/10 mouse. *Journal of Comparative and Physiological Psychology 75,* 258-263.

Belzung, C. & Anderson, J.R. (1986) Social rank and responses to feeding competition in rhesus monkeys. *Behavioural Processes 12,* 307-316.

Benirschke, K. (Ed.) (1986) *Primates: The road to self-sustaining populations.* New York: Springer.

Benirschke, K., Lasley, B. & Ryder, O. (1980) The technology of propagation, In M.E. Soule & B. A. Wilcox (Eds.), *Conservation biology: An evolutionary-ecological perspective* (pp. 225-242). Sunderland MA: Sinauer.

Benjamin, M. (1983) Ethics, animals, and language. In J. de Luce and H.T. Wilder (Eds.), *Language in primates: Perspectives and implications* (pp. 113- 124). New York: Springer.

Bennett, C. L., Davis, R. T., & Miller, J. M. (1983). Demonstration of presbycusis across repeated measures in a nonhuman primate species. *Behavioral Neuroscience, 97,* 602-607.

Berkson, G, Mason, W. & Saxon, S. (1963) Situation and stimulus effects on stereotyped behavior of chimpanzees. *Journal of Comparative and Physiological Psychology 56,* 786-792.

Berman, C.M. (1980). Early agonistic experience and rank acquisition among free-ranging infant rhesus monkeys. *International Journal of Primatology 1,* 153-170.

Berman, C.M. (1982a). The ontogeny of social relationships with group companions among free-ranging infant rhesus monkeys. I. Social networks and differentiation. *Animal Behavior 30,* 149-162.

Berman, C.M. (1982b). The ontogeny of social relationships with group companions among free-ranging infant rhesus monkeys. II. Differentiation and attractiveness. *Animimal Behavior 30,* 163-170.

Bernstein, I. (1964). The integration of rhesus monkeys introduced to a group. *Folia Primatologica 2,* 50-63.

Bernstein, I.S. (1970a). Primate status hierarchies. In L.A. Rosenblum (Ed.) *Primate behavior: Developments in field and laboratory research,* Vol. I (pp. 71-109). New York: Academic Press.

Bernstein, I. (1970b). Some behavioral elements of the cercopithecoidea. In J. R. Napier & P. N. Napier (Eds.), *Old World monkeys: Evolution, systematics and behavior* (pp. 263-297). New York: Academic Press.

Bernstein, I. (1971). Activity profiles of primate groups. In A. M. Schrier & F. Stollnitz (Eds.), *Behavior of nonhuman primates,* Vol. 3 (pp. 69-106). New York: Academic Press.

Bernstein, I.S. (1972). The organization of primate societies: Longitudinal studies of captive groups. In R. Tuttle (Ed.) *The functional and evolutionary biology of primates*. Chicago: Aldine.

Bernstein, I.S. (1976). Activity patterns in a sooty mangabey group. *Folia Primatologia 26*: 185-200.

Bernstein, I.S. (1984) The adaptive value of maladaptive behavior, or you've got to be stupid to be smart. *Ethology and Sociobiology, 5,* 297-303.

Bernstein, I.S. & Ehardt, C. L. (1986) Selective interference in rhesus monkey (*Macaca mulatta*) intragroup agonistic episodes by age-sex class. *Journal of Comparative Psychology, 100,* 380-384.

Bernstein, I., & Gordon, T. (1974). The function of aggression in primate societies. *American Scientist 62,* 304-311.

Bernstein, I. S. & Gordon, T. P. (1977). Behavioral research in breeding colonies of Old World monkeys. *Laboratory Animal Science 27,* 532-540.

Bernstein, I.S., Gordon, T. P. & Rose, R. M. (1974) Factors influencing the expression of aggression during introduction to rhesus monkey groups. In R. L. Holloway (Ed.), *Aggression, territoriality and xenophobia* (pp. 211-239). New York: Academic Press.

Beuving, G. (1980) Corticosteroids in laying hens. In R. Moss (Ed.), *The laying hen and its environment.* The Hague: Martinus Nijhoff.

Birren, J. E. (1959). *Handbook of aging and the individual.* Chicago: University of Chicago Press.

Birren, J. E. & Schaie, K. W. (Eds.). (1977). *Handbook of the psychology of aging.* New York: Van Nostrand Reinhold.

Birren, J. E. & Schaie, K. W. (Eds.). (1985). *Handbook of the psychology of aging* (2nd ed.). New York: Van Nostrand Reinhold.

Bishop, N. (1975). *Social behavior of langur monkeys* (Presbytis entellus) *in a high altitude environment.* Unpublished doctoral dissertation, University of California, Berkeley.

Bishop, N. (1979). Himalayan langurs: Temperate colobines. *Journal of Human Evolution, 8,* 251-281.

Blackshaw, J.K. & McVeigh, J.F. (1984) Stereotype behaviour in sows and gilts housed in stalls, tethers, and groups. In M.W. Fox & L.D. Mickley (Eds.), *Advances in animal welfare science, 1984.* Boston: Martinus Nijhoff.

Bloomsmith, M.A., Alford, P.L. & Maple, T.L. (in press). Successful feeding enrichment for captive chimpanzees. *American Journal of Primatology.*

Bloomstrand, M., Alford, P.L. & Maple, T.L. (1987) *An analysis of feeding enrichment for captive chimpanzees.* Paper presented at meeting of the American Society of Primatologists, Madison, WI.

Bloomstrand, M., Riddle, K., Alford, P. & Maple, T.L. (1986) Objective evaluation of a behavioral enrichment device for captive chimpanzees (*Pan troglodytes*). *Zoo Biology, 5,* 293-300.

Blumenthal, H. T. (1985). Autoimmune reactions in the aging endrocrine system: A comparison between monkeys and human beings. In R. T. Davis & C. W. Leathers (Eds.), *Behavior and pathology of aging in rhesus monkeys* (pp. 233-262). New York: Alan R. Liss.

Boccia, M. L. (1987) Behavior as a predictor of physiological and immunological responses to peer separation. [Abstract]. *American Journal of Primatology, 12,* 330.

Boccia, M. L. (1987) The physiology of grooming: A test of the tension reduction hypothesis. [Abstract]. *American Journal of Primatology, 12,* 330.

Boccia, M.L., Laudenslager, M., & Reite, M. (1984). Spatial distribution of food and dominance related behaviors in Bonnet macaques: A laboratory study. [Abstract]. *American Journal of Primatology, 6,* 399.

Boelkins, C.R. & Wilson, A.P. (1972) Intergroup social dynamics of the Cayo Santiago rhesus. *Primates 13,* 125-140.

Boesch, C. & Boesch, H. (1984). Possible causes of sex differences in the use of natural hammers by wild chimpanzees. *Journal of Human Evolution, 13,* 415-440.

Boggess, J.E. (1976). *Social behavior of the Himalayan langur* (Presbytis entellus) *in eastern Nepal.* Unpublished doctoral dissertation, University of California, Berkeley.

Boinski, S. & Timm, R. M. (1985). Predation by squirrel monkeys and double toothed kites on tent making bats. *American Journal of Primatology, 9,* 121- 127.

Bollhorn, R.W. (1980) Lighting in the animal environment. *Laboratory Animal Science 30,* 440-451.

Bonney, R. C., A. F. Dixson & D. Fleming (1980). Plasma concentrations of oestradial-17 Beta, estrone progesterone and testosterone during the ovarian cycle of the owl monkey (*Aotus trivirgatus*). *Journal of Reproductive Fertility, 60,* 101-107.

Bourliere, F. (1961). Patterns of social grouping among wild primates. In S. L. Washburn (Ed.), *Social life of early man* (pp. 1-10). Chicago: University of Chicago Press.

Bourliere, F., Hunkeler, C., & Bertrand, M. (1970). Ecology and behavior of Lowe's guenon (*Cercopithecus campbelli lowei*) in the Ivory Coast. In J. R. Napier & P. H. Napier (Eds.), *Old World monkeys: Evolution, systematics, and behavior* (pp. 297-351). New York: Academic Press.

Bovard, R. (1959) Social stimulation and the response to stress. *Psychological Review 58,* 267-287.

Bowden, D. M., & Jones, M. L. (1979). Aging research in nonhuman primates. In D.M. Bowden (Ed.), *Aging in nonhuman primates* (pp. 1-13). New York: Van Nostrand Reinhold.

Brambell, M. (1975). Breeding orang-utans. In R. Martin (Ed.), *Breeding endangered species in captivity.* London: Academic Press.

Brambell, M. R. (1977) Reintroduction. *International Zoo Yearbook, 17,* 112- 116.

Bramblett, C.A., Bramblett, S.S., Bishop, D. & Coelho, A.M., Jr. (1982) Longitudinal stability in adult status hierarchies among vervet monkeys (*Cercopithecus aethiops*). *American Journal of Primatology 2,* 10- 19.

Bramblett, C.A. & Coelho, A.M., Jr. (1987) Development of social behavior in vervet monkeys, Sykes' monkeys, and baboons. In E.L. Zucker (Ed.) *Comparative behavior of African monkeys* (pp. 67-7). New York: Alan R. Liss.

Bramblett, C.A., Pejaver, L.D. & Drickman, D.J. (1975) Reproduction in captive vervet and Sykes' monkeys. *Journal of Mammalogy, 56,* 940-946.

Brennan, J. & Anderson, J.R. (1988) Varying responses to feeding competition in a group of rhesus monkeys *(Macaca mulatta). Primates 29,* 353-360.

Breznock, A.w., Harrold, J.B. & Kawakami, T.G. (1977). Successful breeding of the laboratory-housed gibbon (*Hylobates lar*). *Laboratory Animal Science 27*, 222-228.

Brody, E.J. & Brody, A.E. (1974). Breeding Muller's Bornean gibbon *Hylobates lar muelleri. International Zoo Yearbook 14,* 110-113.

Bryant, P. B., & Trabasso, T. (1971) Transitive inferences and memory in young children. *Nature, 232,* 456-458.

Buettner-Janusch, J. (1964). The breeding of *Galagos* in captivity and some notes on their behavior. *Folia Primatologica, 2,* 93-100.

Bunyak, S. C., Harvey, N. C., Rhine, R. J., & Wilson, M. I. (1982). Venipuncture and vaginal swabbing in an enclosure occupied by a mixed-sex group of stump-tailed macaques (*Macaca arctoides*). *American Journal of Primatology 2,* 201-204.

Burdyn, L. E., Jr., & Thomas, R. K. (1984) Conditional discrimination with conceptual simultaneous and successive cues in the squirrel monkey (*Saimiri sciureus*). *Journal of Comparative Psychology, 98,* 405-413.

Calhoun, J.B. (1962). A behavioral sink. In E.L. Bliss (Ed.) *Roots of behavior*, pp. 295-315. New York: Harper Hoeber.

Campbell, G. S. (1977). *An introduction to environmental biophysics* (Heidelburg Science Library). New York: Springer.

Cant, J. G. H., & Temerin, L. A. (1984). A conceptual approach to foraging adaptations in primates. In P. S. Rodman & J. G. H. Cant (Eds.), *Adaptations for foraging in nonhuman primates* (pp. 304-342). New York: Columbia University Press.

Capitanio, J. P. (1988). Behavioral pathology. In G. Mitchell (Ed.), *Comparative primate biology: Vol. 2. Behavior and ecology* (pp. 441-454). New York: Alan R. Liss.

Carder, B. & Berkowitz, K. (1979) Rats' preference for earned in comparison with free food. *Science 167,* 1273-1274.

Carlson, N. R. (1987) *The science of behavior* (2nd ed.) Boston: Allyn & Bacon.

Carpenter, C.R. (1940). A field study in Siam of the behavior and social relations of the gibbon *(Hylobates lar.) Comparative Psychology Monographs 16(5)*, 1-212.

Carroll, M.E. & Meisch, R.A. (1984) Increased drug-reinforced behavior due to food deprivation. In T. Thompson, P.B. Dews, & J.E. Barrett, (Eds), *Advances in behavioral pharmacology,* Vol. 4. (pp. 47-88). New York: Academic Press.

Castell, R., Krohn, H., & Ploog, D. (1969). Ruckenwalzen bei Totenkopfaffen (*Saimiri sciureus*): Korperpflege und soziale Funktion. *Zeitschrift fur Tierpsychologie, 26,* 488-497.

Castell, R. & Wilson, C. (1971) Influence of spatial environment on development of mother-infant interaction in pigtail monkeys. *Behaviour, 39,* 202-211.

Chalmers, N.R. (1979). *Social behaviour in primates*. London: Edward Arnold.

Chamove, A.S. (1979) Three social-preference measures in PKU monkeys. *Journal of Mental Deficiency Research 23,* 265-273.

Chamove, A.S. (1980) Nongenetic induction of acquired levels of aggression. *Journal of Abnormal Psychology 89,* 469-488.

Chamove, A.S. (1984a) Long-term learning deficits of mentally retarded monkeys. *American Journal of Mental Deficiency 88,* 352-368.

Chamove, A.S. (1984b) Role of vision in social interaction in monkeys. *Child Development 55,* 1394-1411.

Chamove, A.S. (1986) Exercise improves behaviour: A rationale for occupational therapy. *British Journal of Occupational Therapy 49,* 83-86.

Chamove, A.S. (in press) Enrichment of captive environments: A review. In H.O. Box (Ed.), *Primate responses to environmental change.* London: Chapman & Hall.

Chamove, A.S. & Anderson, J.R. (1979) Woodchip litter in macaque groups. *Journal of the Institute of Animal Technicians 30,* 69-74.

Chamove, A.S. & Anderson, J.R. (1981) Self-aggression, stereotypy, and self- injurious behaviour in man and monkeys. *Current Psychological Reviews 1,* 245-256.

Chamove, A.S. & Anderson, J.R. (in press) Impact of feeding practices on growth and behavior of stumptailed macaques. In J.E. Fa & C.H. Southwick (Eds.), *The ecology and behavior of food-enhanced primate groups.* New York: Alan R. Liss.

Chamove, A.S., Anderson, J.R., Morgan-Jones, S.C., & Jones, S.P. (1982). Deep woodchip litter: Hygiene, feeding, and behavioral enhancement in eight primate species. *International Journal for the Study of Animal Problems, 3,* 308-318.

Chamove, A.S., Anderson, J.R. & Nash, V.J. (1984) Social and environmental influences on self-aggression in monkeys. *Primates 25,* 319-325.

Chamove, A.S., Bayart, F., Nash, V.J. & Anderson, J.R. (1985) Dominance, physiology, and self-aggression in monkeys. *Aggressive Behavior 11,* 17-26.

Chamove, A.S. & Bowman, R.E. (1978) Rhesus plasma cortisol response at four dominance positions. *Aggressive Behavior 4,* 43-55.

Chamove, A.S., Hosey, J. & Schatzel, P. (in press) Visitors excite primates in zoos. *Zoo Biology.*

Chamove, A.S. & Rohrhuber, B. (in press) Moving callitrichid monkeys from cages to outside areas. *Zoo Biology.*

Chamove, A. S., Rosenblum, L. A. & Harlow, H. F. (1973). Monkeys raised only with peers: A pilot study. *Animal Behaviour, 21,* 316-326.

Chapman, C. (1985) The influence of habitat on behaviour in a group of St. Kitts green monkeys. *Journal of Zoology, London (A) 206,* 311-320.

Charlesworth, E. A. (1982). Music psychology and psychotherapy. *Arts in Psychotherapy, 9,* 191-202.

Charles-Dominique, P. (1977). *Ecology and behavior of nocturnal primates.* London: Duckworth.

Cheney, D., Seyfarth, R.S. & Smuts, B. (1986) Social relationships and social cognition in nonhuman primates. *Science 234,* 1361-1366.

Cherfas, J. (1984) *Zoo 2000.* London: British Broadcasting Corporation.

Cherry, J. A., Izard, M. K., & Simons, E. L. (1987). Description of ultrasonic vocalizations of the mouse lemur (*Microcebus murinus*) and the fat-tailed dwarf lemur (*Cheirogaleus medius*). *American Journal of Primatology 13,* 181-185.

Christian, J.J. (1963). Endocrine adaptive mechanisms and the physiological regulation of population growth. In M.V. Mayer & P.G. Gelder (Eds.),

Physiological Mammalogy, Vol. I (pp. 189-353). New York: Academic Press.

Churchland, P. M. (1984) *Matter and consciousness.* Cambridge, MA: MIT Press.

Clark, A.B. (1978). Sex ratio and local resource competition in a prosimian primate. *Science 201,* 163-165.

Clark, A. B. (1985). Sociality in a nocturnal "solitary" prosimian *Galago crassicaudatus. International Journal of Primatology, 6,* 581-600.

Clark, B. R., Rubin, R. T., Kales, A. & Poland, R. (1970). Comparison of fluorometric method for urinary cortisol with modified Porter-Silber method for 17-OHCS. *Clinical Chimica Acta, 27,* 364.

Clark, B. R., Rubin, R. T., & Poland, R. E. (1971). Modification of new fluorometric method for serum and urine cortisol. *Biochemical Medicine, 5,* 177-179.

Clark, M.M. (1980) Effects of rearing environment on adrenal weights, sexual development, and behavior in gerbils: An examination of Richter's domestication hypothesis. *Journal of Comparative and Physiological Psychology 94,* 857-63.

Clarke, J.R. (1977). Physiological mechanisms of stress. In Institute of Biology (Ed.) *Population control by social behaviour* (pp. 127-155). Exeter, England: A. Wheaton.

Clarke, A. S., Juno, C. J. & Maple, T. L. (1982) Behavioral effects of a change in the physical environment: A pilot study of captive chimpanzees. *Zoo Biology, 1,* 371-380.

Clarke, A. A., Line, S., Ellman, G., & Markowitz, H. (1987) *Hormonal and behavioral responses of rhesus macaques to environmental enrichment apparatus.* Paper presented at the annual meeting of the American Society of Primatology, Madison, WI. (A.A.C.'s address: Health Psychology Program, Sonoma Brain & Behavior Research Center, University of California, San Francisco.)

Cleveland, J. & Snowdon, C. T. (1984) Social development during the first 20 weeks in the cotton-top tamarin, *Saguinus o. oedipus. Animal Behaviour, 32,* 432-444.

Clew, T.H. (1969). Observations on reproduction of squirrel monkeys in captivity. *Journal of Reproduction and Fertility* (Supplement 6), 151-156.

Clew, T. H. & Duvall, W. (1966). Sore tails in squirrel monkeys produced by perches. *Laboratory Primate Newsletter, 5,* 8.

Clutton-Brock, T. H. (1977). Some aspects of intraspecific variation in feeding and ranging behavior in primates. In T. H. Clutton-Brock (Ed.), *Primate ecology: Studies of feeding and ranging behavior in lemurs, monkeys and apes* (pp. 539-556). New York: Academic Press.

Clutton-Brock, T. H. & Harvey, P. H. (1977). Species differences in feeding and ranging behaviors in primates. In T. H. Clutton-Brock (Ed.), *Primate ecology: Studies of feeding and ranging behavior in lemurs, monkeys and apes* (pp. 557-584). New York: Academic Press.

Clutton-Brock, T.H., & Iason, G.R. (1986). Sex ratio variation in mammals. *The Quarterly Review of Biology, 61,* 339-374.

Coe, C. L., Cassayre, P., Levine, S., & Rosenberg, L. T. (1987) Effects of age, sex, and psychological disturbance on immunoglobulin levels in the squirrel monkey. *Developmental Psychobiology, 21,* 161-175.

Coe, C. L., Rosenberg, L. T., Fischer, M. & Levine, S. (1987) Psychological factors capable of preventing the inhibition of antibody responses in separated infant monkeys. *Child Development, 58,* 1420-1431.

Coe, C.L., & Rosenblum, L.A. (1974). Sexual segregation and its ontogeny in squirrel monkey social structure. *Journal of Human Evolution, 3,* 551-561.

Coe, C.L., & Rosenblum, L.A. (1978). Annual reproductive strategy of the squirrel monkey *(Saimiri sciureus). Folia primatologica, 29,* 19-42.

Coe, C. L., Smith, E. R. & Levine S. (1985). The endocrine system of the squirrel monkey. In L. A. Rosenblum & C. L. Coe (Eds.), *Handbook of squirrel monkey research.* New York: Plenum.

Coe, C. L. & Scheffler, J. (in press) Utility of immune measures for evaluating psychological well-being in nonhuman primates. *Zoo Biology.*

Cohen, J. (1960). A coefficient of agreement in nominal scales. *Educational Psychological Measurement 20*:37-46.

Comfort, A. (1959). *Aging: The biology of senescence.* New York: Holt, Rinehart & Winston.

Cooper, D. L. & Markowitz, H. (1979) Handlers' effects on contact comfort behaviors of two trios of juvenile chimpanzees in the zoo. *Psychological Reports, 44,* 1015-1018.

Cooper, R.W. (1968). Squirrel monkey taxonomy and supply. In L.A. Rosenblum, & R.W. Cooper (Eds.), *The squirrel monkey* (pp.1-29). New York: Academic Press.

Cornell, E. H. (1978) Learning to find things: A reinterpretation of object permanence studies. In L. S. Siegel & C. J. Brainerd (Eds.) *Alternatives to Piaget: Critical essays on the theory* (pp. 1-27). New York: Academic Press.

Crandall, L. (1964). *Management of wild animals in captivity.* Chicago: University of Chicago Press..

Crompton, R. H. & P. M. Andau (1987). Ranging, activity rhythm, and sociality in free-ranging *Tarsius bancanus. International Journal of Primatology, 8,* 43-71.

Cross, H. A., & Harlow, H. F. (1965) Prolonged and progressive effects of partial isolation on the behavior of macaque monkeys. *Journal of Experimental Research in Personality, 1,* 39-49.

Curtin, R. (1975). *The socioecology of the common langur* (Presbytis entellus) *in the Nepal Himalaya.* Unpublished doctoral dissertation, University of California, Berkeley.

Curtin, R. (1981). Strategy and tactics in male grey langur competition. *Journal of Human Evolution, 10,* 245-253.

Curtis, S.E. (1985). What constitutes animal well-being? In G.P. Moberg (Ed.), *Animal stress* (pp. 1-14). Bethesda, MD.: American Physiological Society.

Czerny, P., & Thomas, R. K. (1975) Sameness-difference judgments in *Saimiri sciureus* based on volumetric cues. *Animal Learning & Behavior, 3,* 375-379.

Dahl, J.F. (1982). The feasibility of improving the captive environments of the Pongidae. *American Journal of Primatology* (Supplement 1), 77-85.

Dahl, J.F., Bernstein, I.S. & Williams, L.E. (1986). Thermoregulation and social structure of a captive group of rhesus macaques. In D.M. Taub & F.A. King (Eds.), *Current perspectives in primate social dynamics* (pp. 243-251). New York:

Van Nostrand Reinhold.

Dahl, J.F. & Smith, E.O. (1985). Assessing variation in the social behavior of stumptail macaques using thermal criteria. *American Journal of Physical Anthropology 68*, 467-477.

Dainow, E. (1977) Physical effects and motor responses to music. *Journal of Research in Music Education, 25,* 211-221.

Dantzer, R. & Mormede, P. (1983) Stress in farm animals: A need for reevaluation. *Journal of Animal Science 57,* 6-18.

Darrow, K. & Pam, R. (1976). *Appropriate technology sourcebook* (2nd ed.). Stanford, CA: Volunteers in Asia.

Davenport, R. (1979). Some behaviorial disturbances of Great Apes in captivity. In D. Hamburg & E. R. McCown (Eds.) *The great apes* (pp. 341- 356). Menlo Park, CA: Benjamin/Cummings.

Davenport, R. K., & Menzel, E.W., Jr. (1963). Stereotyped behavior of the infant chimpanzee. *Archives of General Psychiatry, 8,* 99-104.

Davenport, R. K., Menzel, E. W., Jr. & Rogers, C. M. (1966). Effect of severe isolation in "normal" juvenile chimpanzees. *Archives of General Psychiatry, 14,* 134-138.

Davenport, R. K., Jr. & Rogers, C. M. (1968). Intellectual performance of differently reared chimpanzees: 1. Delayed response. *American Journal of Mental Deficiencies, 72,* 674-680.

Davenport, R. K., & Rogers, C. M. (1970). Differential rearing of the chimpanzee. In G. H. Bourne (Ed.), *The chimpanzee,* Vol. 3 (pp. 337-360). New York: S. Karger.

Davenport, R. K., Rogers, C. M., & Rumbaugh, D. M. (1973). Long-term cognitive deficits in chimpanzees associated with early impoverished rearing. *Developmental Psychology, 9,* 343-347.

Davis, R. T. (1974). Monkeys as perceivers. In L. A. Rosenblum (Ed.), *Primate behavior: Developments in field and laboratory research* (Vol. 3, whole volume). New York: Academic Press.

Davis, R. T. (1978). Old monkey behavior. *Experimental Gerontology, 13,* 237-250.

Davis, R. T. (1985a). Behavioral history. In R. T. Davis & C. W. Leathers (Eds.), *Behavior and pathology of aging in rhesus monkeys* (pp. 13-56). New York: Alan R. Liss.

Davis, R. T. (1985b). The effects of aging on the behavior of rhesus monkeys. In R. T. Davis & C. W. Leathers (Eds.), *Behavior and pathology of aging in rhesus monkeys* (pp. 57-82). New York: Alan R. Liss.

Davis, R. T., Bennett, C. L., & Weisenburger, W. P. (1982). Repeated measurements of forgetting in rhesus monkeys (*Macaca mulatta*). *Perceptual and Motor Skills, 55,* 703-709.

Davis, R. T., & Leathers, C. W. (Eds.). (1985). *Behavior and pathology of aging in rhesus monkeys.* New York: Alan R. Liss.

Davis, R. T., & Ruggiero, F. T. (1973). Memory in monkeys as a function of preparatory interval and complexity of matrix displays. *American Journal of Physical Anthropology, 38,* 573-578.

Dawkins, M. (1976) Towards an objective method of assessing welfare in domestic fowl. *Applied Animal Ethology 2,* 245-254.
Dawkins, M.S. (1980) *Animal suffering: The science of animal welfare.* London: Chapman and Hall.
Dazey, J., & Erwin, J. (1976). Infant mortality in *Macaca nemestrina*: neonatal and postneonatal mortality at the Regional Primate Research Field Station at University of Washington, 1967-1974. *Theriogeneology 5,* 267-279.
Dazey, J., Kuyk, K., Oswald, M., Mortenson, J., & Erwin, J. (1977). Effects of group composition on agonistic behavior of captive pigtail macaques, *Macaca nemistrina. American Journal of Physical Anthropology 46,* 73-76.
de Waal, F. B. M. (1982) *Chimpanzee politics: Power and sex among apes.* New York: Harper & Row.
de Waal, F. B. M. (1986). The integration of dominance and social bonding in primates. *Quarterly Review of Biology 61,* 459-479.
Denenberg, V.H. (1964) Critical periods, stimulus input, and emotional reactivity: A theory of infantile stimulation. *Psychological Review 71,* 335- 351.
Denenberg, V.H. & Whimbbey, A.E. (1963) Behavior of adult rats is modified by the experiences their mothers had as infants. *Science 142,* 1192-1193.
Dethier, V. G. (1976) *The hungry fly.* Cambridge, MA.: Harvard University Press.
Diener, E. (1984) Subjective well-being. *Psychological Bulletin, 95,* 542- 575.
Dietz, J. M., & Kleiman, D. G. (1986). Reproductive parameters in groups of free-living golden lion tamarins. *Primate Report, 14,* 77.
Dittrich, L. (1976). Food presentation in relation to behavior in ungulates. *International Zoo Yearbook, 16,* 48-54.
Dittus, W.P.J. (1980). The social regulation of primate populations: A synthesis. In D.G. Lindburg (Ed.) *The macaques: Studies in ecology, behavior and evolution* (pp. 263-286). New York: Van Nostrand Reinhold.
Dixson, A. F. & D. Fleming (1981). Parental behaviour and infant development in owl monkeys (*Aotus trivirgatus griseimembra*). *Journal of Zoology, 194,* 25-29.
Dobberstein, J. (1936). Uber die haufigsten todesursachen der in zoologischen garten gehaltenen tiere. *Med. Klinik.* Vol. 32, No. 10.
Dolhinow, P. (Ed.) (1972a) *Primate patterns.* New York: Holt, Reinhart, & Winston.
Dolhinow, P. (1972b). The north Indian langur. In P. Dolhinow (Ed.), *Primate patterns* (pp. 181-238). New York: Holt, Rinehart, and Winston.
Dolhinow, P. (1977). *Caretaking patterns of the Indian Langur monkey.* Paper presented at the annual meeting of the American Association of Physical Anthropology, Seattle, WA.
Dolhinow, P. (1980). An experimental study of mother loss in the Indian langur monkey (*Presbytis entellus*). *Folia primatologica 33,* 77-128.
Dolhinow, P. & DeMay, M. (1982). Adoption: The importance of infant choice. *Journal of Human Evolution, 11,* 391-420.
Dolhinow, P. & Krusko, N. (1984). Langur monkey females and infants: The female's point of view. In M. Small (Ed.), *Female primates: Studies by female primatologists* (pp. 37-57). New York: Alan R. Liss.

Dolhinow, P. & Krusko, N. (1988). *A comparison of colony living langur mothers (*Presbytis entellus*) with own and adopted infants.* Unpublished manuscript.

Dolhinow, P., McKenna, J., & Von der Haas Laws, J. (1979). Rank and reproduction among female langur monkeys: Aging and improvement (They're not just getting older, they're getting better). *Journal of Aggressive Behavior, 5,* 19-30.

Dolhinow, P. & Murphy, G. (1982). Langur monkey (*Presbytis entellus*) development: The first three years of life. *Folia Primatologica, 39,* 305- 331.

Dolhinow, P. & Murphy, G. (1983). Langur monkey mother loss: Profile analysis with multivariate analysis of variants for separation subjects and controls. *Folia primatologica, 40,* 181-196.

Dolhinow, P. & Murphy, G. (1988). *The length of effects of temporary mother- less in langur monkeys (*Presbytis entellus*).* Unpublished manuscript.

Dolhinow, P. & Richenbach, R. (1988). *Weaning in langur monkeys (*Presbytis entellus*).* Unpublished manuscript.

Dolhinow, P. & Taff, M. (1988). *Male caregiving behavior in langurs (*Presbytis entellus*).* Unpublished manuscript.

Doolittle, D.P., Wilson, S.P. & Geisking, D. (1976) Effect of caging variables on body weight and weight gain in mice. *Laboratory Animal Science 26,* 556- 559.

Draper, W.A. & Bernstein, I.S. (1963) Stereotyped behavior and cage size. *Perceptual & Motor Skills 16,* 231-234.

Drickamer, L.C. (1973). Semi-natural and enclosed groups of *Macaca mulatta*: A behavioral comparison. *American Journal of Physical Anthropology 39*: 249- 254.

Dukelow, W.R. (1983). The squirrel monkey *(Saimiri sciureus)*. In F. Hearn (Ed.), *Reproduction in New World primates. New models in medical science* (pp.149-179). Lancaster: MTP Press.

Dukelow, W. R. (1985). Reproductive cyclicity and breeding in the squirrel monkey. In L. A. Rosenblum & C. A. Coe (Eds.), *Handbook of squirrel monkey research.* New York: Plenum.

DuMond, F.V. (1968). The squirrel monkey in a seminatural environment. In L.A. Rosenblum, & R.W. Cooper (Eds.), *The squirrel monkey* (pp. 87- 145). New York: Academic Press.

DuMond, F.V, & Hutchinson, T.C. (1967). Squirrel monkey reproduction: the "fatted" male phenomenon and seasonal spermatogenesis. *Science, 158,* 1067-1070.

Dunbar, R. (1974). Observations on the ecology and social organization of the green monkey (*Cercopithecus sabaeus*) in Senegal. *Primates, 15,* 341-350.

Duncan, I.J.H. (1977) Behavioural wisdom lost? *Applied Animal Ethology 3,* 193-194.

Duncan, I.J.H. (1978) The interpretation of preference tests in animal behaviour. *Applied Animal Ethology 4,* 197-200.

Dutrillaux, B. (1986). The role of chromosomes in evolution: A new interpretation. *Annals of Genetics (Paris), 29*(2), 69-75.

Ehardt-Steward, C. & Bramblett, C. A. (1980). The structure of social space among a captive group of vervet monkeys. *Folia Primatologia 34*: 214-235.

Eisen, E.J. (1966) Comparison of two cage-rearing regimes on reproductive performance and body weight of the laboratory mouse. *Laboratory Animal Care 16,* 447-53.

Eisenberg, J., & Kuehn, R. (1966). Behavior of *Ateles geoffroyi* and related species. *Smithsonian Miscellaneous Collection, 151,* 1-61.

Ekesbo, I. (1981) Some aspects of sow health and housing. In W. Sybesma (Ed.), *The welfare of pigs.* The Hague: Martinus Nijhoff.

Elias, M. G. & Samonds, K. W. (1973). Exploratory behavior of cebus monkeys after having been reared in partial isolation. *Child Development, 44,* 218- 220.

Ellefson, J.O. (1974). A natural history of white-handed gibbons in the Malayan Peninsula. In D.M. Rumbaugh (Ed.), *Gibbon and siamang* (pp. 1-136). Basel: Karger.

Elton, R. H. & Anderson, B. V. (1977). The social behavior of a group of baboons (*Papio anubis*) under artificial crowding. *Primates, 18,* 225-234.

Engle, M. J., & Farrell, P. M. (1985). Lung phospholipids in aging primates. In R. T. Davis & C. W. Leathers (Eds.), *Behavior and pathology of aging in rhesus monkeys* (pp. 275-292). New York: Alan R. Liss.

Epple, G. (1978) Reproductive and social behavior of marmosets with special reference to captive breeding. *Primates in Medicine, 10,* 50-62.

Erwin, J. (1971) Factors influencing survival and development of *Macaca nemestrina* and *Macaca fascicularis* infants in a harem breeding situation. In L.E. Jarrard (Ed.), *Cognitive processes of nonhuman primates.* New York: Academic Press.

Erwin, J. (1977). Factors, influencing aggressive behavior and risk of trauma in the pigtail macaque (*Macaca nemistrina*). *Laboratory Animal Science 27,* 541-547.

Erwin, J. (1978). Factors contributing to intragroup aggression in captive pigtail monkey groups. *Recent Advances in Primatology:* 581-584.

Erwin J. (1979). Aggression in captive macaques: Interaction of social and spatial factors. In J. Erwin, T. Maple, & G. Mitchell (Eds.), *Captivity and behavior: Primates in breeding colonies, laboratories, and zoos* (pp. 139- 171). New York: Van Nostrand Reinhold.

Erwin, J. (1986). Environments for captive propagation of primates: Interaction of social and physical factors. In K. Benirshke (Ed.), *Primates: The road to self-sustaining populations* (pp. 297-305). New York: Springer. Erwin, J. & Deni, R. (1979) Strangers in a strange land: Abnormal behaviors or abnormal environments? In J. Erwin, T.L. Maple & G. Mitchell (Eds.), *Captivity and behavior: Primates in breeding colonies, laboratories and zoos* (pp. 1-28). New York: Van Nostrand Reinhold.

Erwin, J. M., Mitchell, G. & Maple, T. (1973)b Abnormal behavior in non- isolate-reared rhesus monkeys. *Psychological Reports, 33,* 515-523.

Eterovic, V. A. & Ferchmin, P. A. (1985). Interaction with objects by group- living *Saimiri sciureus. Puerto Rico Health Science Journal, 4,* 121-125.

Evans, C. S. (1967). Methods of rearing and social interaction in *Macaca namestrina. Animal Behavior, 15,* 263-266.

Ewbank, R. (1973) Use and abuse of the term "stress" in husbandry and welfare.

Veterinary Record 30, 709-710.

Fa, J.E. (1984). Structure and dynamics of the Barbary macaque population in Gibraltar. In J.E. Fa (Ed.) *The Barbary macaque: A case study in conservation* (pp. 263-306). New York: Plenum.

Fa, J.E. (1986a). *Use of time and resources by provisioned troops of monkeys: Social behaviour, time and energy in the Barbary macaque* (Macaca sylvanus L.) *at Gibraltar.* Contributions to Primatology, No. 23. Basel: Karger.

Fa, J.E. (1986b). Balancing the captive-wild equation: The case of the Barbary macaque (*Macaca sylvanus*). In K. Benirschke (Ed.), *Primates: The road to self-sustaining populations* (pp. 197-211). New York: Springer.

Fa, J.E. (1987). A park for the Barbary macaques at Gibraltar? *Oryx 21*: 242- 245.

Fa, J.E. (in press). Supplemental food as an extranormal stimulus in Barbary macaques (*Macaca sylvanus*) at Gibraltar: Its impact on activity budgets. In J.E. Fa & C.H. Southwick (Eds.) *Ecology and behavior of food-enhanced primate groups.* New York: Alan R. Liss.

Fa, J. E., Ramirez, J. & Valdespino, C. *Impact of zoo visitors on the behavior of a group of green monkeys.* In preparation.

Fa, J.E. & Southwick, C.H. (Eds.) (in press). *Ecology and behavior of food- enhanced primate groups.* New York: Alan R. Liss.

Fagen, R. (1981). *Animal play behavior.* New York: Oxford University Press..

Fairbanks, L. (1974). An analysis of subgroup structure and process in a captive squirrel monkey *(Saimiri sciureus)* colony. *Folia primatologica, 21,* 209-224.

Feistner, A.T.C. & Chamove, A.S. (1986) High motivation toward food increases food-sharing in cotton-top tamarins. *Developmental Psychobiology 19,* 439- 452.

Festing, M.F.W. (1979). *Inbred strains in biomedical research.* New York: Oxford University Press..

Finger, S. & Stein, D. G. (1982). *Brain damage and recovery.* New York: Academic Press.

Fischer, R.B. (1983). Maternal subgrouping in lowland gorillas. *Behavioral Processes 8*:301-306.

Fitts, S. S. (1982). Behavioral stereotypy and social disengagement. *Primates, 23,* 406-415.

Fleagle, J. G., Mittermier, R. A., & Skopec, A. L. (1981). Differential habitat use by *Cebus apella* and *Saimiri sciureus* in central Surinam. *Primates, 22,* 361-367.

Fodor, J. A. (1981) The mind-body problem. *Scientific American, 250,* 132-140.

Foerg, R. (1982). Reproductive behavior in *Varecia variegata. Folia Primatologia 38,* 108-121.

Food Security Act. (1985, December 23). Public Law 99-198, Title 17, Subtitle F: Animal Welfare, Section 1752. (See *Federal Register 52(61)*, 10303, Proposed Rules.)

Fooden, J. (1964). Stomach contents and gastrointestinal proportions in wild- shot Guianan monkeys. *American Journal of Physical Anthropology, 22,* 227- 231.

Fossey, D. (1974). Observations on the home range of one group of mountain gorillas. *Animal Behavior 22*:568-581.

Fossey, D. (1979). Development of the mountain gorilla (*Gorilla gorilla beringei*):

The first thirty-six months. In D.A. Hamburg & E.R. McCoun (Eds.), *The great apes* (139-184). Menlo Park, CA: Benjamin/Cummings.

Fossey, D. (1982). Reproduction among free-living mountain gorillas. *American Journal of Primatology* (Supplement 1): 97-104.

Fossey, D. (1983). *Gorillas in the mist.* Boston: Houghton Mifflin.

Fossey, D., & Harcourt, A. H. (1977). Feeding ecology of the free-ranging mountain gorilla (*Gorilla gorilla berengei*). In T. H. Clutton-Brock (Ed.), *Primate ecology: Studies of feeding and ranging behavior in lemurs, monkeys and apes* (pp. 415-447). New York: Academic Press.

Foster, J.W. (1982). Kin selection and gorilla reproduction. *American Journal of Primatology Supplement 1*:27-35.

Fouts, R. S., Fouts, D. H., & Schoenfeld, D. (1984). Sign language conversational interactions between chimpanzees. *Sign Language Studies, 34,* 1-12.

Fouts, R. S., Hirsch, A. & Fouts, D. H. (1982). Cultural transmission of a human language in a chimpanzee mother/infant relationship. In H. E. Fitzgerald, J. A. Mullins, & P. Page (Eds.), *Psychobiological perspectives: Child nurturance series,* Vol. III (pp. 159-193). New York: Plenum.

Fowles, J. (1970) *The aristos.* Boston: Little, Brown.

Fox, M. W. (1966) Natural environment: Theoretical and practical aspects for breeding and rearing laboratory animals. *Laboratory Animal Care, 16,* 316- 320.

Fox, M. W. (1970) Environmental influences on behavior of domestic and laboratory animals. *Advances in Veterinary Science and Comparative Medicine, 15,* 47-67.

Fox, M. W. (1986a). Handling, socialization, and caretaker effects. In M. Fox (Ed.), *Laboratory animal husbandry: Ethology, welfare and experimental variables* (pp 71-83). Albany: State University of New York Press.

Fox, M. W. (1986b). *Laboratory animal husbandry: Ethology, welfare, and experimental variables.* Albany: State University of New York Press.

Fox, M. W. (1986c, June). The trouble with zoos. *The Animals' Agenda*, pp. 8- 12.

Fragaszy, D. M. (1979). Squirrel and titi monkeys in a novel environment. In J. Erwin, T. Maple, & G. Mitchell (Eds.), *Captivity and behavior.* New York: Holt, Rinehart, & Winston.

Fragaszy, D. M. (1985). Cognition in squirrel monkeys. In L. A. Rosenblum & C. L. Coe (Eds.), *Handbook of squirrel monkey research.* New York: Plenum.

Fragaszy, D. M. & Mason, W. A. (1978). Response to novelty in *Saimiri* and *Callicebus:* Influence of social context. *Primates, 19,* 311-331.

Frankenhaeuser, M. (1982). Challenge-control interaction as reflected in sympathetic-adrenal and pituitary-adrenal activity: Comparison between the sexes. *Scandinavian Journal of Psychology, 1,* 158-164.

French. J. A., Abbott, D. H., Scheffler, G. A., Robinson, J. A. & Goy, R. W. (1983) Cyclic excretion of urinary oestrogens in female tamarins (*Saguinus oedipus*). *Journal of Reproduction and Fertility, 68,* 177-184.

French, J. A., Abbott, D. G. & Snowdon, C. T. (1984) The effect of social environment of estrogen secretion, scent marking and sociosexual behavior in tamarins (*Saguinus oedipus*). *American Journal of Primatology, 6,* 155-167.

French, J. A., & Stribley, J. A. (1985). Patterns of urinary oestrogen excretion in female golden lion tamarins *(Leontopithecus rosalia)*. *Journal of Reproduction and Fertility, 75,* 537-546.

French, J. A., & Stribley, J. A. (1987). Synchronization of ovarian cycles within and between social groups in golden lion tamarins *(Leontopithecus rosalia)*. *American Journal of Primatology, 12,* 469-478.

Fritz, P., & Fritz, J. (1979). Resocialization of chimpanzees. *Journal of Medical Primatology, 8,* 202-221.

Fritz, P. & Nash, L. (1983). Rehabilitation of chimpanzees: Captive population crises. *Laboratory Primate Newsletter, 22,* 4-8.

Galdikas, B. (1978). Orangutans and hominid evolution. In S. Udin (Ed.), *Spectrum* (pp. 287-309). Jakarta: Dian Rakyat.

Galdikas, B. M. F. & Teleki, G. (1981). Variations in subsistence activities of female and male pongids: New perspectives on the origins of hominid labor division. *Current Anthropology, 22,* 241-255 and 316-320.

Gallup, G. G., Jr. (1970) Chimpanzees: Self-recognition. *Science, 167,* 86-87.

Gallup, G. G., Jr. (1982) Self-awareness and the emergence of mind in primates. *American Journal of Primatology, 2,* 237-248.

Gallup, G.G. Jr. (1985) Do minds exist in species other than our own? *Neuroscience Biobehavioral Reviews, 9,* 631-641.

Gallup, G. G., Jr. (1986, July) Self-recognition or the lack thereof in primates. In G. Ettlinger (Chair), *Continuities and discontinuities in cognitive performance within the primates.* Symposium conducted at the Zentrum fur interdisziplinare Forschung, University of Bielefeld, Bielefeld, FRG.

Gallup, G. G., Jr., Boren, J. L., Gagliardi, G. J., & Wallnau, L. B. (1977) A mirror for the mind of man, or Will the chimpanzee create an identity crisis for *Homo sapiens? Journal of Human Evolution, 6,* 303-313.

Gallup, G.G., Jr., Wallnau, L.B. & Suarez, S.D. (1980) Failure to find self-recognition in mother-infant and infant-infant rhesus monkey pairs. *Folia Primatologia 33,* 210-219.

Gamble, M. R. (1982). Sound and its significance for laboratory animals. *Biological Review, 57,* 395-421.

Ganzhorn, J. U. (1986). Feeding behavior of *Lemur catta* and *Lemur fulvus. International Journal of Primatology 7,* 17-30.

Ganzhorn, J. U. (1987). Soil consumption of two groups of semi-free-ranging lemurs (*Lemur catta* and *Lemur fulvus*). *Ethology 74*: 146-154.

Garber, P.A. (1980) Locomotor behavior and feeding ecology of the Panamanian tamarin *(Saguinus oedipus geoffroyi*, Callitrichidae Primates). *International Journal of Primatology 1,* 185-201.

Garber, P. A. (1984). Use of habitat and positional behavior in a Neotropical primate, *Saguinus oedipus.* In P. S. Rodman & J. G. H. Cant (Eds.), *Adaptations for foraging in nonhuman primates* (pp. 112-133). New York: Columbia University Press.

Garber, P.A. (1986) Influence of group size on dietary and foraging patterns in *Saguinus mystax* and *Saguinus fuscicollis* in Amazonian Peru. *Primate Report 14,* 12.

Garber, P.A. & Sussman, R.W. (1984) Ecological distinctions between sympatric species of *Saguinus* and *Sciurus*. *American Journal of Physical Anthropology 65*, 135-146.

Gardner, B. T., & Gardner, R. A. (1971). Two-way communication with an infant chimpanzee. In A. Schrier & F. Stollnitz (Eds.), *Behavior of non-human primates:* Vol 4 (pp. 117-184). New York: Academic Press.

Gardner, R. A., & Gardner, B. T. (1969). Teaching sign language to a chimpanzee. *Science, 165,* 664-672.

Gardner, R. A., & Gardner, B. T. (1978). Comparative psychology and language acquisition. *Annals of the New York Academy of Sciences, 309,* 37-76.

Gardner, R. A., & Gardner, B. T. (1984). A vocabulary test for chimpanzees (*Pan troglodytes*). *Journal of Comparative Psychology, 98,* 381-404.

Garmezy, N. (1982) Children under stress. In R.A. Zucker & A.I. Rabin (Eds.), *Further Exploration in Personality.* New York: Wiley.

Gartlan, J., & Brain, C. (1968). Ecology and social variability in *Cercopithecus aethiops* and *C. mitis*. In P. C. Jay (Ed.), *Primates*. New York: Reinhard and Winston.

Gautier-Hion, A. (1966). The ecology and ethology of the talapoin (*Miopithecus talapoin talapoin*). *Biologica Gabonica, 2,* 311-329.

Gautier-Hion, A. (1968). A study of the annual reproductive cycle of the talapoin (*Miopithecus talapoin*) in its natural environment. *Biologica Gabonica, 4,* 163-173.

Gautier-Hion, A. (1970). Social organization of a band of talapoin (*Miopithecus talapoin*) in northeastern Gabon. *Folia Primatologica, 12,* 116- 141.

Gebo, D. L. (1987). Locomotor diversity in prosimian primates. *American Journal of Primatology 13*: 271-281.

Geller, E., Yuwiler, A. & Zolman, J. R. (1965). Effects of environmental complexity on constituents of the brain and liver. *Journal of Neurochemistry 12,* 949-955.

Gelman, R. (1972) The nature and development of early number concepts. In H. Reese (Ed.), *Advances in child development and behavior,* Vol. 7. New York: Academic Press.

Ghiglieri, M. P. (1984). Feeding ecology and sociality of chimpanzees in Kibale Forest, Uganda, In P. S. Rodman & J. G. H. Cont (Eds.), *Adaptations for foraging in nonhuman primates: Contributions to an organismal biology of prosimians, monkeys and apes* (pp. 161-194). New York: Columbia University Press.

Gillan, D. J. (1981) Reasoning in the chimpanzee: II. Transitive inference. *Journal of Experimental Psychology: Animal Behavior Processes, 7,* 150-164.

Glander, K.E. (in press). Prosimian introduction to a North Carolina forest. In B. Konstant & A. H. Mitchell (Eds.) *Active management for the conservation of wild primates: Rehabilitation, reintroduction, introduction, restocking, and translocation*. SSC-IUCN Publication.

Glander, K. E. & Rabin, D. P. (1983). Food choice from endemic North Carolina tree species by captive prosimians (*Lemur fulvus*). *American Journal of Primatology 5,* 221-229.

Glasersfeld, E., von (1987). *The construction of knowledge.* Seaside, CA: Intersystems.

Glatston, A.R., Geilvoet-Soeteman, E., Hora-Pecek, E. & van Hooff, J.A.R.A.M. (1984) The influence of the zoo environment on social behavior of groups of cotton-topped tamarins, *Saguinus oedipus oedipus. Zoo Biology 3,* 241-253.

Gledhill, L. (1982. May). *Workshop Discussion.* Lion-tailed macaque symposium, sponsored by the Baltimore Zoological Society.

Gluck, J. P. (1979). The intellectual consequences of early social restriction in rhesus monkeys (*Macaca mulatta*). In G. C. Ruppenthal (Ed.), *Nursery care of nonhuman primates.* New York: Plenum.

Goldberg, S.R. (1973) Comparable behavior maintained under fixed-ratio and second-order schedules of food presentation, cocaine injection, or d-amphetamine injection in the squirrel monkey. *Journal of Pharmacology and Experimental Therapeutics, 186,* 18-30.

Goldfoot, D. A. (1977) Rearing conditions which support or inhibit later sexual potential of laboratory-born rhesus monkeys: Hypothesis and diagnostic behavior. *Laboratory Animal Science, 27,* 548-556.

Goldstein, D., Harris, A.H., Izzo, J., Turkkan, J.S., & Keiser, H. (1981) Plasma catecholamines and renin during operant blood pressure conditioning in baboons. *Physiology and Behavior, 26,* 33-37.

Goodall, A. G. (1977). Feeding and ranging behavior of a mountain gorilla group (*Gorilla gorilla beringei*) in the Tshibinda-Kahuzi region (Zaire). In T. H. Clutton-Brock (Ed.), *Primate ecology: Studies of feeding and ranging behavior in lemurs, monkeys and apes* (pp. 449-479). New York: Academic Press.

Goodall, A. G. (1978). On habitat and home range in Eastern gorillas in relation to conservation. In D. J. Chivers and W. Lane-Petters (Eds.), *Recent advances in primatology,* Vol. 2 (pp. 81-84). London: Academic Press.

Goodall, A.G. & Groves, C.P. (1977). The conservation of eastern gorillas. In Prince Rainier & G. Bourne (Eds.), *Primate conservation* (pp 599-637). New York: Academic Press.

Goodall, J. (1965). Chimpanzees of the Gombe Stream Researve. In I. Devore (Ed.), *Primate behavior.* New York: Holt, Rinehart, & Winston.

Goodall, J. van Lawick (1968). The behavior of free-living chimpanzees in the Gombe Stream Reserve. *Animal Behavior Monograph, 1,* 161-311.

Goodall, J. Van Lawick. (1971). *In the shadow of man.* Boston: Houghton Mifflin.

Goodall, J. (1975). Patterns of behavior: The chimpanzee. In J. Goodall (Ed.), *The quest for man.* London: Phaidon Press.

Goodall, J. (1986). *The chimpanzees of Gombe: Patterns of behavior.* Cambridge, MA.: Harvard University Press.

Goosen, C. Fransen, S. & Gommers, M.V.D. (1986) Social aspects of abnormal locomotion stereotypy. *Primate Report 14,* 166.

Gould, E. & Bres, M. (1986) Regurgitation and reingestion in captive gorillas: Description and intervention. *Zoo Biology 5,* 241-250.

Griffin, D. R. (1976/1981) *The question of animal awareness.* New York: The Rockefeller University.

Griffin, D.R. (1983) Prospects for a cognitive ethology. In J. de Luce & H.T. Wilder (Eds.) *Language in primates: Perspectives and implications.* (pp. 159- 186). New York: Springer.

Griffiths, R.R., Lamb, R.J., Ator, N.A., Roache, J.D. & Brady, J.V. (1985). Relative abuse liability of triazolam: Experimental assessment in animals and humans. *Neuroscience and Biobehavioral Reviews, 9,* 133-151.

Griffiths, R.R., Wurster, R.M. & Brady, J.V. (1976). Discrete-trial choice procedure: Effects of naloxone and methadone on choice between food and heroin. *Pharmacological Reviews, 27,* 357-365.

Gross, W. (1972) Effect of social stress on occurrence of Marek's disease in chickens. *American Journal of Veterinary Research 33,* 2275-2279.

Gucwinski, A. (1981). Acclimatization of young lowland gorillas (*Gorilla g. gorilla*) in the Wroclaw Zoological Garden during years 1970-1978. *Zoologische Garten N.F. 51(1)*: 33-51.

Gunnar, M. R. (1980). Control, warning signals, and distress in infancy. *Developmental Psychology, 16,* 281-289.

Haddow, A. J. (1952). Field and laboratory studies on an African monkey (*Cercopithicusa ascanius schmidti*). *Proceedings of the Zoological Society of London, 122,* 297-394.

Hall, E.T. (1969). *The hidden dimension.* Garden City, NY: Doubleday.

Hall, K., & Gartland, J. (1965). Ecology and behavior of the vervet monkey (*Cercopithecus aethiops*), Lolui Island, Lake Victoria. *Proceedings of the Zoological Society of London, 145,* 37-56.

Hamilton, J. H. & Tilson, R. L. (1985). Fishing baboons at desert waterholes. *American Journal of Primatology, 8,* 255-257.

Hampton, J. K., Hampton, S. H. & Landwehr, B. T. (1966). Observations of a successful breeding colony of the marmoset, *Oedipodidas oedipus. Folia Primatologica, 4,* 265-287.

Hannah, A. C., & McGrew, W. C. (1987). Chimpanzees use stones to crack open oil palm nuts in Liberia. *Primates, 28*(1), 31-46.

Hansen, E. W. (1976). Selective responding by recently separated juvenile rhesus monkeys to the calls of their mothers. *Developmental Psychobiology 9*: 83-88.

Hanson, J. P., Larson M. E. & Snowdon, C. T. (1976) The effects of control over high intensity noise on plasma cortisol levels in rhesus monkeys. *Behavioral Biology, 16,* 333-340.

Harcourt, A. H. (1979a). Social relationships between adult male and female mountain gorillas in the wild. *Animal Behaviour, 27,* 325-342.

Harcourt, A.H. (1979b). The social relations and group structure of wild mountain gorillas. In D. A. Hamburg & E. R. McCoun (Eds.), *The great apes* (pp. 187-192). Menlo Park, CA: Benjamin/Cummings.

Harcourt, A. H. (1987). Behaviour of wild gorillas (*Gorilla gorilla*) and their management in captivity. *International Zoo Yearbook 26,* 248-255.

Harcourt, A.H. & Stewart, K.J. (1983) Interactions, relationships and social structure: The great apes. In R.A. Hinde (Ed.), *Primate social relationships: An integrated approach* (pp. 265-279). Sunderland, MA:Sinauer.

Harcourt, A.H., & Stewart, K.J. (1984). Gorillas' time feedings: Aspects of methodology, body size, competition and diet. *African Journal of Ecology, 22,* 207-215.

methodology, body size, competition and diet. *African Journal of Ecology, 22,* 207-215.

Harcourt, A. H., Stewart, K. J. & Fossey, D. (1981). Gorilla reproduction in the wild. In C. E. Graham (Ed.), *Reproductive biology of the great apes: Comparative and biomedical perspectives* (pp. 265-279). New York: Academic Press.

Harcourt, C.S. & Nash, L.T. (1986). Social organization of galagos in Kenyan coastal forests: I. *Galago zanzibaricus. American Journal of Primatology 10*: 339-355.

Haring, D. M. (in press). Natural history and maintenance of sifaka in captivity. *International Zoo Yearbook.* London: Zoological Society of London.

Haring, D. M. & Wright, P.C. (in press). Hand-raising a Philippine tarsier, *Tarsier syrichta. Zoo Biology.*

Haring, D. M., Wright, P.C. & Simons, E.L. (1985). Social behaviors of *Tarsius syrichta* and *Tarsius bancanus. American Journal of Physical Anthropology, 66,* 179.

Harley, D., (1983). *Reproduction in captive female langur monkeys* (Presbytis entellus). Unpublished doctoral dissertation, University of California, Berkeley.

Harley, D. (1985). Birth spacing in langur monkeys (*Presbytis entellus*). *International Journal of Primatology, 6,* 227-242.

Harlow, H. F. (1949) The formation of learning sets. *Psychological Review, 56,* 51-56.

Harlow, H.F. (1950) Learning and satiation of responses in intrinsically motivated complex puzzle performance by monkeys. *Journal of Comparative and Physiological Psychology 43,* 289-294.

Harlow, H.F. (1959) Love in infant monkeys. *Scientific American, 200,* 68-74.

Harlow, H. F. (1962a). Development of affection in primates. In E. L. Bliss (Ed.), *Roots of behavior* (pp. 157-166). New York: Harper.

Harlow, H.F. (1962b) Social deprivation in monkeys. *Scientific American 207,* 136-146.

Harlow, H. F. (1962c) The heterosexual affectional system in monkeys. *American Psychologist, 17,* 1-9.

Harlow, H. F. (1964). Early social deprivation and later behavior in the monkey. In A. Abrams, H. H. Gurner & J. E. P. Tomal (Eds.), *Unfinished tasks in the behavioral sciences* (pp 154-173). Baltimore: Williams & Wilkins.

Harlow, H. F. (1974). Induction and alleviation of depressive states in monkeys. In N.F. White (Ed.), *Ethology and psychiatry* (pp. 197-208). Toronto: University of Toronto Press.

Harlow, H. F. & Harlow, M. K. (1962a). Deprivation in monkeys. *Scientific American, 207,* 136-146.

Harlow, H. F. & Harlow, M. K. (1962b) The effects of rearing conditions on behavior. *Bulletin of the Menninger Clinic, 26(5),* 213-224.

Harlow, H. F. & Harlow, M. K. (1965). The affectional systems. In A. M. Schrier, H. F. Harlow, & F. Stollnitz (Eds.), *Behavior of nonhuman primates,* Vol. 2 (pp. 287-334). New York: Academic Press.

Harlow, H. F. & Harlow, M. K. (1969). Effects of mother infant relationships on rhesus monkey behaviors. In B. M. Foss (Ed.), *Determinants of infant behavior,*

Vol. 4. London: Methuen.

Harlow, H. F. & Harlow, M. K. (1971) Psychopathology in monkeys. In H. D. Kimmel (Ed.), *Experimental psychopathology: Recent research and theory* (pp. 203-229). New York: Academic Press.

Harlow, H. F., Harlow, M. K., Dodsworth, R.S. & Arling, G.L. (1966). Maternal behavior of rhesus monkeys deprived of mothering and peer associations in infancy. *Proceedings of the American Philosophical Society, 40,* 58-66.

Harlow, H. F., Harlow, M. K. & Hansen, E. W. (1963) The maternal affectional system of rhesus monkeys, In H. L. Rheingold (Ed.), *Maternal behavior in mammals* (pp. 254-281). New York: Wiley.

Harlow, H. F., Joslyn, W. D., Senko, M. & Dopp, A. (1966) Behavioral aspects of reproduction in primates. *Journal of Animal Science, 25,* 49-67.

Harlow, H. F. & Mears, C. E. (1977, February 10). The power and passion of play. *New Scientist,* pp. 336-338.

Harris, A.H. & Turkkan, J.S. (1981) Performance characterisitics of conditioned blood pressure elevations in the baboon. *Biofeedback and Self- Regulation, 6,* 11-24.

Harrison, M.J.S. (1983) Age and sex differences in the diet and feeding strategies of the green monkey, *Cercopithecus sabaeus. Animal Behaviour 31,* 969-977.

Harrison, R.M., & Dukelow, R. (1973). Seasonal adaptation of laboratory-maintained squirrel monkeys *(Saimiri sciureus). Journal of Medical Primatology, 2,* 277-283.

Harvey, N. C., & Rhine, R. J. (1983). Some reproductive parameters of stumptailed macaques (*Macaca arctoides*). *Primates 24,* 530-536.

Havelka, J. (1956) Problem-seeking behaviour in rats. *Canadian Journal of Psychology 10,* 91-97.

Hawkins, J. E., Jr., Miller, J. M., Davis, J. A., & Rarey, K. (1985). Inner ear histopathology in aging rhesus monkeys. In R. T. Davis & C. W. Leathers (Eds.), *Behavior and pathology of aging in rhesus monkeys.* New York: Alan R. Liss.

Hayes, K.C., Fay, G., Roach, A., & Stare, F.J. (1972). Breeding New World monkeys in a laboratory environment. In W.I.B. Beveridge (Ed.), *Breeding primates* (pp.122-129). Basel: S.Karger.

Hayes, K.J. & Hayes, C. (1951) The intellectual development of a home-raised chimpanzee. *Proceedings of the American Philosphical Society. 95,* 105-109.

Hearn, J. P., Abbott, D. H., Chambers, P. L., Hodges, J.K. & Lunn, S. F. (1978) Use of the common marmoset, *Callithrix jacchus* in reproductive research. *Primates in Medicine, 10,* 40-49.

Hearn, J. P. & Dixson, A.F. (1984). Assessment of comfort and well-being in New World primates. In *Standards in Laboratory Animal Management, Part 2* (pp. 206-216). Potters Bar: Universities Federation for Animal Welfare.

Hediger, H. (1950/1954). *Wild animals in captivity.* New York: Dover.

Hediger, H. (1968) *The psychology and behavior of animals in zoos and circuses.* New York: Dover.

Hegreberg, G. A., & Hamilton, M. J. (1985). An age series study of skeletal muscle morphology. In R. T. Davis & C. W. Leathers (Eds.), *Behavior and pathology of aging in rhesus monkeys* (pp. 327-334). New York: Alan R. Liss.

Heisenberg, W. (1958) *Physics and philosophy: the revolution in modern science.* New York: Harper & Row.

Hendy-Neely, H., & Rhine, R. J. (1972). Social development of stumptail macaques (*Macaca arctoides*): Momentary touching and other interactions with adult males during the infants' first 60 days of life. *Primates 18*, 589- 600.

Hennesssy, M.B., Mendoza, S., & Kaplan, J. (1982). Behavior and plasma cortisol following brief peer separation in juvenile squirrel monkeys. *American Journal of Primatology, 3*, 143-151.

Henningfield, J.E., Ator, N.A. & Griffiths, R.R. (1981). Establishment and maintenance of oral ethanol self-administration in the baboon. *Drug and Alcohol Dependence, 7,* 113-124.

Henningfield, J.E. & Meisch, R.A. (1976). Drinking device for rhesus monkeys. *Pharmacology, Biochemistry & Behavior, 4,* 609-620.

Herbert, R.T. (1971). Design of cages and diet for metabolic studies using small New World primates. *Laboratory Animals,* 5, 257-265.

Hershkovitz, P. (1977). *Living New World monkeys (Platyrrhini).* Chicago: University of Chicago Press.

Hershkovitz, P. (1984). Taxonomy of squirrel monkeys genus *Saimiri* (Cebidae, Platyrrhini): A preliminary report with description of a hitherto unnamed form. *American Journal of Primatology, 6*, 257-312.

Heymann, E.W. (1987) A field observation of predation on a moustached tamarin *Saguinus mystax* by an anaconda. *International Journal of Primatology 8,* 193-194.

Hienz, R. D., Brady, J. V., Bowers, D. A. & Ator, N.A. (1987) Acquisition of ethanol drinking in baboons: Effects on sensory and motor function. *Proceedings and Abstracts of the Annual Meeting of the Eastern Psychological Association, 58:* 40.

Hill, W. C. O. (1960). *Primates*, Vol. 4, Part A. Edinburgh: Edinburgh University Press.

Hill, W.C. O. (1964). The maintenance of langurs (Colobidae) in captivity: Experiences and some suggestions. *Folia Primatologica, 2*, 222-231.

Hinde, R.A. (1971). Development of social behaviour. In A.M. Shrier & F. Stollnitz (Eds.) *Behaviour of nonhuman primates: Modern research trends,*Vol. III. London: Academic Press.

Hinde, R. A. (Ed.) (1983) *Primate social relationships: An integrated approach.* Oxford: Blackwell.

Hladik, C. M. (1977). Chimpanzees of Gabon and Chimpanzees of Gombe: Some comparative data on the diet. In T. H. Clutton-Brock (Ed.), *Primate ecology: Studies of feeding and ranging behavior in lemurs, monkeys and apes* (pp. 481-501). New York: Academic Press.

Hoage, R. J. (1977) Parental care in *Leontopithecus rosalia*: sex and age differences in carrying behavior and the role of prior experience. In D. G. Kleiman (Ed.) *The biology and conservation of the* Callitrichidae (pp. 293- 305). Washington DC: Smithsonian Institution.

Hoff, M. P., Nadler, R. D. & Maple, T. L. (1981). The development of infant independence in a captive group of lowland gorillas (*Gorilla gorilla gorilla*). *Developmental Psychobiology, 14,* 251-265.

Hoff, M.P., Nadler, R.D. & Maple, T.L. (1983). Maternal transport and infant motor development in a captive group of lowland gorillas. *Primates 24(1):,* 77-85.

Hoffman, R. A. & Stowell, R. E. (1973) Outdoor housing of nonhuman primates. *Laboratory Animal Sciences, 23,* 74-79.

Hoffman, R. & Foerg, R. (1983). Seasonal and daily activity changes in captive *Cheirogaleus medius. Folia Primatologia 38,* 259-268.

Hooff, J. A. R. A. M., van (1967, August). *The care and management of captive chimpanzees, with special emphasis on the ecological aspects.* ARL-TR-67-15, 6571st Aeromedical Research Laboratory, Holloman AFB, N. M.

Hooff, J. A. R. A. M., van (1973). The Arnheim Zoo chimpanzee consortium: An attempt to create an ecologically and socially acceptable habitat. *International Zoo Yearbook, 13,* 195-203.

Hopf, S. (1981). Conditions of failure and recovery of maternal behavior in captive squirrel monkeys (*Saimiri*). *International Journal of Primatology, 2,* 335-349.

Hopf, S., Hartmann-Wiesner, E., Khlmorgen, B., & Mayer, S. (1974). The behavioral repertoire of the squirrel monkey *(Saimiri). Folia Primatologica, 21,* 225-249.

Hopkins, W. D., & Morris, R. (in press). Preliminary evidence of lateralized function for visuospatial processing in two chimpanzees (*Pan troglodytes). Behavioral Neuroscience.*

Hornshaw, S.G. (1981). *A comparative analysis of socio-spatial behavior in a captive and a noncaptive group of* Macaca sylvanus. Ph.D. thesis, University of Toronto.

Hornshaw, S.G. (1984). A comparison of proximity behaviour in two groups of Barbary macaques: Implications for the management of the species in captivity. In J.E. Fa (Ed.) *The Barbary macaque: A case study in conservation* (pp. 221-240). New York: Plenum.

Hosey, G.R. & Druck, P.L. (1987) The influence of zoo visitors on the behaviour of captive primates. *Applied Animal Ethology 18,* 19-29.

Howard, C. F., Jr., & Palotay, J. L. (1985). Fasting metabolites and hormones and changes found in kidneys and islets of Langerhans. In R. T. Davis & C. W. Leathers (Eds.), *Behavior and pathology of aging in rhesus monkeys* (pp. 225-232). New York: Alan R. Liss.

Hrdy, S. B. (1977). *The Langurs of Abu.* Cambridge, MA: Harvard University Press.

Hughes, B.O. (1976) Preference decisions of the domestic hen for wire or litter floors. *Applied Animal Ethology 2,* 155-165.

Hughes, B.O. (1977) Behavioural wisdom and preference tests. *Applied Animal Ethology 3,* 391-392.

Hughes, B.O. (1980) The assessment of behavioural needs. In V. Sybesma (Ed.), *The laying hen and its environment.* The Hague: Martinus Nijhoff.

Hunter, J., Martin, R.D., Dixson, A.F. & Rudder, B.C. (1979). Gestation and interbirth intervals in the owl monkey (*Aotus trivirgatus griseimembra*). *Folia*

Primatologica, 31, 165-175.

Hupp, E.W. (1972). Husbandry procedures in an indoor breeding squirrel monkey colony. In E.I. Goldsmith, & J. Moor-Jankowski (Eds.), *Medical Primatology: Proceedings of the Third Conference on Experimental Medicine and Surgery in Primates, Part 1,* Lyon 1972 (pp. 100-104). Basel: S.Karger.

Inglis, I.R. & Ferguson, N.J.K. (1985) Starlings search for food rather than eat freely-available identical food. *Animal Behaviour 34,* 614-616.

Innis, G.S., Balph, M.H. & Balph, D.F. (1984). On spatial requirements of captive social animals. *Animal Behavior 33(2)*: 680-682.

Ishiyama, T. (1963). Music as a psychotherapeutic tool in the treatment of catatonia. *Psychiatric Quarterly, 37,* 437-461.

Izard, M. K. (1987). Lactation length in three species of *Galago. American Journal of Primatology, 13,* 73-76.

Izard, M. K. & Pereira, M. E. Naturalistic indoor rooms for conservation and behavioral research on captive prosimian primates. In E. Waters, E. Wyers, & E. Menzel (Eds.), *Design and use of indoor naturalistic habitats for behavioral research* (National Science Foundation Conference Proceedings). Stony Brook: State University of New York Press. In preparation.

Izard, M. K. & Rasmussen, D.T. (1985). Reproduction in the slender loris (*Loris tardigradus malabaricus*). *American Journal of Primatology, 8,* 153- 165

Izard, M. K. & Simons, E.L. (1986a). Infant survival and litter size in primigravid and multigravid *Galagos. Journal of Medical Primatology, 15,* 27-35.

Izard, M. K. & Simons, E.L. (1986b). Isolation of females prior to parturition reduces neonatal mortality in *Galago. American Journal of Primatology, 10,* 249-244.

Izard, M.K. & Simons, E. L. (1986c). Management of reproduction in a breeding colony of bushbabies. In J. G. Else & P.C. Lee (Eds.) *Primate ecology and conservation* (pp. 315-324). New York: Cambridge University Press.

Izard, M. K., Wright, P.C. & Simons, E.L. (1985). Gestation length in *Tarsius bancanus. American Journal of Primatology, 9,* 327-331.

Izawa, K. (1976). Group sizes and compositions of monkeys in the upper Amazon basin. *Primates, 17,* 367-399.

James, W.H. (1986). Hormonal control of sex ratio. *Journal of theoretical Biology, 118,* 427-441.

Janson, C. (1985). Aggressive competition and individual food consumption in wild brown capuchin monkeys (*Cebus apella). Behavioral Ecology and Sociobiology, 18,* 125-138.

Jay, P. (1963). *The social behavior of the langur monkey.* Unpublished doctoral dissertation, University of Chicago.

Jay, P. (1965). The common langur of North India In I. DeVore (Ed.), *Primate Behavior: Field studies of monkeys and apes* (pp. 197-247). New York: Holt, Rinehart & Winston.

Jensen, G. D. & Tolman, C. W. (1962) Mother-infant relationship in the monkey, *Macaca nemestrina*: The effect of brief separation and mother-infant specificity. *Journal of Comparative and Physiological Psychology, 55,* 131- 136.

Joffe, J. M., Rawson, R. A. & Mulick, J. A. (1973) Control of their environment reduces emotionality in rats. *Science, 180,* 1383-1384.

Johanson, C.E. (1978). Drugs as reinforcers. In D.E. Sanger & D.J. Sanger (Eds.), *Contemporary research in behavioral pharmacology* (pp. 325-390). New York: Plenum.

John, E.R. (1967). *Mechanisms of memory.* New York: Academic Press.

John, E.R. (1980). A neurophysiological model of purposive behavior. In R.F. Thompson, L.H. Hicks, & V.B. Shvyrkov (Eds.), *Neural mechanisms of goal-directed behavior and learning* (pp.93-115). New York: Academic Press.

Johnson, D. (1981). Atlanta ape fertility conference. *American Association of Zoological Parks and Aquariums annual conference proceedings* (pp.261-263).

Jolly, A. (1986). Lemur survival. In K. Benirschke (Ed.) *Primates: The road to self-sustaining populations* (pp. 71-98). New York: Springer.

Jones, C. & Sabater Pi, J. (1971). Comparative ecology of *Gorilla gorilla* (Savage & Wyman) and *Pan troglodytes* (Blumenbach) in Rio Muni, West Africa. *Bibliotheca Primatologica No. 13.* Basel: S. Karger.

Judge, P. (1982). Reconciliation based on kinship in a captive group of pigtail macaques (*Macaca nemestrina*). [Abstract] *International Journal of Primatology, 3,* 301.

Kalin, N.H., Carnes, M., Barksdale, C.M., Shelton, S.E., Stewart, R.D. & Risch, S.C. (1985) Effects of acute behavioral stress on plasma and cerebrospinal fluid ACTH and B-endorphin in rhesus monkeys. *Neuroendocrinology 40,* 97-101.

Kano, T., & Mulavwa, M. (1984). Feeding ecology of the pygmy chimpanzees (*Pan paniscus*) of Wamba. In R. L. Susman (Ed.), *The pygmy chimpanzee: Evolutionary biology and behavior* (pp. 233-274). New York: Plenum.

Kaplan, J.N. (1977). Breeding and rearing squirrel monkeys (*Saimiri sciureus*) in captivity. *Laboratory Animal Science, 27,* 557-567.

Kaplan, J.R. (1986) Psychological stress and behavior in nonhuman primates. In G. Mitchell, & J. Erwin (Eds.). *Comparative primate biology,* Vol 2, Part A (pp.455-492). New York: Alan R. Liss.

Kaplan, J. R., Manuck, S. B., Clarkson, T. B., Lusso, F. M., Taub, D. M. & Miller, E. W. (1983). Social stress and atheroschlerosis in normocholesterolemic monkeys. *Science, 220,* 733-735.

Kappeler, P.M. (1987a) Reproduction in the crowned lemur (*Lemur coronatus*) in captivity. *American Journal of Primatology 12,* 497-503.

Kappeler, P. M. (1987b). The acquisition of a novel behavior pattern in a group of ring-tailed lemurs (*Lemur catta*). *Primates 28*: 225-228.

Kaufman, I. C. & Rosenblum, L. A. (1967) The reaction to separation in infant monkeys: Anaclitic depression and conservational withdrawal. *Psychosomatic Medicine, 29,* 648-676.

Kawai, M. (1965). Newly acquired pre-cultural behavior of the natural troop of Japanese monkeys on Koshima Islet. *Primates, 6,* 1-30.

Kear, J. (1961) Food selection in finches with special reference to interspecific differences. *Proceedings of the Zoological Society of London 138,* 163-203.

Kelley, T.M. & Bramblett, C.A. (1981) Urine collection from vervet monkeys by instrumental conditioning. *American Journal of Primatology 1,* 95-97.

Kendrick, D.C. (1972) The effects of stimulation and intermittent fasting and feeding on life span in the black-hooded rat. *Developmental Psychobiology 6,* 225-234.

King, F. A. (1986a). Philosophical and practical issues in animal research involving pain and stress. *Annals of the New York Academy of Sciences, 467,* 405-409.

King, F. A. (1986b, September). *Animal research is here to stay.* Keynote address to the National Capitol Area Branch of the American Association for Laboratory Animal Science (AALAS). Washington, D.C.

King, F.A. and Yarbrough, C.J. (1985) Medical and behavioral benefits from primate research. *The Physiologist, 28,* 75-87.

King, M. E., & Wilson, A. C. (1975). Evolution at two levels in humans and chimpanzees. *Science, 188,* 107-116.

Kingdon, J. (1971). *East African mammals: An atlas of evolution in Africa,* Vol. 1 (pp. 201-263). Chicago: University of Chicago Press.

Kinzey, W.G. (1986) Feeding, travel distance and group size in *Callicebus torquatus. Primate Report 14,* 11.

Kirchshofer, R. (1963). Einige bemerkenswerte Verhaltensweisen bei Saimiris im Vergleich zu verwandten Arten. *Zeitschrift fur Morphologie und Anthropologie, 53,* 77-91.

Kirchshofer, R. (1970). Gorillazucht in zoologischen Garden und Forschungsstationen. *Der Zoologische Garten, 38,* 73-96.

Kirchshofer, R. (1979). *International Studbook of the Gorilla (*Gorilla gorilla) *(Savage & Wyman 1847).* Frankfurt: Frankfurt Zoo.

Kirkwood, J.K. (1983) Effects of diet on health, weight and litter-size in captive cotton-top tamarins *Saguinus oedipus oedipus. Primates 24,* 515-520.

Kleiman, D. G. (Ed.) (1977). *The biology and conservation of Callitrichidae.* Washington, D.C.: Smithsonian Institution.

Kleiman, D. G. (1983). The behavior and conservation of the golden lion tamarin *Leontopithecus r. rosalia.* In M. T. deMello (Ed.), *A primatologia no Brasil* (pp. 35-53). Brasilia: Universidade Federale de Brasilia.

Kleiman, D. G., Ballou, J. D., & Evans, R. F. (1982). An analysis of recent reproductive trends in captive golden lion tamarins *Leontopithecus r. rosalia* with comments on their future demographic management. *International Zoo Yearbook, 22,* 94-101.

Kleiman, D. G., Beck, B, B., Dietz, J. M., Dietz, L. A., Ballou, J. D. & Coimbra-Filho, A. F. (1986) Conservation program for the golden lion tamarin: Captive research and management, ecological studies, educational strategies and reintroduction. In K. Benirschke (Ed.), *Primates: The road to self-sustaining populations* (pp. 959-979). New York: Springer.

Knapka, J.J., & Morin, M.L. (1979). Open formula natural ingredient diets for non-human primates. In K.C. Hayes (Ed.), *Primates in nutritional research* (pp.121-138). New York: Academic Press.

Kohler, W. (1927). *The intelligence of apes.* London: Methuen.

Kohler,W. (1959) *The mentality of apes.* New York: Vintage. (First published in English by Routledge & Kegan Paul, London in 1925.)

Kondo, K.T. Minami & Itoigawa, N. Behavioural differences between feral group-reared and mother-reared young Japanese monkeys. In A. B. Chiarelli & R. S. Corruccini (Eds.), *Primate behavior and sociobiology* (pp. 64-71). Berlin: Springer.

Konstant, W. R. & Mittermeier, R. A. (1982) Introduction, reintroduction and translocation of Neotropical primates: Past experiences and future possibilities. *International Zoo Yearbook, 22,* 69-77.

Koomans, P. (1981) Open front piggeries with and without straw. In W. Sybesma (Ed.), *The welfare of pigs.* The Hague: Martinus Nijhoff.

Kortlandt, A. (1966). Chimpanzee ecology and laboratory management. *Laboratory Primate Newsletter, 5,* 1-11.

Kortlandt, A. (1984). Habitat richness, foraching range and diet in chimpanzees and some other primates. In D. J. Chivers, B. A. Wood, and A. Bilsborough (Eds.), *Food acquisition and processing in primates* (pp. 119- 159). New York: Plenum.

Kotelchuk, M. (1982). Child abuse and neglect: prediction and misclassification". In R. J. Starr, Jr. (Ed.), *Child abuse prediction: Policy implications* (pp. 67-104). Cambridge: Ballinger.

Koyama, T., Fujii, H. & Yonekama, F. (1981). Comparative studies of gregariousness and social structure among seven feral *Macaca fuscata* groups. In A.B. Chiarelli & R.S. Corruccini (Eds.) *Primate behavior and sociobiology* (pp. 52-63). New York: Springer.

Koyama, N., Norikoshi, K. & Mano, T. (1975). Population dynamics of Japanese monkeys at Arashiyama. In S. Kondo, M. Kawai & A. Ehara (Eds.), *Contemporary primatology. Proceedings of the 5th International Congress of Primatology* (pp. 411-417). Basel: Karger.

Kreith, F. (1958). *Principles of heat transfer.* Scranton, PA: International Textbook.

Kuhn, T. S. (1962/1970) *The structure of scientific revolutions.*

Chicago: University of Chicago Press.

Kummer, H. (1968). *Social organization of Hamadryas baboons.* Chicago: University of Chicago Press.

Kummer, H. & Kurt, F. (1963). Social units of a free-living population of Hamadryas baboons. *Folia Primatologia 1*: 4-19.

Kupper, J.L. & Britz, W.E. (1972). The squirrel monkey. *Aeromedical Review: Selected Topics in Laboratory Medicine 18*, 1-16. United States Air Force School of Aerospace Medicine.

Lacher, T. E., Da Fonseca, G. A. B., Alves, C., & Magalhaes-Castro, B. (1981). Exudate-eating, scent marking, and territoriality in wild populations of marmosets. *Animal Behaviour, 29*, 306-307.

Landau, V. I. (1987). *The adaptation of New World monkeys to new environmental situations: Food acquisition and food processing behaviors.* Unpublished doctoral dissertation, University of Arizona, Tucson.

Landreth, J. E. & Landreth, H. K. (1974). Effects of music on physiological response. *Journal of Research on Music Education, 22*, 4-12.

Lane-Petter, W. & Pearson, A.E.G. (1971). *The laboratory animal: Principles and practice.* New York: Academic Press.

Laudenslager, M. L. & Reite, M. L. (1984) Loss and separations: Immunological consequences and health implications. In P. Saver (Ed.), *Review of Personality and Social Psychology, 5*, 285-311.

Laudenslager, M. L., Ryan, S. M., Druga, R. C., Hyson, R. L. & Maier, S. F. (1983) Coping and immunosuppression: Inescapable but not escapable shock suppresses lymphocyte proliferation. *Science, 221*, 568-570.

Lave, C.A. & March, J.G. (1975) *An introduction to models in the social sciences.* New York: Harper & Row.

Leary, D. E. (1984) Philosophy, psychology, and reality. *American Psychologist, 39*, 917-919.

Leary, G. A., Bayne, K. A. L., & Bennett, C. L. (1985). The ocular components and senescence. In R. T. Davis & C. W. Leathers (Eds.), *Behavior and pathology of aging in rhesus monkeys* (pp. 155-186). New York: Alan R. Liss.

Lee, P.C. (1983). Caretaking of infants and mother-infant relationships. In R. A. Hinde (Ed.), *Primate social relationships: An integrated approach* (pp 146-151). Sunderland, MA., Sinauer.

Lehner, N.D.M., Bullock, B.C., Feldner, M.A., & Clarkson, T.B. (1967). Observations on reproduction of laboratory-maintained squirrel monkeys. *Laboratory Primate Newsletter, 6*, 1-3.

Lester D., & Freed, E. X. (1973). Criteria for an animal model of alcoholism. *Pharmacology, Biochemistry & Behavior, 1*, 103-107.

Levitt, M. (1985) Dysesthesias and self-mutilation in humans and subhumans: A review of clinical and experimental studies. *Brain*

Research Reviews, 10, 247-290.

Lewin, R. (1984). DNA reveals surprises in human family tree. *Science, 226*, 1179-1183.

Lewis, M. & Goldberg, S. (1969). Perceptual-cognitive development in infancy: A generalized expectancy model as a function of mother-infant interaction. *Merrill Palmer Quarterly, 15*, 81-85.

Lewontin, R.C., Rose, S. & Kamin, L.J. (1984) *Not in our genes*. New York: Pantheon. Lieberman, J. N. (1965). Playfulness and divergent thinking: An investigation of their relationship at the kindergarten level. *Journal of Genetic Psychology, 107*, 219-224.

Line, S. W. (1987). Environmental enrichment for laboratory primates. *Journal of the American Veterinary Medical Association, 190*, 854-859.

Levine, S., Johnson, D.F., & Gonzales, C.A. (1985). Behavioral and hormonal responses to separation in rhesus monkeys and mothers. *Behavioral Neurosciences, 99*, 399-410.

Long, J.O., & Cooper, R.W. (1968). Physical growth and dental eruption in captive-bred squirrel monkeys, *Saimiri sciureus* (Leticia, Colombia). In L.A. Rosenblum, & R.W. Cooper (Eds.), *The squirrel monkey* (pp.193-205). New York: Academic Press.

Lovelace, W. E., & Davis, R. T. (1963). Visual capacity of rhesus monkeys as a function of aging and whole-body radiation with X-rays. *Journal of Genetic Psychology, 103*, 251-257.

Loy, J. (in press). Effects of supplementary feeding on maturation and fertility of primate groups. In J.E. Fa & C.H. Southwick (Eds.) *Ecology and behavior of food-enhanced primate groups*. New York: Alan R. Liss.

Loy, J.D., Keifer, G., Conaway, C., & Kingston, R.I. (1984). *The behavior of gonadectomized rhesus monkeys*. Basel: Karger.

Lukas, S.E., Griffiths, R.R., Bradford, L.D., Brady, J.V., Daley, L. & Delorenzo, R. (1982). A tethering system for intravenous and intragastric drug administration in the baboon. *Pharmacology, Biochemistry & Behavior, 17*, 823-829.

Macedonia, J. M. (1986). Individuality in a contact call of the ringtailed lemur (*Lemur catta*). *American Journal of Primatology 11*: 163-179.

Macedonia, J.M. (1987). Effects of housing differences upon activity budgets in captive sifakas (*Propithecus verreauxi*). *Zoo Biology 6*: 55-67.

Macedonia, J. M. & Taylor, L. L. (1985). Subspecific divergence in a loud call of the ruffed lemur (*Varecia variegata*). *American Journal of Primatology* 9:295-303.

MacKinnon, J. (1974). The behavior and ecology of wild orang-utans (*Pongo pygmaeus). Animal Behavior, 22*, 3-74.

MacLean, P.D. (1959). The limbic system with respect to two basic life principles. In M.A.B. Brazier (Ed.), *Transactions of the Second Conference on the Central Nervous System and Behavior* (pp.31-118). New York: Josiah Macy Jr. Foundation.

MacLean, P.D. (1972). Implications of microelectrode findings on exteroceptive inputs to the limbic cortex. In C.H. Hockman (Ed.), *Limbic system mechanisms and autonomic function* (pp.115-130). Springfield, IL: C.C. Thomas.

MacRae, L., & Salzen, E.A. (1978). A laboratory colony of *Saimiri sciureus* in the Department of Psychology, University of Aberdeen, Scotland. *Primate Information, 5,* 8.

Mager, W. (1981). Stimulating maternal behavior in the lowland gorilla (*Gorilla g. gorilla*) at Apeldoorn. *International Zoo Yearbook 21*:138-142.

Major zoos: Attendance figures during 1984. (1987). *International Zoo Yearbook 26*: 321-394.

Maki, S., Alford, P., Bloomstrand, M. & Franklin, J. A simulated "termite- fishing" food puzzle device for captive chimpanzees (*Pan troglodytes*). [Abstract]. *American Journal of Primatology, 14,* 431.

Maki, S. & Bloomsmith, M.A. (in press). Uprooted trees facilitate the psychological well-being of captive chimpanzees. *Zoo Biology.*

Manicas, P. T., & Secord, P. F. (1983) Implications for psychology of the new philosophy of science. *American Psychologist, 38,* 399-413.

Maple, T.L. (1979). Great apes in captivity: The good, the bad, and the ugly. In J. Erwin, T. Maple & G. Mitchell (Eds.) *Captivity and behavior: Primates in breeding colonies, laboratories and zoos* (pp. 239-272). New York: Van Nostrand Reinhold.

Maple, T. L. (1980). *Chimpanzee reproduction, rearing, and rehabilitation in captivity.* Report submitted to the Ad Hoc Task Force, National Chimpanzee Breeding Program. Atlanta: Georgia Institute of Technology Publications.

Maple, T. L. & Bloomstrand, M. (1983). *Feeding, foraging and mental health.* Paper presented at the Dr. Scholl Conference on the Nutrition of Captive Wild Animals, Chicago, IL.

Maple, T. L., & Finlay, T. W. (1986). Evaluating the environments of captive nonhuman primates. In K. Benirschke (Ed.), *Primates: The road to self- sustaining populations* (pp.479-488). New York: Springer.

Maple, T. L. & Hoff, M. (1982). *Gorilla behavior.* New York: Van Nostrand Reinhold.

Maple, T., & Stine, W. (1982). Environmental variables and great ape husbandry. *American Journal of Primatology, 1,* 67-76.

Maple, T.L. & Warren-Leubecker, A. (1983). Variability in the parental conduct of captive great apes and some generalizations to humankind. In M. Reite & N. G. Caine (Eds.), *Child abuse: The nonhuman primate data.* (pp. 120-136). Monographs in

Primatology, Vol. 1. New York: Alan R. Liss.
Maple, T.L. & Wren Stine, W.M. (1982). Environmental variables and great ape husbandry. *American Journal of Primatology* (Supplement 1), 67-76.
Maples, T.C. & Finlay, T.W. (1986). Evaluating the environments of captive nonhuman primates. In K. Benirshke (Ed.), *Primates: The road to self- sustaining populations* (pp. 479-488). New York: Springer.
Maranto, G. (1986). Will the guenons make a monkey out of Darwin? *Discover*, pp. 87-101.
Markowitz, H. (1977) Effects of environmental enrichment and engineering on behavior of gibbons, diana monkeys, and mandrills. In J. Erwin (Ed.), *Environmental influences on behavior of nonhuman primates*. New York: Van Nostrand.
Markowitz, H. (1979) Environmental enrichment and behavior engineering for captive primates. In J. Erwin, T.L. Maple & G. Mitchell (Eds.), *Captivity and behavior: Primates in breeding colonies, laboratories and zoos* (pp. 217-238). New York: Van Nostrand Reinhold.
Markowitz, H. (1982). *Behavioral enrichment in the zoo.* New York: Van Nostrand Rheinhold.
Markowitz, H. & Stevens, V. J. (Eds.) (1978) *Behavior of captive wild animals*. Chicago: Nelson Hall.
Markowitz, H., Stevens, V.J., Mellen, J.D. & Barrow, B.C. (1982). Performance of a mandrill (*Mandrillus sphinx*) in competition with zoo visitors and computer on a reaction-time game. *Acta Zoologica Pathologica Antwerp 76*: 169-180.
Markowitz, H., & Spinelli, J. (1986a). *The interface between research and environmental enrichment.* Paper presented at the annual meeting of the American Psychological Association, Washington D.C.
Markowitz, H., & Spinelli, J. (1986b). Environmental engineering for primates. In K. Benirschke (Ed.), *Primates: the road to self-sustaining populations* (pp. 489-498). New York: Springer.
Marriott, B.M., & Salzen, E.A. (1979). Food-storing behavior in captive squirrel monkeys *(Saimiri sciureus)*. *Primates, 20,* 307-311.
Martin, P., & Bateson, P. (1986). *Measuring behaviour.* Cambridge, England: Cambridge University Press.
Mason, J. W., Maher, J. T., Hartley, L. H., Mougey, E. H., Perlow, M. J. & Jones, L. G. (1976). Selectivity of corticosteroid and catecholamine responses to various natural stimuli. In G. Serban (Ed.), *Psychopathology of human adaptation* (pp. 147-171). New York: Plenum.
Mason, W. (1965a). Determinants of social behavior in young

chimpanzees. In A. M. Schrier, H. F. Harlow, & F. Stollnitz (Eds.), *Behavior of nonhuman primates,* Vol. 2 (pp. 335-364). New York: Academic Press.

Mason, W. A. (1965b). The social development of monkeys and apes. In I. DeVore (Ed.), *Primate behavior* (pp. 514-543). New York: Holt Rinehart & Winston.

Mason, W. A. (1976) Windows on other minds: Review of *The question of animal awareness. Science, 194,* 930-931.

Mason, W.A. (1986). Can primate political traits be identified? In J.G. Else & P.C. Lee (Eds.), *Primate ontogeny, cognition and social behavior* (pp. 3-11). Cambridge, England: Cambridge University Press.

Matthews, L.H. (1946-1947). Notes on the genital anatomy and physiology of the gibbon (*Hylobates*). *Proceedings of the Zoological Society of London, 116,* 339-364.

Max Lang, C. (1968). The laboratory care and clinical management of *Saimiri* (Squirrel monkeys). In L.A. Rosenblum, & R.W. Cooper (Eds.), *The squirrel monkey* (pp.393-416). New York: Academic Press.

McClure, H.M. (1984) Nonhuman primate models for human disease. *Advances in Veterinary Science and Comparative Medicine, 28,* 267-304.

McConnell, P. B. & Snowdon, C. T. (1986) Vocal interactions between unfamiliar groups of captive, cotton-top tamarins. *Behaviour, 97,* 273-296.

McFarland, M.J. (1986) Food competition and foraging group size in the black spider monkey, *Ateles paniscus. Primate Report 14,* 12.

McFarland, R. A. (1985). Relationship of skin temperature changes to the emotions accompanying music. *Biofeedback and Self Regulation, 10,* 255-267.

McGonigle, B. O., & Chalmers, M. (1977) Are monkeys logical? *Nature, 267,* 694- 696.

McGrath, V.M. (1986). *Male-infant care among captive gorillas.* MA Thesis, University of Calgary, Calgary, Alberta.

McGrew, W. C. (1974). Tool use by wild chimpanzees in feeding upon driver ants. *Journal of Human Evolution, 3,* 501-508.

McGrew, W.C. (1981) Social and cognitive capabilities of nonhuman primates: Lessons from the wild to captivity. *International Journal for the Study of Animal Problems 2,* 138-149.

McGrew, W. C. (1983). Animal foods in the diets of wild chimpanzees (*Pan troglodytes*): Why cross-cultural variation? *Journal of Ethology, 1,* 46-61.

McGrew, W.C., Brennan, J.A. & Russell, J. (1986) An artificial "gum tree" for marmosets *(Callithrix j. jacchus). Zoo Biology 5,* 45-50.

McGrew, W. C. & McLuckie, E. C. (1986) Philopatry and dispersion in the cotton-top tamarin, *Saguinus (o.) oedipus*: An attempted laboratory simulation. *International Journal of Primatology, 7,* 401-422.

McKenzie, S. M., Chamove, A. S. & Feistner, A. T. C. (1986). Floor coverings and hanging screens alter arboreal monkey behavior. *Zoo Biology, 5,* 339-348.

McKim, W.A. (1986). *Drugs and behavior: An introduction to behavioral pharmacology.* Englewood Cliffs, NJ: Prentice-Hall.

McNab, B. & Wright, P. C. (1987). Temperature regulation and oxygen consumption in the Phillipine tarsier *Tarsius syrichta*. *Physiological Zoology 60*: 596-600.

Mears, C. E. & Harlow, H. F. (1975) Play early and eternal. *Proceedings of the National Academy of Sciences, 72 (5),* 1878-1882.

Medin, D. L. (1969). Form perception and pattern reproduction by monkeys. *Journal of Comparative and Physiological Psychology, 68,* 412-419.

Medin, D. L. (1972). Evidence for short- and long-term memory in monkeys. *American Journal of Psychology, 85,* 117-120.

Medin, D. L., O'Neal, P., Smeltz, E., & Davis, R. T. (1974). Age differences in retention of concurrent discrimination problems in monkeys. *Journal of Gerontology, 28,* 63-67.

Meisch, R.A. (1977). Ethanol self-administration: Infrahuman studies. In T. Thompson & P.B. Dews (Eds.), *Advances in behavioral pharmacology,* Vol. 1, (pp. 35-84). New York: Academic Press.

Mendoza, S.P., Lowe, E.L., Davidson, J.M., & Levine, S. (1978). Annual cyclicity in the squirrel monkey *(Saimiri sciureus):* The relationship between testosterone, "fatting," and sexual behavior. *Hormones and Behavior, 11,* 295-303.

Menzel, E. W., Jr. (1963). The effects of cumulative experience on response to novel objects in young isolation-reared chimpanzees. *Behaviour, 21,* 1-12.

Menzel, E. W., Jr. (1964). Patterns of responsiveness in chimpanzees reared through infancy under conditions of environmental restriction. *Psychologische Forschung, 27,* 337-365.

Menzel, E. W., Jr. (1969) Responsiveness to food and signs of food chimpanzee discrimination learning. *Journal of Comparative and Physiological Psychology, 68,* 484-489.

Menzel, E. W., Jr. (1987). *Searching for the problem.* Manuscript submitted for publication.

Menzel, E. W., Jr., Davenport, R. K., Jr. & Rogers, C. M. (1963a). Effects of environmental restriction upon the chimpanzee's responsiveness in novel situations. *Journal of Comparative Physiology and Psychology, 56,* 329-334.

Menzel, E. W., Jr., Davenport, R. K., Jr. & Rogers, C. M. (1963b). The effects of environmental restriction upon the chimpanzee's responsiveness to objects. *Journal of Comparative Physiology and Psychology, 56,* 78-85.

Menzel, E. W., Davenport, R. K., Jr. & Rogers, C. M. (1970). The development of tool-using in wild-born and restriction-reared chimpanzees. *Folia Primatologica, 12,* 273-283.

Meyer, J. S., Novak, M. A., Bowman, R. E. & Harlow, H. F. (1975). Behavioral and hormonal effects of attachment object separation in surrogate-peer- reared and mother-reared infant monkeys. *Developmental Psychobiology, 8,* 425-436.

Michael, R. P. & Zumpe, D. (1978). Potency in male rhesus monkeys: Effects of continuously receptive females. *Science, 200,* 451-453.

Miles, R. C. & Meyer, D. R. (1956) Learning sets in marmosets. *Journal of Comparative and Physiological Psychology, 49,* 219-222.

Miller, L. C. & Nadler, R. D. (1981). Mother-infant relations and infant development in captive chimpanzees and orang-utans. *International Journal of Primatology, 2,* 247-261.

Miller-Schroeder, P.L. (1985). *Infant abuse in captive gorillas.* MA Thesis, University of Calgary, Calgary, Alberta.

Milton, K. (1980). *The foraging strategy of howler monkeys.* New York: Columbia University Press.

Milton, K. (1981) Distribution patterns of tropical plant foods as an evolutionary stimulus to primate mental development. *American Anthropologist 83,* 534-548.

Mineka S. & Cook, M. (1986) Immunization against the observational conditioning of snake fear in rhesus monkeys. *Journal of Abnormal Psychology, 95,* 307-318

Mineka, S. & Cook, M. *Observational conditioning of snake fear in rhesus monkeys using videotaped models.* Manuscript in preparation.

Mineka, S., Davidson, M., Cook, M. & Keir, R. (1984) Observational conditioning of snake fear in rhesus monkeys. *Journal of Abnormal Psychology, 93,* 355-372.

Mineka, S., Gunnar, M. & Champoux, M. (1986) Control and early socioemotional development: Infant rhesus monkeys reared in controllable versus uncontrollable environments. *Child Development 57,* 1241-1256.

Mineka, S., Keir, R. & Price, V. (1980) Fear of snakes in wild- and lab-reared rhesus monkeys. *Animal Learning and Behavior, 8,* 653-663.

Mineka, S. & Keir, R. (1983) The effects of flooding on reducing snake fear in monkeys: Six month follow-up and further flooding. *Behavioral Research and Therapy, 21,* 527-535.

Mitcham, J. (1986) Notes for a prayer in June. *The Georgia Review, 39,* 239- 241.

Mitchell, G. D. (1967) Long-term effects of maternal punishment on the behavior of monkeys. *Psychonomic Science, 8,* 209-210.

Mitchell, G. D. (1970) Abnormal behavior in primates. In L. Rosenblum (Ed.), *Primate behavior: Developments in field and laboratory research,* Vol. 1 (pp. 195-249). New York: Academic Press.

Mitchell, G. (1988) In defense of primate psychology. *The American Biology Teacher, 50,* 86-90.

Mitchell, G. D. & Clark, D. L. (1968) Long-term effects of social isolation on non-socially adapted rhesus monkeys. *Journal of Genetic Psychology, 113,* 117-128.

Mitchell, D., Herbert, H., Widogsky, H., & Morris, M. (1980) Noninvasive oscillometric measurement of blood pressure in baboons *(Papio cynocephalus). Laboratory Animal Science, 30,* 666-672.

Mitchell, G., Maple, T., & Erwin, J. (1979). Development of social attachment potential in captive rhesus monkeys. In J. Erwin, T. Maple, & G. Mitchell (Eds.), *Captivity and behavior: Primates in breeding colonies, laboratories and zoos* (pp. 55-111). New York: Van Nostrand Reinhold.

Moberg, G.P. (1985). Biological response to stress: Key to assessment of animal well-being? In G.P. Moberg (Ed.), *Animal stress* (pp. 27-49). Bethesda, MD : American Physiological Society,

Mohnot, S.M. (1971a). Some aspects of social changes and infant-killing in Hanuman langur, *Presbytis entellus* (Primates: Cercopithecidae) in western India. *Mammalia, 35,* 175-198.

Mohnot, S.M. (1971b). Social organisation in all-male groups of Hanuman langur, *Presbytis entellus* (Primates: Cercopithecidae), around Jodhpur, western India. *Abstracts of the Twelfth International Ethological Conference* (Edinburgh, 1971), pp. 69.

Monkeys agree - Noise is upsetting. (1973, Spring). *Primate Record* (pp. 3- 6). Madison, WI: University of Wisconsin Regional Primate Center. Moore, E. O. (1981) A prison environment's effect on health care service demands. *Journal of Environmental Systems II, 1,* 17.

Moorhouse, K. (1980). *Social integration and the genesis of male-male conflict in a group of captive langur monkeys (*Presbytis entellus). Unpublished doctoral dissertation, University of California, Berkeley.

Morris, D. (1964) The response of animals to a restricted environment. *Symposia of the Zoological Society of London, 13,* 99-118.

Morse, W.H. & Kelleher, R.T. (1977) Schedules as fundamental determinants of behavior. In W.N. Schoenfeld (Ed.), *The theory of reinforcement schedules* (pp. 139-185) New York:

Appleton-Century-Crofts.

Morton, D. B. & Griffiths, P. H. M. (1985). Guidelines on the recognition of pain, distress, and discomfort in experimental animals and an hypothesis for assessment. *Veterinary Record, 116*, 431-436.

Muncer, S. J. (1983) "Conservations" with a chimpanzee. *Developmental Psychology, 16,* 1-11.

Murer, B. (1939) *Pathologisch-Anatomische Untersuchungen an Gefangengehaltenen Wilden Tieren des Basler Zoologischen Gartens.* Basel.

Nadler, R. D. (1975). Determinants of variability in maternal behavior of captive female gorillas. In S. Kondo, M. Kawai, A. Ehara and S. Kawamura (Eds.), *Proceedings of the Symposium of the 5th Congress of the International Primatological Society* (pp. 207-216). Tokyo: Japan Science Press.

Nadler, R. D. (1976). Three gorillas born at Yerkes in one month. *Yerkes Newsletter, 13(2)*, 15-19.

Nadler, R. D. (1980). Child abuse: Evidence from nonhuman primates. *Developmental Psychobiology, 13*, 507-512.

Nadler, R. D. (1984). Biological contributions to maternal behavior of great apes. In M. Lewis (Ed.), *Genesis of behavior, Vol. 4: Beyond the dyad* (pp. 109-128). New York: Plenum.

Nadler R.D., Collins, D.C., Miller, L.C. & Graham, C.E. (1983). Menstrual cycle patterns of hormones and sexual behavior in gorillas. *Hormones and Behavior, 17*, 1-17.

Nadler, R. D. & Green, S. (1975). Separation and reunion of a gorilla infant and mother. *International Zoo Yearbook, 15*, 198-201.

Nadler, R.D., Herndon, J.G. & Wallis, J. (1986) Adult sexual behavior: hormones and reproduction. In G. Mitchell and J. Erwin, (Eds.), *Behavior, conservation, and ecology* (pp. 363-407). New York: Alan R. Liss.

Nadler, R.D. & Miller, L.C. (1982). Influence of male aggression on mating of gorillas in the laboratory. *Folia Primatologia, 38*, 233-239.

Napier, J., & Napier, P. (1985). *Natural history of the primates*. MIT Press.

Napier, J.R. & Napier, P.H. (1967). *A handbook of living primates.* London/New York: Academic Press.

Nash, L. T. & Chilton, S. (1986). Space or novelty?: Effects of altered cage size on galago behavior. *American Journal of Primatology, 10*, 37-49.

Nash, L. T. & Harcourt, C.S. (1986). Social organization of Galagos in Kenyan coastal forests. II. *Galago garnettii. American Journal of Primatology, 10,* 357-369.

Nash, V.J. (1982) Tool use by captive chimpanzees at an artificial termite mound. *Zoo Biology 1,* 211-221.

Natale, F., Antinucci, F., Spinozzi, G., & Poti, P. (1986) Stage 6 object concept in nonhuman primate cognition: A comparison between gorilla (*Gorilla gorilla gorilla*) and Japanese macaque (*Macaca fuscata*). *Journal of Comparative Psychology, 100,* 335-339.

National Institutes of Health. (1985). *Guide for the care of laboratory animals.* (DHHS/NIH Publication No. USPHS 85-23). Washington, DC: U. S. Government Printing Office.

National Institutes of Health. (1986). *Public Health Service policy on humane care and use of laboratory animals.* (DHHS/NIH Publication.) Bethesda, MD: Office for Protection from Research Risks, National Institutes of Health. Natsoulas, T. (1978) Toward a model for consciousness in the light of B. F. Skinner's contribution. *Behaviorism, 6,* 139-175.

Natsoulas, T. (1983) Concepts of consciousness. *The Journal of Mind and Behavior, 4,* 13-39.

Natsoulas, T. (1985) An introduction to the perceptual kind of conception of direct (reflective) consciousness. *The Journal of Mind and Behavior, 6,* 333- 356.

Neuringer, A. J. (1969). Animals respond for food in the presence of free food. *Science, 166,* 399-401.

Newell-Morris, L. & Fahrenbruch, C.E. (1985) Practical and evolutionary consideration for use of the nonhuman primate model in prenatal research. In E.S. Watts (Ed.), *Nonhuman primate models for human growth and development* (pp. 9-40). New York: Alan R. Liss.

Newman, J. D. & Symmes, D. (1982) Inheritance and experience in the acquisition of primate acoustic behavior. In C. T. Snowdon, C. H. Brown, & M. R. Petersen (Eds.) *Primate communication* (pp. 259-278). New York: Cambridge University Press.

Neyman, P. F. (1977) Aspects of the ecology and social organization of free- ranging cotton-top tamarin (*Saguinus oedipus*) and the conservation status of the species. In D. G. Kleiman (Ed.), *The biology and conservation of the Callitrichidae* (pp. 39-71). Washington DC: Smithsonian Institution.

Neyman, P.F. (1980) *Ecology and social organization of the cotton-top tamarin.* PhD dissertation, University of California, Berkeley.

Nicolosi, R.J., & Hunt, R.D. (1979). Dietary allowances for nutrients in nonhuman primates. In K.C. Hayes (Ed.), *Primates in nutritional research* (pp.11-37). New York: Academic Press.

Nie, N.H., Hull, C.H., Jenkins, J.G., Steinbrenner, K. & Bent, D.H. (1975). *SPSS: Statistical package for the social sciences* (2nd ed.). New York: McGraw Hill.

Niemitz, C. (1979). Outline of the behaviour of *Tarsius bancanus.* In G. A. Doyle & R. D. Martin (Eds.) *The study of prosimian behavior* (pp. 631-660). London: Academic Press.

Nieuwenhuijsen, K. & de Waal, F. B. M. (1982). Effects of spatial crowding on social behavior in a chimpanzee colony. *Zoo Biology, 1*, 5-28.

Nishida, T. (1979). The social structure of chimpanzees in the Mahale Mountains. In D. A. Hamburg and E. R. McCown (Eds.), *The great apes*. Menlo Park, CA.: Benjamin/Cummings.

Nishida, T., Wrangham, R. W., Goodall, J., & Uehara, S. (1983). Local differences in plant-feeding habits of chimpanzees between the Mahale Mountains and Gombe National Park, Tanzania. *Journal of Human Evolution, 12*, 467-480.

Nissen, H. W. (1958) Axes of behavioral comparison. In A. Roe & G. G. Simpson (Eds.) *Behavior and evolution* (pp. 183-205). New Haven, CT: Yale University Press.

Novak, M. A. (1979). Social recovery of monkeys isolated for the first year of life: II. Long-term assessment. *Developmental Psychology, 15*, 50-61.

Novak, M. A. & Harlow, H. F. (1975). Social recovery of monkeys isolated for the first year of life. I. Rehabilitation and therapy. *Developmental Psychology, 11*, 453-465.

Nowak, R. M. & S. L. Paradiso (1983). *Walker's mammals of the world* (pp. 382- 384). Baltimore: The Johns Hopkins University Press.

Oates, J.F. (1987) Food distribution and foraging behavior. In B.B. Smuts, D.L. Cheney, R.M. Seyfarth, R.W. Wrangham & T.T. Struhsaker (Eds.), *Primate societies*. Chicago: University of Chicago Press.

Ogburn, C. E., & Martin, G. M. (1985). Age-related declines in the replicative potentials of aortic cells from rhesus monkeys: Evidence from primary cloning and organoid culture techniques. In R. T. Davis & C. W. Leathers (Eds.), *Behavior and pathology of aging in rhesus monkeys* (pp. 101-106). New York: Alan R. Liss.

Oliver, C. (1959) Transfusion. *Astounding Science Fiction 63(4)*, 44-77.

Oliver, J.I. & Lee, P. C. (1978). Comparative aspects of the behaviour of juveniles in two species of baboon in Tanzania. In D. J. Chivers & J. Herbert (Eds.), *Recent advances in primatology* 1 (pp. 151-153). New York: Academic Press.

Olson, T. R. (1979). *Studies on aspects of the morphology of the genus* Otolemur, *Coquerel, 1859.* PhD dissertation, University of London.

Oswald, M., & Erwin, J. (1976). Control of intragroup aggression by male pigtail monkeys (*Macaca nemestrina*). *Nature 262*, 686-688.

Overmier, J. B. (1981) Interference with coping: An animal model. *Academic Psychology Bulletin, 3*, 105-118.

O'Neill, P. L. (1987) Enrichment techniques for confined primates. *Scientists Center Newsletter 9(4)*, 5-8.

O'Neill, P. L. (1988, May/June). Developing effective social and environmental enrichment strategies for macaques in captive groups. *Lab Animal*, pp. 23- 24, 27-28, & 31-35.

Parrish, J. A. (1983). *The effect of ultraviolet radiation on the immune system*. Skillman, NJ: Johnson & Johnson.

Pasnak, R. (1979). Acquisition of prerequisites to conservation by macaques. *Journal of Experimental Psychology: Animal Behavior Processes, 5,* 194-210.

Patterson, F.G. (1981) Ape language. *Science, 211,* 86-87.

Patterson, J.D. (1986). Shape as a factor in primate thermoregulation. In D.M. Taub & F.A. King (Eds.), *Current perspectives in primate social dynamics* (pp. 228-242). New York: Van Nostrand Reinhold.

Paulk, H., Dienske, H. & Ribbens, L. (1977) Abnormal behavior in relation to cage size in rhesus monkeys. *Journal of Abnormal Psychology, 86,* 87-92.

Pearson, E., Williams, M., & Allison, L. (1988). *Proximity patterns in the first two years of life of a colony of langur monkeys (*Presbytis entellus). Unpublished manuscript.

Pereira, M. E. (1986). Maternal recognition of juvenile offspring coo vocalizations in Japanese macaques. *Animal Behavior 34,*. 935-937.

Pereira, M. E. (1988a). Effects of age and sex on social association in juvenile savannah baboons *(Papio cynocephalus cynocephalus)*. *Animal Behavior 36*: 184-204.

Pereira, M. E. (in press). Agonistic interactions in juvenile savanna baboons. I. Fundamental features. *Ethology.*

Pereira, M. E. & Altmann, S. A. (1983, July 12). *Statement for Institute of Laboratory Animal Resources' public meeting.* Rosemont, IL. Copies available from M.E.P.

Pereira, M. E. & Altmann, J. (1985). Development of social behavior in free- living nonhuman primates. In E. S. Watts (Ed.), *Nonhuman primate models for human growth and development* (pp. 217-309). New York: Alan R. Liss.

Pereira, M. E., Klepper, A., & Simons, E. L. (1987). Tactics of care for young infants in ruffed lemurs (*Varecia variegata variegata*): Ground nests, "parking," and biparental guarding. *American Journal of Primatology 13*: 129-144.

Pereira, M. E., Seeligson, M. L. & Macedonia, J. M. (in press). The behavioral repertoire of ruffed lemurs (Lemuridae: *Varecia variegata variegata*). *Folia Primatologia.*

Pereira, M. E. & Simons, E. L. (1986). Sexually-differentiated responses to potential immigrant males in semi-free-ranging ringtailed lemurs. *Primate Report 14*: 113.

Peres, C. A. (1986). Consequences of territorial defense in wild golden lion tamarins, *Leontopithecus rosalia. Primate Report, 14,* 234.

Petersen, M. R., Beecher, M. D., Zoloth, S. R., Green, S., Marler, P., Moody, D. B. & Stebbins, W. C. (1984). Neural lateralization of vocalizations by Japanese macaques: Communicative significance is more important than acoustic structure. *Behavioral Neurosciences 98,* 779-790.

Peterson, E. A. (1980). Noise and laboratory animals. *Laboratory Animal Science, 30,* 422-439.

Peterson, J. I. (1987). Laboratory experimentation with a personal computer. *Science, 235,* 162.

Pfeiffer, A. J., & Koebner, L. J. (1978). The resocialization of single caged chimpanzees and the establishment of an island colony. *Journal of Medical Primatology, 7,* 70-81.

Phillips, J. L., Jr. (1975) *The origins of intellect: Piaget's theory* (2nd ed.). San Francisco: W. H. Freeman.

Piaget, J. (1954) *The construction of reality in the child.* New York: Basic Books. (Original work published 1937)

Piaget, J. (1971) *Biology and knowledge.* Chicago: University of Chicago Press. (Original work published 1967)

Piaget, J., & Inhelder, B. (1969). *The psychology of the child.* New York: Basic Books. (Original work published 1966)

Platt, M. & Thompson, R.L. (1984). Mirror responses in a Japanese macaque troop (Arashiyama West). *American Journal of Primatology 6,* 417-418.

Plimpton, E. & Rosenblum, L. (1983). The ecological context of infant maltreatment in primates. In M. Reite and N. Caine (Eds.), *Child abuse: The nonhuman primate data* (pp. 103-117). New York: Alan R. Liss.

Plimpton, E. H., Swartz, K. B. & Rosenblum, L. A. (1981). The effects of foraging demand on social interactions in a laboratory group of bonnet macaques. *International Journal of Primatology 2,* 175-185.

Ploog, D., Blitz, J., & Ploog, F. (1963). Studies on social and sexual behavior of the squirrel monkey *(Saimiri sciureus). Folia Primatologica, 1,* 29-66.

Ploog, D., Hopf, S., & Winter, P. (1967). Ontogenese des Verhaltens von Totenfopfaffen *(Saimiri sciureus). Psychologische Forschung, 31,* 1-141.

Ploog, D.W., & MacLean, P.D. (1963). Display of penile erection in squirrel monkey *(Saimiri sciureus). Animal Behaviour, 11,* 32-39.

Poirier, F. E. & Hu, H. (1983). *Macaca mulatta* and *Rhinopithecus* in China: Preliminary results. *Current Anthropology 24*: 387-388.

Poirier, F., & Smith, E. (1974). Socializing functions of primate play. *American Zoologist, 14,* 275-287.

Pollock, J. I. (1986a). Primates and conservation priorities in Madagascar. *Oryx 20*: 209-216.

Pollock, J. I. (1986b). The management of prosimians in captivity for conservation and research. In K. Benirschke (Ed.) *Primates: The road to self-sustaining populations* (pp. 269-288). New York: Springer.

Pollock, J. I. & Mullin, R. J. (1987). Vitamin C biosynthesis in prosimians: Evidence for the anthropoid affinity of *Tarsius. American Journal of Physical Anthropology 73,* 65-70.

Pook, A. G., & Pook, G. (1981). A field study of the socio-ecology of the Goeldi's monkeys (*Callimico goeldii*) in northern Bolivia. *Folia Primatologica, 35,* 288-312.

Porter, W.P. & Gates, D.M. (1969). Thermodynamic equilibria of animals with environment. *Ecological Monographs, 39,* 245-270.

Portman, O.W. (1970). Nutritional requirements of squirrel and woolly monkeys. In R.S. Harris (Ed.), *Feeding and nutrition of non-human primates* (pp.159-173). New York: Academic Press.

Portman, O.W., Alexander, M., Tanaka, N., & Osuga, T. (1979). Role of diet in normal biliary physiology and gallstone formation. In K.C. Hayes (Ed.), *Primates in nutritional research* (pp.139-180). New York: Academic Press.

Preilowski, B., Roger, W., & Ngele, H. (in press). Combining scientific instrumentation with species-adequate housing in laboratory studies of non-human primates: A pilot study with rhesus monkeys. *American Journal of Primatology.*

Premack, D. (1983) The codes of man and beast. *The Behavioral and Brain Sciences, 6,* 125-166.

Premack, D. & Premack, A. J. (1983) *The mind of an ape.* New York: W. W. Norton.

Premack, D. & Woodruff, G. (1978). Does the chimpanzee have a theory of mind? *The Behavioral and Brain Sciences 4,* 515-526.

Preston, D. G., Baker, R.P. & Seay, B. M. (1970). Mother-infant separation in the patas monkey. *Developmental Psychology, 3,* 298-306.

Price, D. L., Cork, L. C., Struble, R. G., Kitt, C. A., Price, D. L., Jr., & Hedreen, J. C. (1985). Neuropathological, neurochemical, and behavioral studies of the aging nonhuman primate. In R. T. Davis & C. W. Leathers (Eds.), *Behavior and pathology of aging in rhesus monkeys* (pp. 113- 136). New York: Alan R. Liss.

Prost, J.H. (1965). Definitional system for the classification of primate locomotion. *American Anthropologist 67,* 1198-1214.

Przybeck, T. R. (1985). Histomorphology of the rib: Bone mass and cortical remodeling. In R. T. Davis & C. W. Leathers (Eds.), *Behavior and pathology of aging in rhesus monkeys* (pp. 303-326). New York: Alan R. Liss.

Putnam, H. (1981) Reductionism and the nature of psychology. In J. Jaugeland (Ed.) *Mind design* (pp. 205-219). Cambridge, MA: Bradford.

Quick, D.L.F. (1984). An integrative approach to environmental engineering in zoos. *Zoo Biology 3(1)* :65-77.

Ralls, K., & Ballou, J. (1982). Effects of inbreeding on infant mortality in captive primates. *International Journal of Primatology, 3,* 491-505.

Ralls, K., & Ballou, J. (1986). Captive breeding programs for populations with a small number of founders. *Trends in Ecology and Evolution, 1,* 19- 22.

Rasmussen, D. T. (1985). A comparative study of breeding seasonality and litter size in eleven taxa of captive lemurs (*Lemur* and *Varecia*). *International Journal of Primatology 6*: 501-517.

Rasmussen, D. T. (1986). *Behavior and life-history of slow lorises and slender lorises: Implications for the lorisine-galagine divergence.* PhD thesis, Duke University, Durham, NC.

Rasmussen, D. T. & Izard, M. K. (in press). Scaling of growth and life- history traits relative to body size, brain size, and metabolic rate in lorises and galagos (Lorisidae, Primates). *American Journal of Physical Anthropology.*

Rasmussen, K.M., Ausman, L.M., & Hayes, K.C. (1980). Vital statistics from a laboratory breeding colony of squirrel monkeys *(Saimiri sciureus). Laboratory Animal Science, 30,* 99-106.

Rawlins, R. G. & Kessler, M. (1986). *The Cayo Santiago macaques: History, behavior and biology.* Albany: State University of New York.

Redshaw, M. (1978) Cognitive development in human and gorilla infants. *Journal of*

Human Evolution, 7, 133-141.

Reidy, T.J. (1980). The aggressive characteristics of abused and neglected children. In G.J. Williams and J. Money (Eds.), *Traumatic abuse and neglect of children at home* (pp. 262-269). Baltimore: The John Hopkins University Press.

Reinhardt, V., Cowley, D., Eisele, S.G., Vertein, R., & Houser, W.D. (1987). Preliminary comments on pairing unfamiliar adult female rhesus monkeys for the purpose of environmental enrichment. *Laboratory Primate Newsletter, 26*(2), 5-8.

Reinhardt, V., Cowley, D., Eisele, S., Vertein, R., & Houser, D. (1988) Pairing compatible female rhesus monkeys for cage enrichment has no negative impact on body weight. *Laboratory Primate Newsletter, 27(1),* 13-15.

Reinhardt, V., Houser, W. D., Cowley, D. & Champoux, M. (1987). Preliminary comments on environmental enrichment with branches for individually caged rhesus monkeys. *Laboratory Primate Newsletter, 26,* 1-3.

Reinhardt, V., Houser, W. D. Eisele, S. G. & Champoux, M. (1987) Social enrichment of the environment for singly caged adult rhesus monkeys. *Zoo Biology, 6,* 365-371.

Reinhardt, V., Reinhardt, A. & Houser, W.D. (1986) Hair pulling and eating in captive rhesus monkey troops. *Folia Primatologica 47,* 158-164.

Reinhardt, V., Reinhardt, A., Eisele, S., Houser, D. & Wolf, J. (1987) Control of excessive aggressive disturbance in a heterogeneous troop of rhesus monkeys. *Applied Animal Behaviour Science, 18,* 371-377.

Reiser, K. M., Rucker, R. B., & Last, J. A. (1985). Age-related changes in connective tissue cross-linking. In R. T. Davis & C. W. Leathers (Eds.), *Behavior and pathology of aging in rhesus monkeys* (pp. 293-300). New York: Alan R. Liss.

Reite, M. & Capitanio, J. P. (1985). On the nature of social separation and attachment. In M. R. Reite & T. Field (Eds.), *Psychobiology of attachment* (pp. 223-258). New York: Academic Press.

Reite, M. & Field, T. (Eds) (1985). *The psychobiology of attachment and separation.* New York: Academic Press.

Reite, M., Harbeck, R., & Hoffman, A. (1981). Altered cellular immune response following peer separation. *Life Sciences, 29,* 1133-1136

Reite, M., Seiler, C., Crowley, T., Hydinger-MacDonald, M., & Short, R. (1982). Circadian rhythm changes following maternal separation. *Chronobiologia, 9,* 1-11.

Reite, M. & Short, R. (1978). Nocturnal sleep in separated monkey infants. *Archives of General Psychiatry, 35,* 1247-1253.

Reite, M., Short, R., Kaufman, I.C., Stynes, A.J., & Pauley, J.D., (1978). Heart rate and body temperature in separated monkey infants. *Biological Psychiatry, 13,* 91-105

Renner, M.J. & Rosenzweig, M.R. (1987) *Enriched and impoverished environments.* London: Springer.

Renquist, D.M. & Judge, F.J. (1984) Use of nylon balls as behavioral modifier for caged primates. *Laboratory Primate Newsletter 24,* 4.

Reynolds, V., & Reynolds, F. (1965). Chimpanzees of the Budonga Forest. In I.

Devore (Ed.), *Primate behavior.* New York: Holt, Rinehart, & Winston.

Rhine, R. J. (1972). Changes in the social structure of two groups of stumptail macaques (*Macaca arctoides*). *Primates 13*, 181-194.

Rhine, R. J. (1973). Variation and consistency in the social behavior of two groups of stumptail macaques (*Macaca arctoides*). *Primates 14*, 21-35.

Rhine, R. J., & Ender, P. E. (1983). Comparability of methods used in the sampling of primate behavior. *American Journal of Primatology, 5*, 1-15.

Rhine, R. J., & Hendy-Neely, H. (1978a). Social development of stumptail macaques (*Macaca arctoides*): Momentary touching, play, and other inter actions with aunts and immatures during the infants' first 60 days of life. *Primates, 19*, 115-123.

Rhine, R. J., Hendy-Neely, H. (1978b). Social development of stumptail macaques (*Macaca arctoides*): Synchrony of changes in mother-infant interactions and individual behaviors during the first 60 days of life. *Primates 19*, 681-692.

Rhine, R. J., Hopper, J. S., Harvey, N. C., & Bunyak, S. C. (1981). Meat- eating and possible stalking in a colony group of *Macaca arctoides. Laboratory Primate Newsletter 20*, 5-7.

Rhine, R. J. & Kronenwetter, C. (1972). Interaction patterns of two newly formed groups of stumptailed macaques (*Macaca arctoides*). *Primates 13*, 19- 33.

Rhine, R. J., & Linville, A. K. (1980). Properties of one-zero scores in observational studies of primate social behavior: The effect of assumptions on empirical analyses. *Primates 21*, 111-122.

Rhoads & Goldsworthy, R.J. (1979). The effects of zoo environments on public attitudes toward endangered wildlife. *Internernational Journal of Environmental Studies 13*: 283-287.

Rice, D.C. (1979) Operant conditioning of infant monkeys (*Macaca fascicularis*) for toxicity testing. *Neurobehavioral Toxicology, 1* (Supplement 1), 85-92.

Rice, D.C. (1983) Central nervous effects of perinatal exposure to lead or methylmercury in the monkey. In T.W. Clarkson, G.F. Nordberg & P.R. Sayer (Eds.), *Reproductive and developmental toxicity of metals* (pp. 517-539). New York: Plenum.

Richard, A. F. (1978). *Behavioral variation: Case study of a Malagasy lemur.* Lewisburg, PA: Bucknell University Press.

Richard, A.F. (1981) Changing assumptions in primate ecology. *American Anthropologist 83*, 517-533.

Richard, A.F. (1982). The world's endangered primate species: A case study on the lemur fauna of Madagascar. In R. A. Mittermeier & M. J. Plotkin (Eds.), *Primates and the tropical forest* (pp. 23-31). World Wildlife Fund - U.S.A. & L.S.B. Leakey Foundation.

Richard, A.F. (1985). *Primates in nature.* New York: W.H. Freeman.

Rickenbach, F. (1988) *Langur monkey* (Presbytis entellus) *early development: A captive study.* Unpublished manuscript

Rijksen, H.D. (1978). *A field study on Sumatran orangutans* (Pongo pygmaeus abelii, *Lesson 1827): Ecology, behaviour and conservation.* Wageningen, The Netherlands: H. Veenman & Zonen.

Riley, V. (1981) Psychoneuroendocrine influence on immunocompetence and neoplasia. *Science 212,* 465-467.

Ripley, S. (1965). *The ecology and social behavior of the Ceylon gray langur,* Presbytis entellus thersites. Unpublished doctoral dissertation, University of California, Berkeley.

Ripley, S. (1967). Intertroop encounters among Ceylon gray langurs (*Presbytis entellus thersites*).In S.A. Altmann (Ed.), *Social communication among primates* (pp. 237-253). Chicago: University of Chicago Press.

Rissler, L., Wasser, S. K. & Sackett, G. P. (1987). Measurement of excreted steroids in *Macaca nemestrina. American Journal of Primatology, 12,* 91-100.

Robert, S. (1986) Ontogeny of mirror behavior in two species of great apes. *American Journal of Primatology, 10,* 109-117.

Robinson, J. G., Wright, P.C. & Kinzey, W.G. (1987). Monogamous cebids and their relatives: Intergroup calls and spacing. In B. B. Smuts, D. L. Cheney, R. M. Seyfarth, R. Wrangham & T. T. Struhsaker (Eds.), *Primate societies.* Chicago: University of Chicago Press.

Rodman, P. S. (1977). Feeding behavior of orang-utans of the Kutai Nature Reserve, East Kalimantan. In T. H. Clutton-Brock (Ed.), *Primate ecology: Studies of feeding and ranging behavior in lemurs, monkeys and apes* (pp. 383-413). New York: Academic Press.

Rodman, P. S. (1979). Individual activity profiles and the solitary nature of orang-utans. In D. A. Hamburg and E. R. McCown (Eds.), *The great apes* (pp. 234-255). Menlo Park, CA: Benjamin/Cummings.

Rodman, P. S. (1984). Foraging and social systems of orangutans and chimpanzees. In P. S. Rodman and J. G. H. Cant (Eds.), *Adaptations for foraging in nonhuman primates: Contributions to an organismal biology of prosimians, monkeys and apes* (pp. 134-160). New York: Columbia University Press.

Rodriguez, L. E. (1987) Rehabilitation program for the spider monkey (*Ateles geoffroyi*), first results. [Abstract] *American Journal of Primatology, 12,* 357.

Rondinelli, R. & Klein, L. L. (1976). An analysis of adult social spacing tendencies and related social interactions in a colony of spider monkeys (*Ateles geoffroyi*) at San Francisco Zoo. *Folia Primatologia 25-26*: 122-142.

Roonwal, M. & Mohnot, S.M. (1977). *Primates of South Asia.* Cambridge, MA: Harvard University Press.

Rosen, C. (1974). The effects of sociodramatic play on problem-solving behavior among culturally disadvantaged preschool children. *Child Development, 45,* 920-927.

Rosenblum, L.A. & Smiley, J. (1984) Therapeutic effects of an imposed foraging task in disturbed monkeys. *Journal of Child Psychology & Psychiatry & Allied Disciplines 25,* 485-497.

Rosenson, L. M. (1972). Interactions between infant greater bushbabies (*Galago crassicaudatus crassicaudatus*) and adults other than their mothers under experimental conditions. *Zeitschrift fur Tierpsychologie, 31,* 240- 269.

Rowell, T. E. & Hinde, R.A. (1964). Responses of rhesus monkeys to mildly stressful situations. *Animal Behavior, 11,* 235-243.

Rowell, T.E. (1967). A quantitative comparison of the behaviour of a wild and a caged baboon group. *Animal Behavior 15*: 499-509.

Rowell, T. E. (1972). *The social behaviour of monkeys.* Baltimore: Penguin.

Roy, M. A., Hennig, C. W., Hill, E. M. & Walker, M. L. (1981). Subspecies differences in curiosity and general activity for developing squirrel monkeys. *Developmental Psychobiology, 14*, 55-58.

Roy, M. A., Wolf, R. H., Martin, L. N., Rangan, S. R. S. & Allen, W. P. (1978) Social and reproductive behaviors in surrogate reared squirrel monkeys (*Saimiri sciureus*). *Laboratory Animal Science, 28*, 417-421.

Rueter, A. (1984) Animals in research: Politics and polemics. *The Research News 35(10-11)*, 11-18

Rumbaugh, D. M. (1975). The learning skills of the rhesus monkey. In G. H. Bourne (Ed.), *The rhesus monkey.* New York: Academic Press.

Rumbaugh, D. M. (1977). *Language learning by a chimpanzee: The Lana project.* New York: Academic Press.

Rumbaugh, D. M. (1987, April) *Chimpanzee skills with language and numbers.* Invited address at the annual meeting of the Southern Society for Philosophy and Psychology, Atlanta, GA.

Rumbaugh, D. M. (1988). Cage design attenuates male chimpanzees' displays. *Zoo Biology, 7(2)*, 177-180.

Rumbaugh, D. M., Richardson, W. K., Washburn, D. A., Savage-Rumbaugh, E. S. & Hopkins, W. D. (in press). Rhesus monkeys (*Macaca mulatta*) learn complex video tasks: Implications for stimulus-response spatial contiguity. *Journal of Comparative Psychology.*

Rumbaugh, D. M, Savage-Rumbaugh, E. S., Hopkins, W. D., Washburn, D. A., & Runfeldt, S. A. (1987, August). *Lana chimpanzee (Pan troglodytes) counts by Numath.* Paper presented at the annual meeting of the American Psychological Association, New York, NY.

Russell, W.M.S. & Burch, R.L. (1959). *The principles of humane experimental technique.* London: Methuen.

Sabater-Pi, P. J. (1977). Contribution to the study of alimentation of lowland gorillas in the natural state, in Rio Muni, Republic of Equatorial Guinea (West Africa). *Primates, 18*, 183-204.

Sabater-Pi, P. J. (1979). Feeding behavior and diet of chimpanzees (*Pan troglodytes*) in the Okorobiko Mountains of Rio Muni (West Africa). *Zeitschrift fur Tierpsychologie, 50*, 265-281.

Sackett, D., Oswald, M., & Erwin, J. (1975). Aggression among captive female pigtail monkeys in all-female and harem groups. *Journal of Biological Psychology 17*, 17-20.

Sackett, G.P. (1966) Monkeys reared in isolation with pictures as visual input: Evidence for an innate releasing mechanism. *Science 154*, 1468-1470.

Sackett, G. (1968a) Abnormal behavior in laboratory-reared rhesus monkeys. In M. Fox (Ed.), *Abnormal behavior of animals* (pp. 293-331). Philadelphia: Saunders.

Sackett, G. P. (1968b) The persistence of abnormal behavior in monkeys following isolation rearing. In R. Porter (Ed.), *The role of rearing in psychotherapy.* London: Churchill.

Sackett, G.P. (1970) Innate mechanisms, rearing conditions, and a theory of early experience effects in primates. In R.M. Jones, (Ed.), *Miami symposium on the prediction of behavior: Early experience.* Miami: University of Miami Press.

Sackett, G. P., Gunderson, V., & Baldwin, D. (1982). Studying the ontogeny of primate behavior. In J. L. Fobes & J. E. King (Eds.), *Primate behavior* (pp. 135-169). New York: Academic Press.

Sackett, G. P., Holm, R. A., Ruppenthal, G. C., & Fahrenbruch, C. E. (1976). The effects of total isolation rearing on the behavior of rhesus and pigtail macaques. In N. E. Walsh & W. Greenough (Eds.), *Environments as therapy for brain dysfunction.* New York: Plenum.

Sauer, E. G. F. & E. M. Sauer (1963). The South West African bushbaby of the *Galago senegalensis* group. *Journal of the South West African Scientific Society, Windhock, 16,* 5-36.

Savage, A., Dronzek, L. & Snowdon, C. T. (1987) Color discrimination and its relation to the feeding ecology of the cotton-top tamarin (*Saguinus o. oedipus*). *Folia Primatologica, 49,* 57-69.

Savage, A., Ziegler, T. E. & Snowdon, C. T. (1988). Sociosexual development, pairbond formation and mechanisms of ferility suppression in female cotton- top tamarins (*Saguinus oedipus oedipus*). *American Journal of Primatology, 14,* 345-359.

Savage-Rumbaugh, E. S. (1986). *Ape language: From conditioned response to symbol.* New York: Columbia University Press.

Schaie, K. W., & Willis, S. L. (1986). *Adult development and aging.* Boston: Little, Brown.

Schaik, C.P. van, & Noordwijk, M.A. van. (1983). Social stress and sex ratio of neonates and infants among non-human primates. *Netherlands Journal of Zoology,* 33, 249-265.

Schaller, G. (1963). *The mountain gorilla: Ecology and behavior.* Chicago: University of Chicago Press.

Schapiro, S.J. & Mitchell, G. (1983). Infant-directed abuse in a seminatural environment: Precipitating factors. In M. Reite & N. G. Caine (Eds.), *Child abuse: The nonhuman primate data. Monographs in Primatology,* Vol. 1 (pp.29-48). New York, Alan R. Liss.

Schlottmann, R. S. & Seay, B. M. (1972) Mother-infant separation in the Java monkey (*Macaca irus*). *Journal of Comparative and Psychological Psychology, 79,* 334-340.

Schmidt, M.J. & Markowitz, H. (1977) Behavioral engineering as an aid in the maintenance of healthy zoo animals. *Journal of the American Veterinary Medical Association 171,* 966-969.

Schneeburg, N. G. (1970). *Essentials of clinical endocrinology.* St. Louis: C. V. Mosby.

Schullian, P. M. & Schoen, M. (1948). *Music and medicine.* Freeport, NY: Books for Library Press.

Schultz, A.H. (1933). Observations on the growth, classification and evolutionary specialization of gibbons and siamangs. *Human Biology 5*, 212- 255.

Schuster, D.H., & Schuster, L. (1972). Speculative mechanisms affecting sex ratio. *Journal of Genetic Psycholology, 121,* 245-254.

Seay, B. (1962) Mother-infant separation in monkeys. *Journal of Child Psychology and Psychiatry and Allied Disciplines, 3,* 123-132.

Segal, E. F. (1976) Mind-body: What is the question? *Philosophy Forum, 14,* 325-350.

Segal, E. F. (1987, October). *Lessons taught and questions posed by an educated macaque.* Paper presented at the 9th Congress of the Mexican Society of Behaviour Analysis, Puebla, Mexico.

Seiden, L.S. & Dykstra, L.A. (1977). *Psychopharmacology: A biochemical and behavioral approach.* New York: Van Nostrand Reinhold.

Seligman, M. E. P. (1975) *Helplessness: On depression, development, and death.* San Francisco: W. H. Freeman.

Selye, H. (1974). *Stress without distress.* Philadelphia: J. B. Lippincott.

Senner, J. W. (1980) Inbreeding depression and the survival of zoo populations. In M. E. Soule & B. A. Wilcox (Eds.), *Conservation biology: An evolutionary-ecological perspective* (pp. 209-224). Sunderland MA: Sinauer. Senner, W. W. (1962). *Saturation discrimination by rhesus monkeys.* Unpublished Master's thesis, University of South Dakota, Vermillion.

Seyfarth, R. M. & Cheney, D. L. (1986) Vocal development in vervet monkeys. *Animal Behaviour 34,* 1640-1658.

Sharon, I. M., Fellner, R. P. & Burney, S. W. (1977). The effects of lights of different spectra on caries incidence in the golden hamster. *Archives of Oral Biology, 16,* 1427-1432.

Siegel, H.S. (1980) Physiological stress in birds. *Bioscience 30,* 529-534.

Simons, E. L. (1986). Lemurs in a natural habitat in North Carolina: A crucial step in reintroduction to the wild. *Primate Conservation 7*: 60-62.

Singh, M., Akram, N. & Pirta, R.S. (1984). Evolution of demographic patterns in the bonnet monkey (*Macaca radiata*). In M.L. Roonwal, S.M. Mohnot & N.S. Rathore (Eds.) *Current primate researches* (pp. 7-16). Jodhpur: University of Jodhpur Press.

Singh, S.D. (1969). Urban monkeys. *Scientific American 221,* 108-115.

Siple, P.A. & Passel, C.F. (1945). Measurements of dry atmospheric cooling in subfreezing temperatures. *Proceedings of the American Philosophical Society 89,* 177-199.

Smith, C. A. & Morris, L. W. (1976). Effects of stimulative and sedative music on cognitive and emotional components of anxiety. *Psychological Reports, 38,* 1187-1193.

Smith, E.C. (1987). Deception and evolutionary biology. *Cultural Anthropology 2(1),* 50-64.

Smith, H. J., King, J. E., Witt, E. D., & Rickel, J. E. (1975). Sameness- difference matching from sample by chimpanzees. *Bulletin of the Psychonomic Society, 6,* 469-471.

Smuts, B. B. (1985). *Sex and friendship in baboons.* Hawthorn, NY: Aldine.

Smuts, B. B. (1987). Introduction. In B.B. Smuts, D.L. Cheney, R. M. Seyfarth, R. W. Wrangham, & T. T. Struhsaker (Eds.), *Primate societies* (pp 1-9). Chicago: University of Chicago Press.

Snowdon, C. T. (in press). The criteria for successful propagation of endangered primates, *Zoo Biology.*

Snowdon, C. T., Savage, A. & McConnell, P. B. (1985). A breeding colony of cotton-top tamarins (*Saguinus oedipus*). *Laboratory Animal Science 35,* 477- 481.

Southwick, C. (1967). An experimental study of intragroup agonistic behavior in rhesus monkeys (*Macaca mulatta*). *Behaviour 28,* 182-209.

Southwick, C.H., Beg, M.A. & Siddiqi, M.R. (1965). Rhesus monkeys in North India. In I. Devore (Ed.), *Primate behavior: Field studies of monkeys and apes* (pp. 111-159). New York: Holt, Rinehart & Winston.

Southwick, C. & Siddiqi, M. F. (1985). Rhesus monkey's fall from grace. *Natural History, 2/85,* 62-70.

Southwick, C. H., Siddiqi, M. F., Farooqui, M. Y. and Pal, B. C. (1974). Xenophobia among free-ranging rhesus groups in India. In R. L. Holloway (Ed.), *Primate aggression, territoriality and xenophobia* (pp. 185-210). New York: Academic Press.

Spencer-Booth, Y. & Hinde, R. A. (1967) The effects of separating rhesus monkey infants from their mothers for six days. *Journal of Child Psychology and Psychiatry and Allied Disciplines, 7,* 179-197.

Spinelli, J.S. & Markowitz, H. (1985) Prevention of cage-associated distress. *Lab Animal, 14,* 19-28.

Spitz, R. A. (1946). Anaclitic depression. *Psychoanalytic Study of the Child, 2,* 313-342.

Stewart, K.J. (1984). Parturition in wild gorillas: Behavior of mothers, neonates and others. *Folia Primatologia 42:*62-69.

Stoskopf, M.K. (1983). The physiological effects of psychological stress. *Zoo Biology 2(3):* 179-190.

Stolba, A. & Wood-Gush, D.G.M. (1980) Arousal and exploration in growing pigs in different environments. *Applied Animal Ethology 6,* 382-383.

Stribley, J. A. (1984). *The reproductive cycle and social behavior of the golden lion tamarin,* Leontopithecus rosalia rosalia. Unpublished MA thesis, University of Nebraska at Omaha.

Strickland, H. L. & Clarkson, T. B. (1985). Use of squirrel monkeys in cardiovascular research. In L. A. Rosenblum & C. A. Coe (Eds.), *Handbook of squirrel monkey research.* New York: Plenum.

Stroebel, C. F. (1969) Biologic rhythm correlates of disturbed behavior in rhesus monkeys. *Bibliotheca Primatologica, 9,* 91-105.

Struhsaker, T. (1967). Behavior of vervet monkeys and other *Cercopithecus. Science, 156,* 1197-1203.

Struhsaker, T. (1970). Phylogenetic implications of some vocalizations of *Cercopithecus* monkeys. In J. R. Napier & P. H. Napier (Eds.), *Old World monkeys* (pp. 344-365). New York: Academic Press.

Struhsaker, T. (1971). Social behavior of mother and infant vervet monkeys (*Cercopithecus aethiops*). *Animal Behaviour, 19,* 233-250.

Struhsaker, T. (1978). Infanticide and social organization in the red tail monkey (*Cercopithecus ascanius schmidti*) in the Kabale forest, Uganda. In D. Chivers & J. Herbert (Eds.), *Recent advances in primatology: Behavior,* Vol. 1. New York: Academic Press.

Struhsaker, T. (1980). Comparison of the behavior and ecology of Red Colobus and redtail monkeys in the Kibale Forest, Uganda. *African Journal of Ecology, 18,* 33-51.

Sugiyama, Y. (1964). Group composition, population density and some sociological observations of Hanuman langurs (*Presbytis entellus*). *Primates, 6,* 213-247.

Sugiyama, Y. (1965a). Behavioral development and social structure in two troops of Hanuman langurs (*Presbytis entellus*). *Primates, 5,* 7-27.

Sugiyama, Y. (1965b). On the social change of Hanuman langurs (*Presbytis entellus*) in their natural condition. *Primates, 6,* 381-418.

Sugiyama, Y. (1966). An artificial social change in a Hanuman langur troop (*Presbytis entellus*). *Primates, 7,* 41-72.

Sugiyama, Y. (1967). Social organisation of Hanuman langurs. In S.A. Altmann (Ed.), *Social communication among primates* (pp. 221-236). Chicago: University of Chicago Press.

Sugiyama, Y., Yoshiba, K., & Parthasarathy, M.D. (1965). Home range, mating season, male group and inter-troop relations in Hanuman langurs (*Presbytis entellus*). *Primates, 6,* 73-106.

Suomi, S. J. (1986) Anxiety-like disorders in young nonhuman primates. In R.Gitelman (Ed.), *Anxiety disorders of childhood* (pp. 1-23). New York: Guilford Press.

Suomi, S. J. (1986) Behavioral aspects of successful reproduction in primates. In K. Benirschke (Ed.), *Primates: The road to self-sustaining populations* (pp. 331-340). New York: Springer.

Suomi, S. J. & Harlow, H. F. (1972). Social rehabilitation of isolate-reared monkeys. *Developmental Psychology, 6,* 487-496.

Suomi, S. J., Harlow, H. F. & Kimball, S. D. (1971). Behavioral effects of prolonged partial social isolation in the rhesus monkey. *Psychological Reports, 29,* 1171-1177.

Suomi, S. J., Harlow, H. F., & Novak, M. A. (1974). Reversal of social deficits produced by isolation rearing in monkeys. *Journal of Human Evolution, 3,* 527-534.

Suomi, S.J. & C. Ripp (1983). A history of motherless monkey mothering at the University of Wisconsin primate laboratory. In M. Reite & N.G. Caine (Eds.), *Child abuse: The nonhuman primate data. Monographs in Primatology,* Vol. 1 (pp. 49-77). New York, Alan R. Liss.

Sussman, R.W. & Kinzey, W.G. (1984) The ecological role of the Callitrichidae: A review. *American Journal of Physical Anthropology 64,* 419-449.

Taff, M.A. (1988). *Social dynamics within a captive, all-male group of langurs.* In preparation.

Taff, M.A. & Geary, S. (1988) *Play patterns in captive langurs (*Presbytis entellus). Unpublished manuscript.

Tardif, S. D., Richter, C. B. & Carson, R. L. (1984a) Effects of sibling- rearing experience on future reproductive success in two species of Callitrichidae. *American Journal of Primatology 6,* 377-380.

Tardif, S. D., Richter, C. B. & Carson, R.L. (1984b) Reproductive performance of three species of Callitrichidae. *Laboratory Animal Science 34,* 272-275.

Tattersall, I. (1982). *The primates of Madagascar.* New York: Columbia University Press..

Taub, D.M. (1980). Age at first pregnancy and reproductive outcome among colony-born squirrel monkeys *(Saimiri sciureus, Brazilian). Folia Primatologica, 33,* 262-272.

Taub, D.M., Adams, M.R., & Auerbach, K.G. (1978). Reproductive performance in a breeding colony of Brazilian squirrel monkeys *(Saimiri sciureus). Laboratory Animal Science, 28,* 562-566.

Taylor, L. L. (1986). *Kinship, dominance, and social organization in a semi-free-ranging group of ringtailed lemurs (*Lemur catta). PhD thesis, Washington University, St. Louis, MO.

Taylor, L. L. & Sussman, R. W. (1985). A preliminary study of kinship and social organization in a semi-free-ranging group of *Lemur catta. International Journal of Primatology 6*: 601-614.

Teas, J., Richie, T.L., Taylor, H.G., Siddiqi, M.F. & Southwick, C.H. (1981). Natural regulation of rhesus monkey populations in Kathmandu, Nepal. *Folia Primatologia 35*: 117-123.

Teas, J., Richie, T., Taylor, H. & Southwick, C. H. (1980). Population patterns and behavioral ecology of rhesus monkeys (*Macaca mulatta*) in Nepal. In D. G. Lindburg (Ed.) *The macaques: Studies in ecology, behavior and evolution* (pp. 247-262). New York: Van Nostrand Reinhold.

Terborgh, J. (1985). The ecology of Amazonian primates. In G. T. Prance & T. E. Lovejoy (Eds.), *Key environments: Amazonia* (pp. 284-304).

Terrace, H.S. (1985) On the nature of animal thinking. *Neuroscience & Biobehavior Reviews 9,* 643-652.

Terry, R. (1970). Primate grooming as a tension reduction mechanism. *The Journal of Psychology, 76,* 129-136.

Thomas, R. K. (1980). Evolution of intelligence: An approach to its assessment. *Brain, Behavior and Evolution, 17,* 454-472.

Thomas, R. K., & Peay, L. (1976). Length judgments by squirrel monkeys: Evidence for conservation? *Developmental Psychology, 12,* 349-352.

Thomas, R. K., & Walden, E. L. (1985). The assessment of cognitive development in nonhuman primates. In E. S. Watts (Ed.), *Nonhuman primate models for human growth and development* (pp. 187-215). New York: Alan R. Liss.

Thomas, W. (1986). Housing and furniture. In K. Benirschke (Ed.), *Primates: The road to self-sustaining populations* (pp. 463-464). New York: Springer.

Thorington, L. (1985). Spectral irradiance and temporal aspects of natural and artificial light. In R. J. Wurtman, M. J. Baum and J. T. Potts (Eds.) *The medical and biological effects of light* (pp. 28-54). New York: New York Academy of Sciences.

Thorington, R.W. (1968). Observations of squirrel monkeys in a Colombian forest. In L.A. Rosenblum & R.W. Cooper (Eds.), *The squirrel monkey* (pp.69-85). New York: Academic Press.

Thorington, R. W. Jr. (1985). The taxonomy and distribution of squirrel monkeys (*Saimiri*). In L. A. Rosenblum & C. A. Coe (Eds.), *Handbook of squirrel monkey research.* New York: Plenum.

Tilford, B. L. (1982). *Mate selection by females in free-ranging rhesus monkeys.* Doctoral dissertation, University of California, Riverside.

Titus, K., Mosher, J.A. & Williams, B.K. (1984). Chance corrected classification for use in discriminant analysis: Ecological applications. *The American Midland Naturalist 3(1):* 1-7.

Todt, D., Bruser, E., Hultsch, H., & Lange, R. (1982). Nocturnal actions and interactions of new-born monkeys. *Journal of Human Evolution, 11,* 383-389.

Traylor-Holzer, K., & Fritz, P. (1985). Utilization of space by adult and juvenile groups of captive chimpanzees (*Pan troglodytes*). *Zoo Biology, 4,* 115-127.

Tripp, J.K. (1985) Increasing activity in captive orangutans: Provision of manipulable and edible materials. *Zoo Biology 4,* 225-234.

Troisi, A., D'Amato, F.R., Aureli, F., & Piovesan, P. (1987) Infant abuse by a macaque mother who experienced neonatal abandonment and foster rearing. [Abstract] *American Journal of Primatology, 12,* 375.

Truelove, J.F., Gilbert, S.G. & Rice, D.C. (1985) Effect of diet on blood lead concentration in the cynomolgus monkey. *Fundamental Applied Toxicology, 5,* 588-596.

Turckheim, G. de & Merz, E. (1984). Breeding Barbary macaques in outdoor open enclosures. In J.E. Fa (Ed.), *The Barbary macaque: A case study in conservation* (pp. 241-262). New York: Plenum.

Turkkan, J.S. & Harris, A.H. (1981). Shaping blood pressure elevations: An examination of acquisition. *Behaviour Analysis Letters, 1,* 97-106.

Turkkan, J.S. & Harris, A.H. (1984). Hyperdipsia in the baboon during operant conditioning of blood pressure elevations. *Behavioral Neuroscience, 98,* 631-639.

Turnbull, P.C.B. & Snoyenbos, G.H. (1973). The roles of ammonia, water activity, and pH in the salmonellacidal effect of long-used poultry litter. *Poultry Science 12,* 72-86.

Turnquist, J. E. (1983). Influence of age, sex, and caging on joint mobility in the patas monkey. *American Journal of Physical Anthropology, 61,* 211- 220.

Turnquist, J.E. (1985). Passive joint mobility in patas monkeys (*Erythrocebus patas*): Rehabilitation of caged animals after release into a free-ranging environment. *American Journal of Physical Anthropology 67,* 1-5.

Tuttle, R. H. (1986). *Apes of the world: Their social behavior, communication, mentality and ecology.* Park Ridge, NJ: Noyes.

Ullrey, D.E. (1986). Nutrition of primates in captivity. In K. Benirshke (Ed.), *Primates: The road to self-sustaining populations* (pp. 823-835). New York:

Springer.

Ulrich, R. (1979). Visual landscapes and psychological well-being. *Landscape Research, 4(1),* 17.

Ulrich, R. (1984). View through a window may influence recovery from surgery. *Science, 224,* 420.

Ulrich, R. & Simons, R. (1986). Recovery from stress during exposure to everyday outdoor environments. *Proceedings of the Seventeenth Annual Conference of the Environmental Design Research Association.* Washington, DC: EDRA.

Urbain, A. & Nouvel, J. (1954, October). On keeping wild animals in zoos. *Endeavour,* pp. 184-189

Ursin, H. (1982). The search for stress markers. *Scandinavian Journal of Psychology,* Supplement 1, 165-169.

van Hooff, J.A.R.A.M. *See Hooff, J.A.R.A.M., van.*

Veit, P.G. (1982). Gorilla society. *Natural History 93(3):* 48-59.

Vestergaard, K. (1981) Influence of fixation on the behaviour of sows. In W. Sybesma (Ed.), *The welfare of pigs.* The Hague: Martinus Nijhoff.

Vincent (1969). *Contribution a l'etude des prosimians Africains. Le Galago de Demidoff.* PhD dissertation. Faculty of Sciences, Paris.

Visalberghi, E. & Antinucci, F. (1986) Tool use in the exploitation of food resources in *Cebus apella.* In J.G. Else & P.C. Lee (Eds.), *Primate ecology and conservation,* Vol 2. London: Cambridge University Press.

Vogel, C. (1971). Behavioral differences of *Presbytis entellus* in two different habitats. *Proceedings of the Third International Congress of Primatology, 3,* 41-47.

Vogel, C. (1973). Acoustical communication among free-ranging common Indian langurs (*Presbytis entellus*) in two different habitats of North India. *American Journal of Anthropology, 38,* 469-479.

Vogel, C. (1976). Ekologie, lebensweise und sozialverhalten der grauen languren in verschiedenen biotropen. *Advances in Ethology No. 17,* supplement to *Zeitshrift fur Tierpsychologie.* Berlin: Paul Parey.

Vogt, J. L. (1978). The social behavior of a marmoset (*Saguinus fuscicollis*) group. II: Behavior patterns and social interaction. *Primates, 19,* 287-300.

von Glasersfeld, E. *See Glasersfeld, E., von.*

Walsberg, G.E., Campbell, G.S. & King, J.R. (1978). Animal coat color and radiative heat gain: A re-evaluation. *Journal of Comparative Physiology 126,* 211-222.

Waser, P. (1986) Interactions among primate species. In B.B. Smuts, D.L. Cheney, R.M. Seyfarth, R.W. Wrangham & T.T. Struhsaker (Eds.), *Primate societies.* Chicago: University of Chicago Press.

Watson, J. S. (1979). Perception of contingency as a determinant of social responsiveness. In E. Thoman (Ed.), *Origins of infant's social responsiveness* (pp. 33-64). New York: Lawrence Erlbaum.

Watts, D. P. (1985). Observations on the ontogeny of feeding behavior in mountain gorillas (*Gorilla gorilla berengei*). *American Journal of Primatology 8*: 1-10.

Webster's Seventh New Collegiate Dictionary (1967). Springfield, MA: G. & C. Merriam.

Webster's Ninth New Collegiate Dictionary. (1987). Springfield, MA: G. & C.

Merriam.

Weick, B. G., & Ritter, S. (1985). Adrenergic binding in cardiac ventricular membranes of aging rhesus monkeys. In R. T. Davis & C. W. Leathers (Eds.), *Behavior and pathology of aging in rhesus monkeys*. New York: Alan R. Liss.

Weigel, R.M. (1980). Dyadic spatial relationships in pigtail and stumptail macaques: A multiple regression analysis. *International Journal of Primatology 1*: 287-321.

Weisenburger, W. P. (1979). *Hearing loss in old monkeys*. Unpublished master's thesis, Washington State University, Pullman.

Welch, B. L. (1964). Psychophysiological response to the mean level of environmental stimulation: A theory of environmental integration. *Symposium on the medical aspects of stress in a military climate* (pp. 39- 99). Washington, DC: Walter Reed Army Institute of Research.

Welch, B. L. & Welch, A. S. (1966). Differential effect of chronic grouping and isolation on the metabolism of brain biogenic amines. *Federation of American Societies of Experimental Biology: Federation Proceedings 25*: 623.

Welch, B. L. & Welch, A. S. (1966). Graded effect of social stimulation upon d-amphetamine toxicity, aggressiveness, and heart and adrenal weight. *Journal of Pharmacology and Experimental Therapeutics 141*, 331-338.

Welch, B. L. & Welch, A. S. (1968). Greater lowering of brain and adrenal catecholamines in group-housed than in individually housed mice administered dl-a-methyl-tyrosine. *Journal of Pharmacology 20*: 244-246.

Wemelsfelder, F. (1985) Animal boredom: Is a scientific study of the subjective experiences of animals possible? In M.W. Fox & L.D. Mickley (Eds.), *Advances in animal welfare science, 1984*. Boston: Martinus Nijhoff.

Westergaard, G. C. & Fragaszy, D. M. (1985). Effects of manipulatable objects on the activity of captive capuchin monkeys (*Cebus apella*). *Zoo Biology, 4*, 317-327.

Westergaard, G. C., & Fragaszy, D. M. (1987). The manufacture and use of tools by capuchin monkeys. *Journal of Comparative Psychology, 10*, 159-168.

Westergaard, G. C. & Lindquist, T. (1987). Manipulation of objects in a captive group of lion-tailed macaques (*Macaca silenus*). *American Journal of Primatology, 12*, 231-234.

Whelton, P.K., Thompson, S.G., Barnes, G.R., & Miall, W.E. (1983). Evaluation of the Vita-Stat automatic pressure recorder: A comparison with the random- zero sphygmomanometer. *American Journal of Epidemiology, 117*, 46-54.

Whitehead, J. M. (1986). Development of feeding selectivity in mantled howling monkeys (*Alouatta palliata*). In J. G. Else & P. C. Lee (Eds.) *Primate ontogeny, cognition and social behaviour* (pp. 105-117). New York: Cambridge University Press.

Willes, R.F., Kressler, P., Truelove, J. (1977). Nursery rearing of infant monkeys (*Macaca fascicularis*) for toxicity studies. *Laboratory Animal Science, 27*, 90-98.

Williams. G.J. (1980). Child abuse and neglect: Problems of definition and incidence. In G.J. Willliams & J. Money (Eds.), *Traumatic abuse and neglect of children at home* (pp. 1-6). Baltimore: The John Hopkins University Press.

Wilson, S. F. (1982). Environmental influences on the activity of captive apes. *Zoo*

Biology, 1, 201-209.

Winkler, P. (1984). The adaptive capacities of the Hanuman langur and the categorizing of diet. In D.J. Chivers, B.A. Wood and A. Bilsborough (Eds.), *Food acquisition and processing in primates* (pp. 161-168). New York: Plenum.

Wolfheim, J.H. (1983). *Primates of the world: Distribution, abundance, and conservation.* Seattle: University of Washington Press.

Wood, S., Moriarity, K. M., Gardner, B. T., & Gardner, R. A. (1980) Object permanence in child and chimpanzee. *Animal Learning & Behavior, 8,* 3-9.

Woodruff, G., Premack, D., & Kennel, K. (1978) Conservation of liquid and solid quantity by the chimpanzee. *Science, 202,* 991-994.

Wood-Gush, D.G.M. (1973) Animal welfare in modern agriculture. *British Veterinary Journal 129,* 164-174.

Wrangham, R. W. (1977). Feeding behavior of chimpanzees in Gombe National Park, Tanzania. In T. H. Clutton-Brock (Ed.), *Primate ecology: Studies of feeding and ranging behavior in lemurs, monkeys and apes* (pp. 503--538). New York: Academic Press.

Wright, P. C. (1981). The night monkeys, genus *Aotus.* In R. A. Mittermeier & A. F. Coimbra-Filho (Eds.), *Ecology and behavior of neotropical primates* (pp. 211-240). Rio de Janeiro, Brazilian Academy of Sciences.

Wright, P. C. (1984). Biparental care in *Aotus trivirgatus* and *Callicebus moloch.* In M. Small (Ed.), *Female primates: Studies by women primatologists* (pp. 59-750). New York: Alan R. Liss.

Wright, P. C. (1985). *Costs and benefits of nocturnality to the night monkey* Aotus trivirgatus. PhD dissertation. City University of New York, New York.

Wright, P. C. (1986). Ecological correlates of monogamy in *Aotus* and *Callicebus moloch.* In J. Else & P. C. Lee (Eds.), *Primate behavior and evolution* (pp. 159-167). London: Cambridge University Press.

Wright, P. C., Haring, D.M., Simons, E.L. & Andau, P. 1987). Tarsiers: A conservation perspective. *Primate Conservation, 8,* 51-54.

Wright, P.C., Izard, M.K. & Simons, E.L. (1986b). Reproductive cycles in *Tarsius bancanus. American Journal of Primatology, 11,* 207-215.

Wright, P.C., Izard, M. K. & Simons, E. L. (in press) Reproductive cycles of the Bornean tarsier (*Tarsius bancanus*). *American Journal of Primatology.*

Wright, P. C., Toyoma, L. & Simons, E.L. (1986a). Courtship and copulation in *Tarsius bancanus. Folia Primatologica, 46,* 142-148.

Wurtman, R. J. & Weisel, J. (1969). Environmental lighting and neuroendocrine function: Relationship between spectrum of light source and gonadal growth. *Endocrinology, 85,* 1218-1221.

Yamagiwa, J. (1983). Diachronic changes in two eastern lowland gorilla groups (*Gorilla gorilla graueri*) in the Mt. Kahuzi region, Zaire. *Primates 24(2):* 174-183.

Yanagita, T., Deneau, G.A. & Seevers, M.H. (1965). Evaluation of pharmacological agents in the monkey by long-term intravenous self- or programmed administration. *Excerpta Medica International Congress Series, 87,* 453-457.

Yerkes, R. M. (1925) *Almost human.* London: Jonathan Cape

Yerkes, R.M. (1943). *Chimpanzees: A laboratory colony.* New Haven: Yale University Press.

Yerkes, R.M. & Yerkes, A.W. (1929). *The great apes: A study of anthropoid life.* New Haven, CT: Yale University Press.

Yoneda, M. (1984). Comparative studies on vertical separation, foraging behavior and traveling mode of saddle-backed tamarins (*Saguinus fuscicollis*) and red-chested moustached tamarins (*Saguinus labiatus*) in northern Bolivia. *Primates, 25,* 414-422.

Yoshiba, K. (1967). An ecological study of Hanuman langurs. *Primates, 8,* 127-154.

Yoshiba, K. (1968). Local and intertroop variability in ecology and social behaviour of common Indian langurs. In P.C. Jay (Ed.), *Primates: Studies in adaptation and variability* (pp. 217-242). New York: Holt, Reinhart, & Winston.

Yoshino, M.M. (1975). *Climate in a small area: An introduction to local meteorology.* Tokyo: University of Tokyo Press.

Zabel, J. (1980) *Object/image discrimination by vervet monkeys.* Unpublished student project.

Ziegler, T. E., Bridson, W. E., Snowdon, C. T. & Eman, S. (1987a) Urinary gonadotropin and estrogen excretion during the postpartum estrus, conception, and pregnancy in the cotton-top tamarin (*Saguinus oedipus oedipus*). *American Journal of Primatology 12,* 127-140.

Ziegler, T. E., Savage, A., Scheffler, G. & Snowdon, C.T. (1987b) The endocrinology of puberty and reproductive functioning in female cotton-top tamarins (*Saguinus oedipus*) under varying social conditions. *Biology of Reproduction, 37,* 618-627.

Index